T4-AVD-347

Evolutionary Biology

VOLUME 7

Evolutionary Biology

VOLUME 7

Edited by

THEODOSIUS DOBZHANSKY
Department of Genetics
University of California
Davis, California

MAX K. HECHT
Queens College
Flushing, New York

and

WILLIAM C. STEERE
New York Botanical Garden
Bronx, New York

PLENUM PRESS • NEW YORK AND LONDON

The Library of Congress cataloged the first volume of this title as follows:

Evolutionary biology. v. 1- 1967-
New York, Appleton-Century-Crofts.

v. illus. 24 cm. annual.

Editors: 1967- T. Dobzhansky and others.

1. Evolution — Period. 2. Biology — Period. I. Dobzhansky, Theodosius Grigorievich, 1900-

QH366.A1E9 575'.005 67—11961

Library of Congress [67n7]

Library of Congress Catalog Card Number 67-11961
ISBN 0-306-35407-1

A Division of Plenum Publishing Corporation
227 West 17th Street, New York, N.Y. 10011

United Kingdom edition published by Plenum Press, London
A Division of Plenum Publishing Company, Ltd.
4a Lower John Street, London, W1R 3PD, England

Printed in the United States of America

Contributors

CLAUDIO BARIGOZZI, *Institute of Genetics, University of Milan, Milan, Italy*

PETER A. CORNING, *Institute of Political Studies, Stanford University, Stanford, California*

NILES ELDREDGE, *Department of Invertebrate Paleontology, American Museum of Natural History, New York, N.Y.*

STEPHEN J. GOULD, *Museum of Comparative Zoology, Harvard University, Cambridge, Masachusetts*

MAX K. HECHT, *Department of Biology, Queens College, Flushing, New York*

HAROLD W. KERSTER, *Department of Environmental Studies, California State University, Sacramento, California*

H. B. LÉJOHN, *Department of Microbiology, University of Manitoba, Winnipeg, Manitoba, Canada*

DONALD A. LEVIN, *Department of Botany, University of Texas, Austin, Texas*

LYNN MARGULIS, *Department of Biology, Boston University, Boston, Massachusetts*

P. K. K. NAIR, *National Botanic Gardens, Lucknow, India*

J. WILLIAM SCHOPF, *Department of Geology, University of California, Los Angeles, California*

Contents

1

Paleobiology of the Precambrian: The Age of Blue-Green Algae

J. WILLIAM SCHOPF
Department of Geology
University of California
Los Angeles, California

INTRODUCTION

Depending one one's predilections and the geological time in question, the Eras of the Phanerozoic can be properly (if informally) referred to as "the age of mammals," "the age of angiosperms," "the age of reptiles," and so forth. In like manner and with equal force, the Precambrian Era, encompassing the earliest seven-eighths of geological history, can be termed "the age of blue-green algae"; microscopic cyanophytes formed the dominant component of the earth's earliest biota (Schopf, 1970*a*).

Until quite recently, the Precambrian fossil record, evidencing the antiquity and early dominance of the Cyanophyta, was virtually unknown. Despite this lack of fossil evidence, however, biologists have long regarded these microorganisms as being among the most simple and presumably primitive of living systems. Moreover, mound-shaped stromatolites, thought possibly to be of blue-green algal origin, have been known for several decades to be particularly prevalent in Precambrian strata. Thus the occurrence of cyanophytic and cyanophyte-like microfossils in Precambrian sediments, first reported near the turn of the century (Peach *et al.*, 1907; Walcott, 1914; Moore, 1918; Gruner, 1925) but not firmly established until 1954 (Tyler and Barghoorn, 1954), was not wholly unexpected. What is remarkable, however, is the fidelity of cellular preservation and the biotic diversity exhibited by certain of the Precambrian blue-green algal assemblages discovered in recent years. These microbiological communities have provided exceptional insight into the antiquity, composition, and evolutionary development of Precambrian life.

Of the fossil cyanophytes now known, most occur in microcrystalline cherts, preserved by silica permineralization as three-dimensional, structurally intact, organic residues. In such deposits, clear-cut evidence of ecological setting, growth habit, general morphology, detailed cellular anatomy, and mode of reproduction is rather commonly present. Such features provide a sound basis for comparison of living and fossil taxa. Thus, since classification of cyanophytes is based primarily on morphology, the fossil record can provide some indication of the temporal distribution of various types of blue-green algae (especially at the familial level where systematic categories are morphologically quite distinct); if judiciously applied, these data can be used to infer possible phylogenetic relationships. Obviously, the "probable correctness" (or, perhaps more accurately, the "plausibility") of any such postulated phylogeny will be influenced by many factors, the most important of which are the *quantity, quality*, and *geological distribution* of the fossil evidence available. In the following discussion, therefore, a primary task will be to appraise these three aspects of the known cyanophytic fossil record. Following this assessment, tentative inferences will be drawn regarding the timing and nature of major events in blue-green algal evolution. As is true of almost all biological groups, the known fossil record of the Cyanophyta leaves much to be desired. Nevertheless, this treatment should serve to focus attention on the early evolutionary importance of these primitive microorganisms and will, I hope, illustrate both the actual and the potential contribution that fossil evidence can make toward deciphering the course of thallophytic evolution.

MORPHOLOGY AND CLASSIFICATION OF CYANOPHYTES

Together with bacteria, blue-green algae are distinguished from all other cellular organisms by their prokaryotic organization. Like bacteria, cyanophytes are microscopic in size and unicellular, colonial, or filamentous in form; their prokaryotic cells lack an endoplasmic reticulum, true vacuoles, and membrane-bound organelles (e.g., plastids, mitochondria, and a nucleus), their cell walls are characteristically composed of peptidoglycans (mucopeptides), and their histone-free genetic material is distributed to daughter cells via a direct process such as "binary fission" rather than by the more complex mechanisms of eukaryotic mitosis. All prokaryotes are nonmeiotic, almost all are entirely asexual, and in the few bacteria in which genetic transfer has been shown to occur the recipient cell is not genetically equivalent to a diploid eukaryotic zygote (Stanier *et al.*, 1970, p. 44).

Unlike bacteria, however, a large majority of blue-green algae are obligate photoautotrophs, obtaining carbon and producing gaseous oxygen by energy-yielding mechanisms similar to those of higher plants.

The division Cyanophyta consists of one class, the Cyanophyceae (sometimes referred to as the Myxophyceae, "slime algae," or the Schizophyceae, "fission algae"), and five orders which can be grouped conveniently into two "tribes": (1) the "Coccogoneae," including three orders (Chroococcales, Chamaesiphonales, and Pleurocapsales) composed of unicellular and colonial genera and (2) the "Hormogoneae," including the remaining two orders of filamentous genera (Nostocales and Stigonematales). Of the 1500–2000 known species of extant cyanophytes, about 70% are nostocaleans and 20% chroococcaleans; remaining species are distributed more or less equally among the other three orders. The two major extant orders are also dominant in the fossil record; about 60% of reported fossil cyanophytes are referrable to the Nostocales and about 35% to the Chroococcales.

Almost all chroococcaleans are included in a single family, the Chroococcaceae, members of which are spheroidal to ellipsoidal in shape, are commonly enveloped by a prominent mucilagenous sheath, and occur singly or in colonies of a few to many cells (Fig. 1). Cell division produces two equal unicells that become separated by the gelatinous matrix; reproduction of colonies is by fragmentation. Unlike other members of the "Coccogoneae," chroococcaleans are commonly planktonic in habit, and their cells are not specialized for attachment to substrates.

The unit of organization in the Nostocales is the trichome, a uniseriate, unbranched row of cylindrical, discoidal, or spheroidal cells (Fig. 1). A nostocalean filament is composed of one or more such trichomes, enclosed by a tubular, mucilagenous investment. All nostocaleans have the capability of reproducing by liberation of hormogones, short trichomic segments that, like many mature trichomes, move by gliding motility; in the Oscillatoriaceae, the largest of the four major families of the order, no other means of reproduction is known. Trichomes of the Nostocaceae, the second largest nostocalean family, are composed predominantly of barrel-shaped or nearly spherical vegetative cells. However, like most other "hormogoneans" apart from the Oscillatoriaceae, nostocacean trichomes also commonly contain specialized, somewhat enlarged cells. Such cells include akinetes, a type of elongate resting spore, and heterocysts, a spheroidal, thick-walled, unpigmented cell type that performs a reproductive function and is the principal site of nitrogen fixation in species having that capability

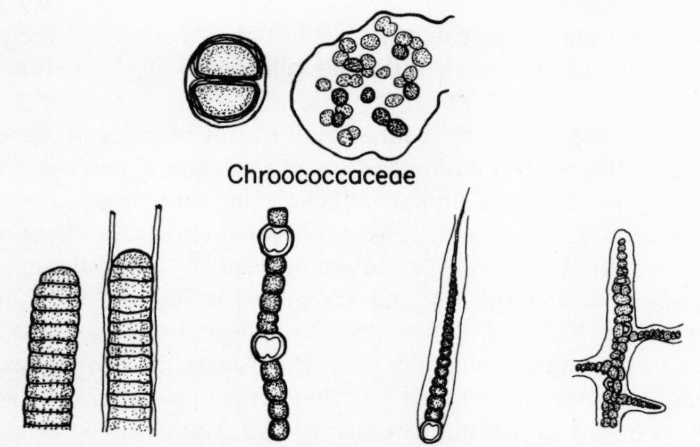

FIG. 1. Representative modern blue-green algae. "Coccogoneae" (above): Chroococcaceae—sheath-enclosed cell pair (*Chroococcus*) and irregular, mucilage-embedded colony (*Aphanocapsa*). "Hormogoneae" (below): Oscillatoriaceae—uniseriate trichome without evident sheath (*Oscillatoria*) and filament with prominent sheath (*Lyngbya*); Nostocaceae—filament with enlarged, intercalary heterocysts (*Nostoc*); Rivulariaceae—short, tapered filament with terminal heterocyst (*Calothrix*); Stigonemataceae—true-branching, heterocystous filament (*Stigonema*).

(Stewart *et al.*, 1969). Members of the Rivulariaceae and Scytonemataceae, the remaining major families of the Nostocales, are generally rather heavily ensheathed and commonly exhibit false branching; typically, rivulariacean trichomes are markedly attenuated toward one end, terminating in a fine cellular hair. Among all "Hormogoneae," true trichomic branching is exhibited only by members of the Stigonematales, commonly regarded as the most advanced order of the class.

ASSESSMENT OF THE CYANOPHYTIC FOSSIL RECORD

Quantity of Fossil Evidence

As is true of most types of unmineralized microorganisms (i.e., those lacking "hard parts" such as siliceous and calcareous tests), relatively few occurrences of cellularly preserved cyanophytes are known. In part, this apparently limited fossil record reflects the delicate nature of these microscopic algae; unless rapidly permineralized in chemical sediments (e.g., calcareous coal balls, many cherts, and some limestones), they seem rather unlikely to be preserved. However, such chemical sediments, many of which contain permineralized spores and axes of higher plants, are not uncommon in the geological record. Moreover, most of these plant-bearing units were deposited in peat bogs and coal swamps, environments in which blue-green algae would be expected to have thrived. Why then the apparent paucity of fossil cyanophytes? This scarcity and the general scarcity of reported fossil thallophytes appear to be a prime result of the emphasis on megascopic organisms traditional to paleobiology; although fossil prokaryotes, fungi, and unmineralized algae are no doubt of rather widespread occurrence, they commonly have been overlooked or ignored in studies chiefly concerned with remains of tracheophytes and metazoans. Recent increased interest in the phylogeny of thallophytes (Sagan, 1967; Klein and Cronquist, 1967; Whittaker, 1969; Raven, 1970) and in the origin of the vascular flora (Banks, 1968; Chaloner, 1970) should spur the search for new fossil evidence of lower plants. To date, it is only in deposits of Precambrian age, essentially devoid of traces of higher organisms, that blue-green algae have received detailed attention (Schopf, 1968, 1970*a*; Schopf and Blacic, 1971), and there only a bare beginning has been made.

As is shown in Fig. 2, approximately 185 occurrences of cellularly preserved cyanophytes have been reported since the 1850s. Of these, more than half have been described during the past 5 years and more than 75% since 1950. About two-thirds of all occurrences have been reported from the Precambrian (e.g., cyanophytes of the Gunflint, Bitter Springs and Beck Spring microfloras); it is evident that recent expansion of the known cyanophytic fossil record has resulted directly from the marked increase of interest in Precambrian life witnessed during the past two decades. In fact, the cumulative curve of Precambrian occurrences shown in Fig. 2 rather accurately parallels the historical development of Precambrian paleobiology as a recognized subdiscipline of paleontological science—its halting beginnings in the early 1900s with the studies of Walcott and Gruner, its renaissance during the 1950s, and its subsequent meteoric rise to its current highly active status (reviewed in Schopf, 1970*a*).

Since about 1965, it has become increasingly apparent that most and perhaps all major events in cyanophytic evolution occurred prior to the beginning of the Phanerozoic. Thus although the quantity of available fossil evidence seems rather limited, the most important phase in blue-green algal evolution, that occurring during the Middle and Late Precambrian, is relatively well documented.

Quality and Geological Distribution of Fossil Evidence

The quality of preservation of reported fossil cyanophytes is highly variable, being primarily a function of the environment in which preservation occurred and the postdepositional geological history of the fos-

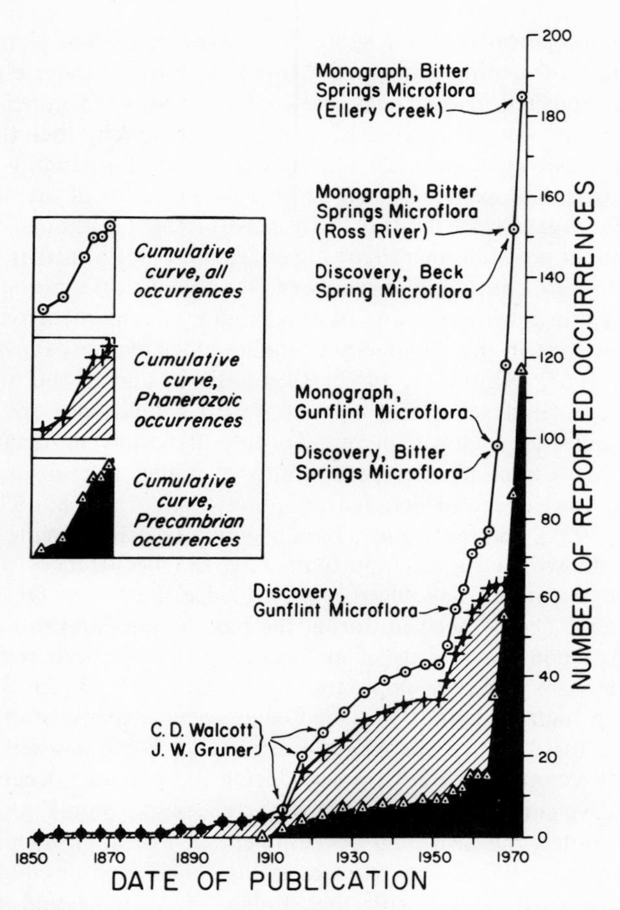

FIG. 2. Reported occurrences of fossil blue-green algae. Cumulative curves showing date of publication and number of published occurrences of cellularly preserved blue-green algae (excluding stromatolites and similar organosedimentary structures) reported from Precambrian to Phanerozoic deposits since 1855. References to items noted: Walcott (1914) and Gruner (1925); Gunflint microflora (Tyler and Barghoorn, 1954; Barghoorn and Tyler, 1965; Cloud, 1965); Bitter Springs microflora (Schopf, 1968; Schopf and Blacic, 1971; Barghoorn and Schopf, 1965); Beck Spring microflora (Cloud *et al.*, 1969; Gutstadt and Schopf, 1969; Licari, 1971).

siliferous unit. Thus highly oxidized facies, such as redbeds and ferruginous cherts, coarse-grained (e.g., sandstones) or highly compacted clastic deposits (e.g., many shales), and sediments that have been subject to recrystallization (e.g., many limestones) rarely contain well-preserved cyanophytes. In addition to such environmental and early diagenetic factors, subsequent geological history plays an important role in determining fidelity of preservation. It is not surprising, therefore, that in general there is an inverse correlation between geological age and both the quality and quantity of the preserved fossil record—the older the sedimentary unit, the greater the probability that it will have been metamorphosed, recycled by normal geological processes, or simply buried by younger materials and thus be unavailable for study. This relationship between geological age and number of reported occurrences of cellularly preserved blue-green algae is well illustrated by the data summarized in Fig. 3.

As a rule, anatomical and morphological details are discernible only in blue-green algae that have become embedded (i.e., permineralized) in chemically deposited cherts and similar fine-grained sediments. The exact mechanisms involved in this type of preservation are poorly defined and no doubt vary depending on the mineral emplaced and local environmental conditions. However, recent laboratory experiments, designed to produce

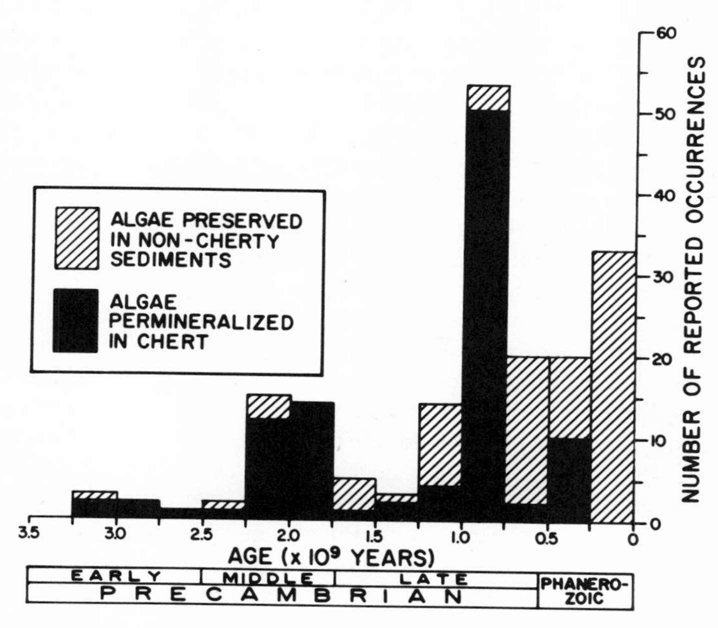

FIG. 3. Geological distribution of fossil blue-green algae. Histogram showing known geological distribution of cellularly preserved cyanophytic and cyanophyte-like microfossils reported from cherty and noncherty Precambrian and Phanerozoic sediments.

synthetic "fossiliferous chert" containing artificially silicified blue-green algae, have provided insight into some aspects of this complex process (Oehler and Schopf, 1971). Like their naturally occurring counterparts, the experimentally produced "artificial microfossils" are three-dimensional in form, are composed entirely of organic matter, and are thoroughly embedded in the sileceous matrix, being roughly comparable in appearance to algae that have been embedded in paraffin, epoxy, or some similar medium in preparation for microtomy. Under both laboratory and natural conditions, silica infills cell lumina, intracellular areas, and the intermicellar spaces of cell walls and encompassing sheaths. During this process, cytoplasmic monomers and absorbed, adsorbed, and intermicellar water (J. M. Schopf, 1971) are generally displaced; however, resilient long-chained or cross-linked polymers, such as those of the cell walls and sheaths, become impregnated by the embedding material and remain structurally intact.

In nature, the initial stages of permineralization occur rather rapidly; cells are occasionally preserved as if "frozen" during a dynamic process such as cell division. And, because primary bedded cherts are formed at the sediment–water interface, mat-building blue-green algal communities are especially susceptible to *in situ* preservation of this type. Although many such deposits are of limited areal distribution, some fossiliferous cherty horizons, particularly those of Precambrian age, are of substantial extent; the Middle Precambrian Gunflint cherts apparently were deposited over an area of nearly 2500 km^2 and those of the Late Precambrian Bitter Springs Formation over an area of more than 1500 km^2. When one considers that these two cherty horizons have an average thickness in excess of 10 cm, that they contain, on average, about 100 well-preserved microorganisms per cubic centimeter (with a density for the Gunflint cherts of 2 or 3 orders of magnitude greater in highly fossiliferous regions), and that 70–90% of these microfossils are of blue-green algal affinities, it becomes apparent that such deposits can provide an astonishing wealth of information regarding the early evolutionary status of cyanophytes.

It should be recognized, of course, that the Gunflint and Bitter Springs horizons are quite unusual. Few fossiliferous cherts approach the microbiological density of the Gunflint deposit, and the permineralized Bitter Springs assemblage is exceptional in both biological diversity and fidelity of preservation. Nevertheless, the ratio of reported occurrences of fossil cyanophytes in cherts of all ages to those described from all noncherty sediments should give some indication of the overall quality of available fossil evidence. As is shown in Fig. 3, this ratio is about 1.2:1—more than half of all known occurrences of cellular fossil cyanophytes have been reported from fine-grained cherts. Moreover, since cherts tend to be much more fossiliferous than noncherty sediments, commonly containing many representatives of each species, a very large majority of individual cyanophytic microfossils have been described from cherty facies. And, as is also evident in Fig. 3, cherts containing permineralized blue-green algae are stratigraphically well distributed, spanning a time range from the Early Precambrian to the mid-Phanerozoic. In general, therefore, much of the currently available fossil evidence can be expected to be of relatively high quality.

The fidelity of preservation obtained under favorable conditions is perhaps best illustrated by comparing the morphology of fossil cyanophytes with that of closely related living forms. As a specific example, one of many that might be chosen, permineralized filaments of the genus *Schizothrichites* (from cherts of the Ordovician of Poland; Starmach, 1963) can be compared with *Schizothrix*, one of the most common blue-green algae of the modern flora. In Fig. 4 are shown optical photomicrographs and scanning electron micrographs of the modern cyanophyte; the uniseriate, undifferentiated trichomes (Fig. 4A,B), encompassing, lamellated, mucilagenous sheaths (Fig. 4C–F), and rounded terminal cells (Fig. 4G) are all features typical of the Oscillatoriaceae. *Schizothrix* reproduces asexually by liberation of the short trichomic hormogones shown in Fig. 4(H–K). In Fig. 5 are shown specimens of *Schizothrichites* in petrographic thin sections (Fig. 5A,D) and in hydrofluoric acid–resistant residues (Fig. 5B,C,E–I) of the Ordovician chert. The oscillatoriacean organization of *Schizothrichites* is evident. Like those of modern *Schizothrix*, the trichomes of the fossil alga have rounded terminal cells, are uniseriate and undifferentiated, and are encompassed by a well-developed organic sheath. Although diagenetically somewhat altered, the fossil sheaths (Fig. 5F–I) are multilayered like those of *Schizothrix*. To complete the comparison, *Schizothrichites* evidently reproduced by liberation of hormogones (Fig 5E) morphologically quite similar to those of the modern alga. Thus in all important morphological characteristics, and apparently also in mode of reproduction, the fossil and modern taxa (separated by some 480 million years of earth history) seem closely comparable.

Conclusions

The following general conclusions can be drawn from the foregoing discussion:

1. Although the known cyanophytic fossil record is quantitatively rather unimpressive, the most active phase in blue-green algal evolution, occurring during the Middle and Late Precambrian, is relatively well documented.
2. The majority of known fossil cyanophytes have been preserved by

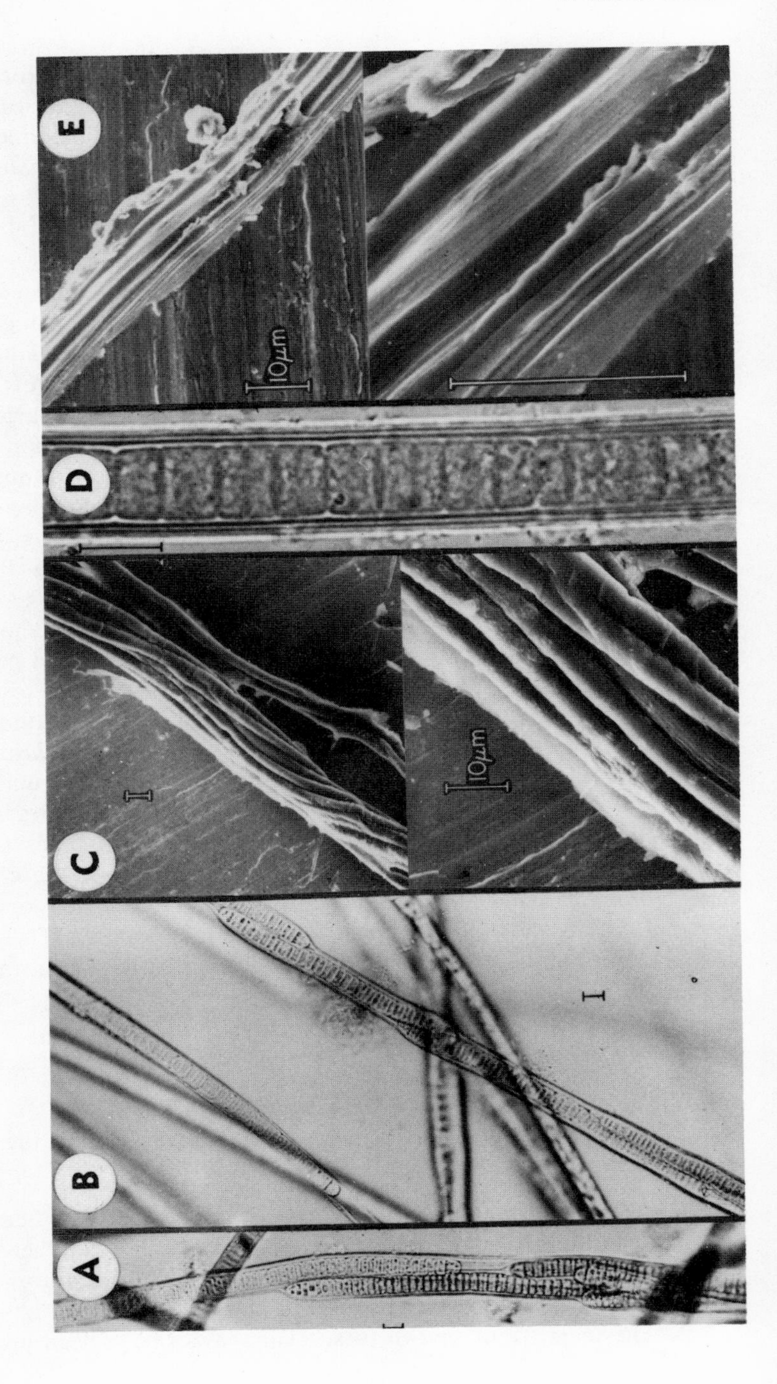
A
B
C
10μm
D
E
10μm

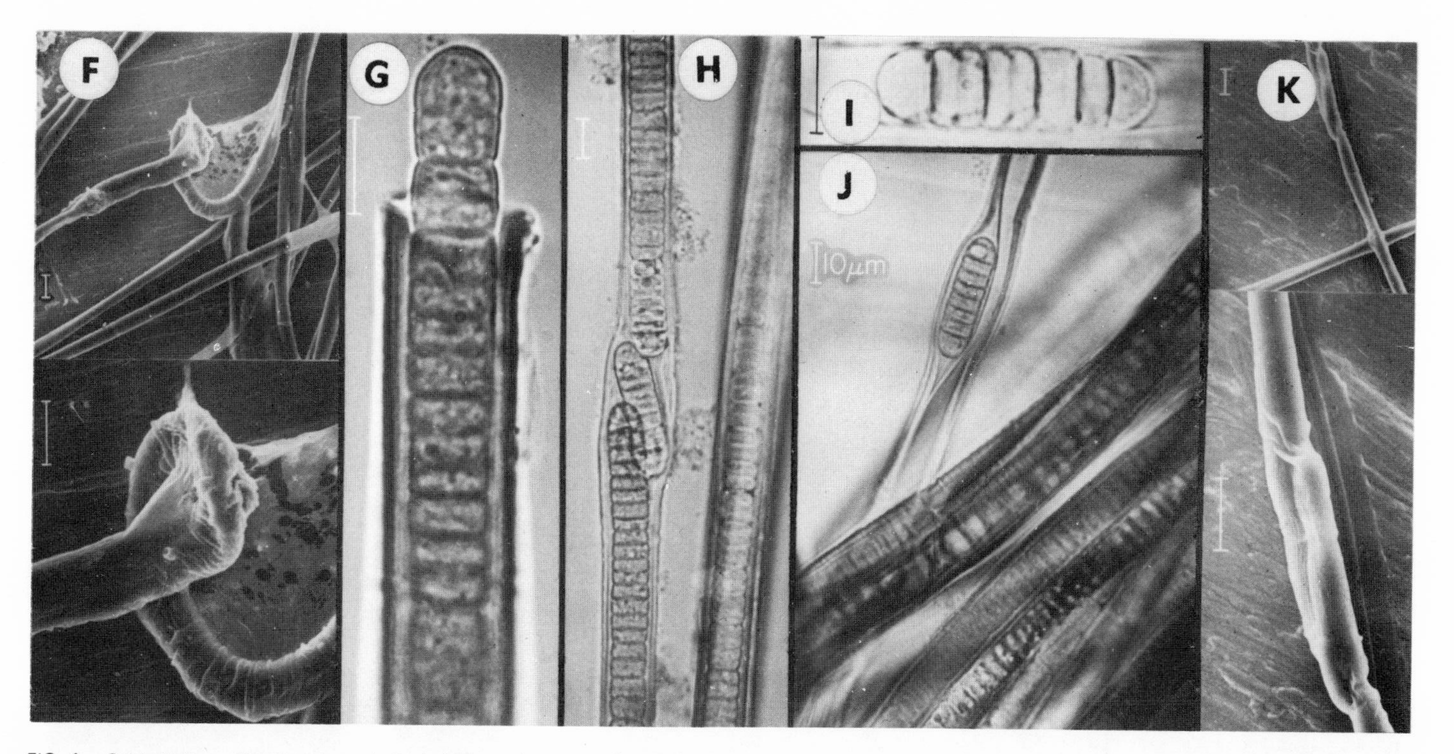

FIG. 4. *Schizothrix* sp. (Oscillatoriaceae). Optical photomicrographs (A,B,D,G–J) and scanning electron micrographs (C,E,F,K) showing filaments, sheaths (E,F), and reproductive hormogones (H–J) of this widespread, modern alga; lines for scale represent 10 μm.

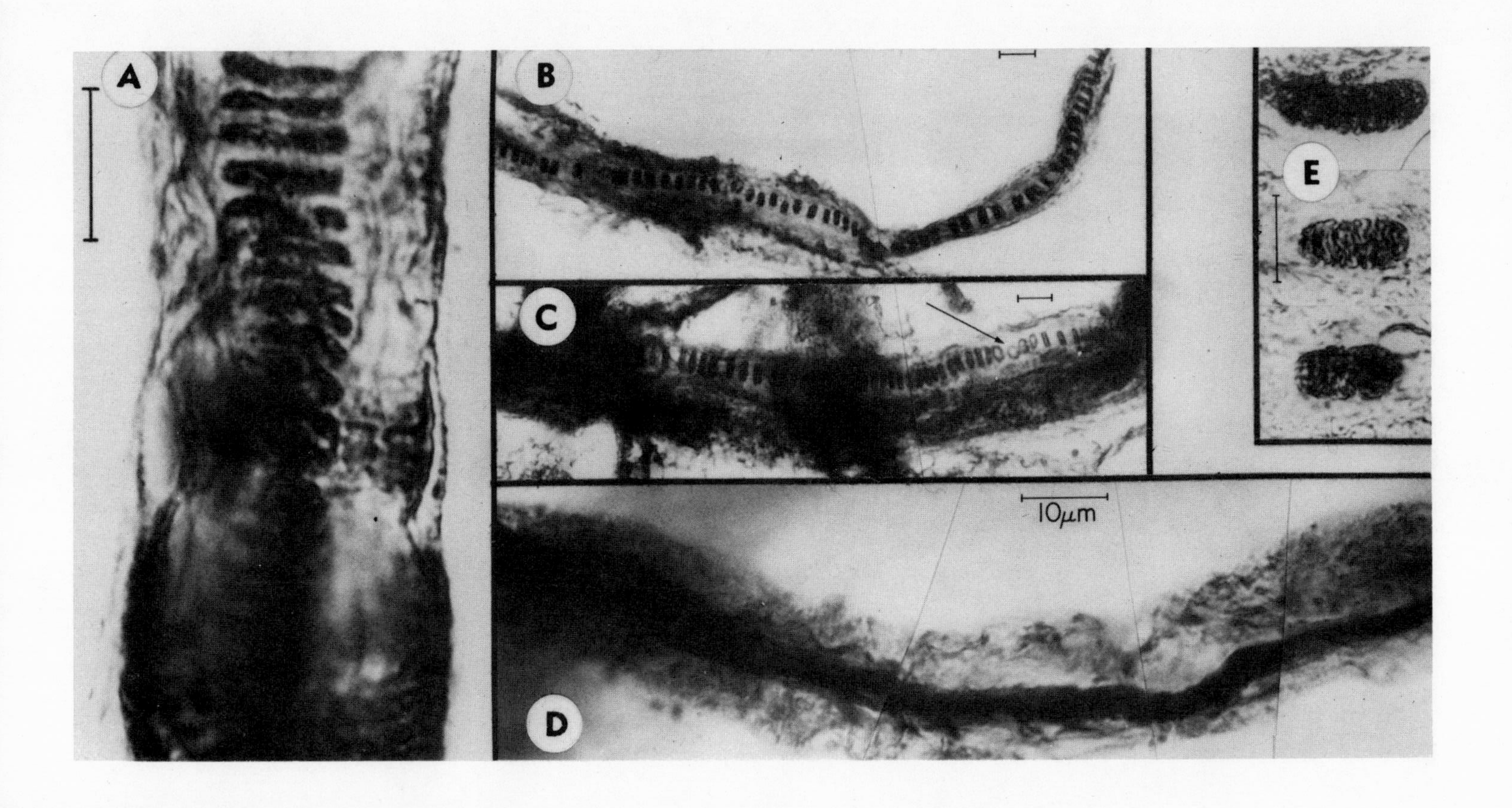
A
B
C
D
E
10μm

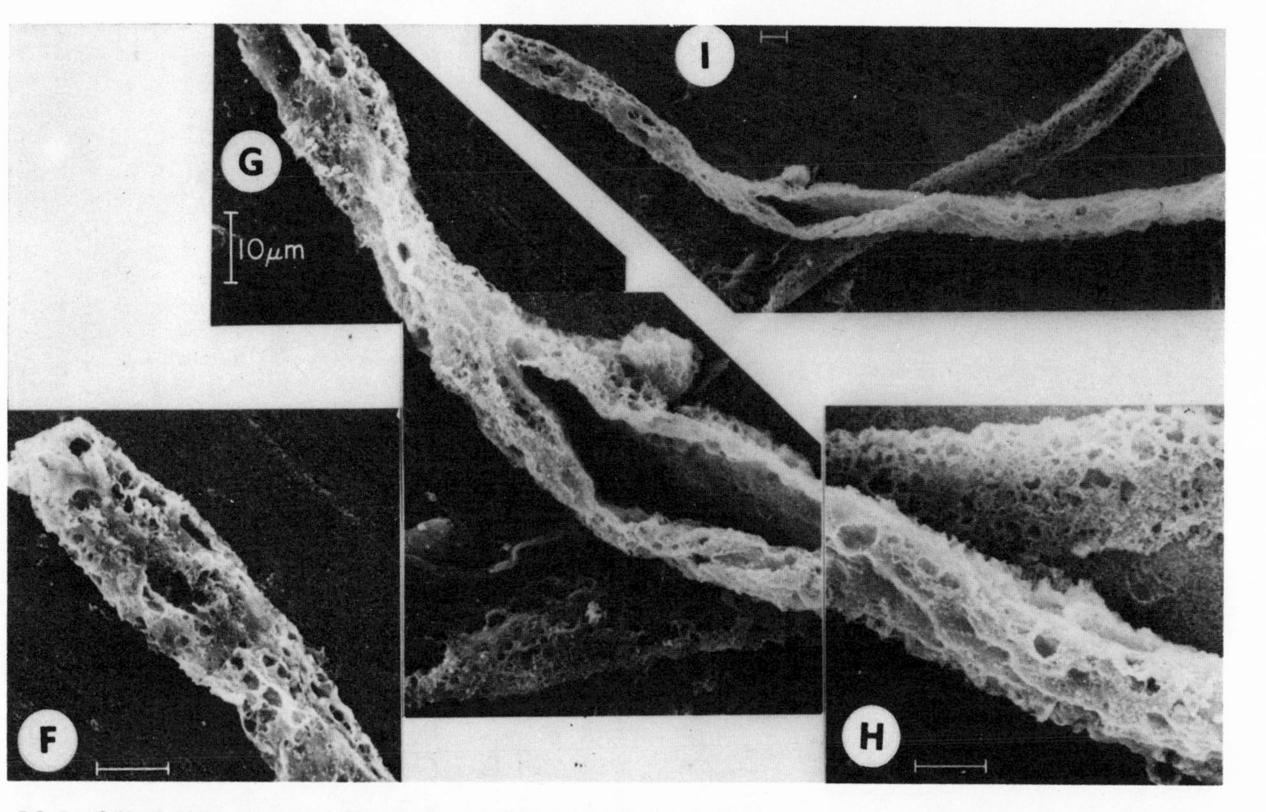

FIG. 5. *Schizothrichites ordoviciensis* (Oscillatoriaceae). Optical photomicrographs (A–E) and scanning electron micrographs (F–I) showing filaments, sheaths (F–I), and hormogones (E) of this fossil alga in petrographic thin section (A,D) and in hydrofluoric acid-resistant residues (B,C,E–H) of chert from the lower Ordovician of Poland (*cf.* Starmach, 1963); lines for scale represent 10 μm.

silica permineralization in fine-grained cherts. In such deposits, preservation is commonly of excellent quality, providing a basis for detailed morphological comparison of extant and fossil taxa.

3. There is, in general, an inverse correlation between geological age and both the quality and quantity of available fossil evidence. Nevertheless, cyanophyte-like microfossils are known from sediments more than 3 billion years old, and the fossil record of the group extends from the Early Precambrian to the Recent.
4. Thus in quantity, quality, and geological distribution the available fossil evidence appears to provide a promising basis for phylogenetic speculation. Although the data are far from ideal, especially in terms of quantity (a situation that appears certain to improve markedly in the near future), they seem sufficient for tentative identification of major events in the evolution of the Cyanophyta.

ORIGIN OF BLUE-GREEN ALGAE

Mode of Origin

As is discussed above, blue-green algae and bacteria share many fundamental features of cellular organization and differ significantly from all other biological systems; it seems evident that these two prokaryotic groups are closely related. Among the physiologically diverse varieties of extant bacteria, the photosynthetic types seem most obviously allied with blue-green algae. Photosynthetic bacteria and cyanophytes both contain chlorophylls, carotenoids, and photosynthetic lamellae, both are primarily photoautotrophic, and both are similar in cell wall ultrastructure and chemistry to gram-negative bacteria (Brock, 1970). However, the two groups differ substantially in their photosynthetic mechanisms. Photosynthetic bacteria contain bacteriochlorophylls, use H_2S, other sulfur compounds, or organic materials as a source of reducing power, and apparently exhibit only one photochemical reaction (Case *et al.*, 1970; however, see Olson, 1970).

In cyanophytes, chlorophyll *a* serves as the light-absorbing molecule, hydroxyl ions from water are used as electron donors (resulting in production of oxygen as a by-product), and two photoacts linked in series are involved in the photosynthetic process. Thus, in contrast to bacterial photosynthesis, blue-green algal photosynthesis is quite comparable to that occurring in eukaryotic algae and higher plants. Furthermore, cyanophytes, like higher plants, are generally aerobic; bacterial photosynthesis is an anaerobic process and, with the exception of certain of the Athiorhodaceae (nonsulfur purple bacteria, some of which are heterotrophic in the presence of oxygen), photosynthetic bacteria are obligate anaerobes. Since it seems quite likely that the earliest living systems were heterotrophic (Wald, 1964), photoautotrophic bacteria and blue-green algae must both be regarded as derived groups. Of the two, cyanophytes appear to be relatively more advanced, as evidenced by their aerobiosis and their more elaborate morphology. As a first approximation, therefore, blue-green algae can be viewed as representing an early stage in biological evolution, intermediate and apparently transitional between anaerobic, bacterial precursors and aerobic, eukaryotic descendants (or eukaryotic chloroplasts, if one accepts the symbiotic theory of eukaryotic origins; Margulis 1970).

The precise course of evolutionary history between the origin of life and the appearance of the first cyanophytes is unknown. It seems apparent, however, that the prime feature of this sequence was the progressive modification of the structure and function of metalloporphyrin-containing compounds, i.e., the "biochemical evolution" of porphyrin-based metabolism that resulted in formation of cytochromes and chlorophylls and ultimately in development of the complex apparatus of oxygen-producing photosynthesis. The following scenario outlines a possible course for this evolutionary progression. Initially, biological systems probably lacked porphyrin-containing molecules (*cf.* Margulis, 1970, p. 94); these most primitive organisms were anaerobic heterotrophs, entirely dependent for a source of carbon and chemical energy on substrate-level phosphorylation (fermentation) of organic compounds abiologically synthesized in the essentially anoxic environment. As these heterotrophs progressively depleted the supply of relatively simple organic materials available in their aqueous environment, organisms first appeared that possessed iron-containing porphyrins functionally similar to modern cytochromes. This new biological stock was able to carry out oxidative phosphorylation anaerobically (using terminal electron acceptors other than oxygen such as SO_4^{2-}, CO_2, NO_3^-, or organic compounds such as fumarate), and thus could utilize a variety of nutrients that could not be broken down by simple fermentation (Brock, 1970, p. 498).

Further evolution resulted in the appearance of enzymes that extended the previously established biosynthetic pathway to metalloporphyrin. The new pathway, yielding porphyrins complexed with magnesium rather than iron, resulted in formation of chlorophylls and the concomitant "discovery" of anaerobic photoheterotrophy, the use of chemical energy generated by light-driven reactions to "photoassimilate" exogenous organic matter (Olson, 1970). With the addition of CO_2-fixing pathways similar to those of extant autotrophs, these photoassimilators gave rise to anaerobic photoautotrophs. Of the various steps postulated in this hypothetical sequence, this last development, the origin of autotrophy, was especially crucial since it freed the primitive biota from its dependence on "ready-made" organic materials and thus surmounted the "carbon crisis" brought about by the previous steady depletion of available foodstuffs. Although the earliest photoautotrophs were probably rather similar to modern photosynthetic bacteria, their porphyrin-containing pigments may have more closely resembled chlorophyll *a* than the bacteriochlorophylls (Olson, 1970). Finally, with the development of phycobilins and the establishment of serial linkage between two photoacts of the type occurring in green-plant photosynthesis, hydroxyl ions from water could be used as electron donors; the resulting oxygen-producing photoautotrophs were the earliest blue-green algae.

Compared to their photoautotrophic bacterial precursors, primitive cyanophytes had several selective advantages, not the least of which was their substantially more efficient use of light energy for the production of cellular chemical energy. Moreover, during daylight, they would have been enshrouded by a continuously renewed "protective envelope" of gaseous and dissolved oxygen, toxic to contemporary anaerobes. For this reason, and perhaps also because blue-green algae and photosynthetic bacteria absorb light in complementary regions of the visible and near infrared spectra (although this complementarity could be a secondary adaptation of the bacteria, necessitated by the appearance of oxygen-producing cyanophytes), competition for photosynthetic space would have been minimal. Probably, the earliest blue-green algae did not respire aerobically. Photosynthetic oxygen production has been shown to have an inhibitory effect on the pathways necessary for aerobic respiration in modern cyanophytes (Carr and Hallaway, 1965; Holm-Hansen, 1968), and at night, when respiration might be expected to have occurred, local oxygen concentrations would have rapidly decreased to a minimal level as oxygen evaded into the essentially anoxic environment and was consumed in the oxidation of various inorganic substrates (e.g., ferrous iron and sulfides). Thus, at night, the primitive cyanophytes may have reverted to anaerobic heterotrophy, a capability exhibited by certain modern blue-green algae (Holm-Hansen, 1968: Fay, 1965), or they may have simply generated chemical energy from breakdown of photosynthetically manufactured cellular components.

In the absence of competitors, cyanophytes would be expected to have become widespread rather rapidly. As is discussed below, however, the earliest forms appear to have been planktonic unicells, and since an ultraviolet-absorbing atmospheric ozone layer had yet to become established, their dispersal may well have been hindered by the high ultraviolet flux that would have penetrated to the earth's surface. On the other hand, this factor may not have been of overriding importance; modern cyanophytes are highly resistant to ultraviolet light (Gerloff *et al.*, 1950) and, like bacteria, have unusually effective genetic repair mechanisms (Witkin, 1966).

It is perhaps more probable that the availability of critical metabolic requirements, such as usable nitrogen (the capacity for nitrogen fixation appears to have been a later "invention"), may have been more directly limiting. Within a moderate period of geological time, as increasing quantities of photosynthetically produced oxygen were added to the environment, the inorganic "oxygen sinks" approached saturation and the concentration of atmospheric oxygen began to markedly increase. With oxygen now available for use as an electron acceptor, cyanophytes developed the process of aerobic respiration, a highly efficient means of energy production involving the complete oxidation of photosynthetically produced carbohydrates. With the appearance of this process, the Cyanophyta had become essentially "modern" in physiology; subsequent evolutionary innovations were for the most part morphological rather than biochemical in nature.

Paleobiological Evidence

The postulated sequence outlined above is based in large measure on the physiological characteristics of extant microorganisms. Although it may therefore seem plausible, in view of such neobiological considerations, its paleobiological relevance remains to be established, for it can be assumed neither that the biosynthetic mechanisms of modern microbes faithfully reflect, in detail, those of their Precambrian precursors nor that all evolutionarily important prokaryotic groups, especially primitive anaerobic varieties, have survived to the present. On the contrary, microbial stocks have no doubt waxed and waned during the geological past; the modern microbiota represents a selected, derivative, and very limited sample of earlier assemblages. In principle, therefore, critical evidence relating to the origin of the Cyanophyta can be provided only by the geological record.

What may be inferred from the fossil record regarding the timing and nature of these early evolutionary events? As outlined above, two key developments appear to have occurred during early biological diversification: *viz.*, the development of autotrophy (i.e., anaerobic photosynthesis) and the appearance of oxygen-producing photoautotrophs (i.e., blue-green algae). Organic geochemical investigations of early sediments have yielded data relating to the first of these events, whereas the second, the origin of the Cyanophyta, is primarily reflected in the morphological fossil record.

Chlorophylls (including bacteriochlorophylls) are composed of two subunits, a magnesium-chelated tetrapyrrole nucleus and an isoprenoid, phytyl alcohol side-chain. Geochemical degradation of chlorophyll results in cleavage of the molecule, yielding metalloporphyin complexes and isoprenoid hydrocarbons (Bendoraitis *et al.*, 1962; McCarthy and Calvin, 1967; Ikan and Bortinger, 1971). Thus the occurrence in ancient sediments of these geochemically stable derivatives of chlorophyll can be regarded as evidence consistent with and suggestive of the existence of chlorophyllous organisms. As is discussed above, however, such organisms would include anaerobic photoheterotrophs; for this and other reasons (Schopf, 1970*a*), the reported occurrence of porphyrins (Kvenvolden *et al.*, 1968; Kvenvolden and Hodgson, 1969) and of isoprenoid hydrocarbons (Hoering, 1967; Oró and Nooner, 1967; Han and Calvin, 1969; Meinschein, 1967; MacLeod, 1968; Calvin, 1969) in Early Precambrian sediments may not necessarily reflect the existence of autotrophy.

The most obvious innovation accompanying the appearance of autotrophic organization was the development of biochemical pathways necessary for fixation of CO_2. In modern photoautotrophs, these pathways result in a fractionation of the stable isotopes of carbon, C^{12} and C^{13}, such that the photosynthetically produced organic matter is substantially enriched in the lighter isotope, C^{12}, as compared to the inorganic-source carbon (Park and Epstein, 1960, 1961). Oehler *et al.* (1972) have shown that the carbon isotopic composition of Precambrian organic matter younger than about 3.2 billion years is essentially identical to Phanerozoic carbon of known photosynthetic origin. Similarly, the isotopic composition of inorganic carbon reservoirs, as reflected in carbonate sediments, appears to have remained virtually constant since the Early Precambrian (Degens, 1969; Becker, 1971; Brooks, 1971, Schopf *et al.*, 1971). These data suggest that pathways resulting in autotrophic CO_2 fixation and concomitant isotopic fractionation were probably established as early as 3.2 billion years ago. In contrast, organic matter indigenous to certain of the oldest sedimentary units now known (the lowest chert horizons of the Onverwacht Group of South Africa, with an age of approximately 3.4×10^9 years) is distinctly anomalous in carbon isotopic composition (Oehler *et al.*, 1972); several lines of evidence (MacLeod, 1968; Calvin, 1969; Scott *et al.*, 1970) suggest (but by no means prove) that this isotopically unusual organic matter might be of heterotrophic or abiological origin. Thus it seems possible that photoautotrophic organization, presumably of the bacterial variety, may have first appeared about 3.3 billion years ago.

Morphological fossil evidence similarly suggests that photoautotrophy was an early evolutionary innovation. The oldest *bona fide* fossils now known occur in black cherts of the approximately 3.1 billion-year-old Fig Tree Group of South Africa, the geological unit immediately overlying the Onverwacht deposits. Numerous "bacterium-like" (Barghoorn and Schopf,

1966) and "alga-like" (Schopf and Barghoorn, 1967; Pflug, 1967; Pflug *et al.*, 1969) microfossils have been detected in the Fig Tree cherts. Somewhat similar carbonaceous bodies have been described from upper units of the Onverwacht Group (Engel *et al.*, 1968; Nagy and Nagy, 1969; Brooks and Muir, 1971); these bodies, however, exhibit a high degree of morphological variability, and their origin—whether biological or inorganic—remains to be established. As is discussed above, early microbial evolution primarily involved diversification at a biochemical level, changes that need not have been reflected in morphology. Moreover, many physiologically diverse types of modern microbes are morphologically quite similar. Thus the physiological characteristics of the Fig Tree microorganisms cannot be inferred on the basis of morphology alone; comparisons with extant forms, while seemingly suggestive of biological affinities (and, for the "alga-like" forms, of photoautotrophic metabolism), could give a faulty view of early metabolic capabilities. Fortunately, the Early Precambrian fossil record includes additional evidence that bears more directly on this problem.

In 1935, mound-shaped sedimentary structures comprised of alternating carbonaceous and calcareous laminae were discovered by A. M. Macgregor in Early Precambrian limestones of the Bulawayan Group of Rhodesia (Macgregor, 1940). The gross morphology and finely laminated organization of these structures led Macgregor to compare them with Paleozoic and modern bioherms (stromatolites) formed through the activities of microbiological communities. The great geological age of the Bulawayan limestones, reported to be in excess of 2.6 billion years (Holmes, 1954), was not fully appreciated until the mid-1950s, by which time much of the deposits had been destroyed by quarrying. Although stromatolites have been shown during the past two decades to be widely distributed in Middle Precambrian and younger sediments, they have been reported only rarely from units older than about 2.3 billion years. For this reason, and because the Bulawayan material was thought to be unavailable for further detailed study, an element of skepticism has existed as to both the reported age and the presumed biogenicity of these structures. To resolve the question of biogenicity, additional material from the original Rhodesian locality has recently been investigated (Fig. 6). These studies (Schopf *et al.*, 1971), based on both morphological and carbon isotopic analyses, fully confirmed the biological origin initially proposed by Macgregor. Furthermore, new stratigraphic and radiometric evidence appears to indicate that the Bulawayan Group has an age in excess of 2.8 or possibly 3.0 billion years (Vail and Dodson, 1969). Thus the Bulawayan limestones, containing the oldest stromatolitic structures now known, are apparently only slightly younger than the fossiliferous Fig Tree cherts.

These Early Precambrian stromatolites seem to provide important

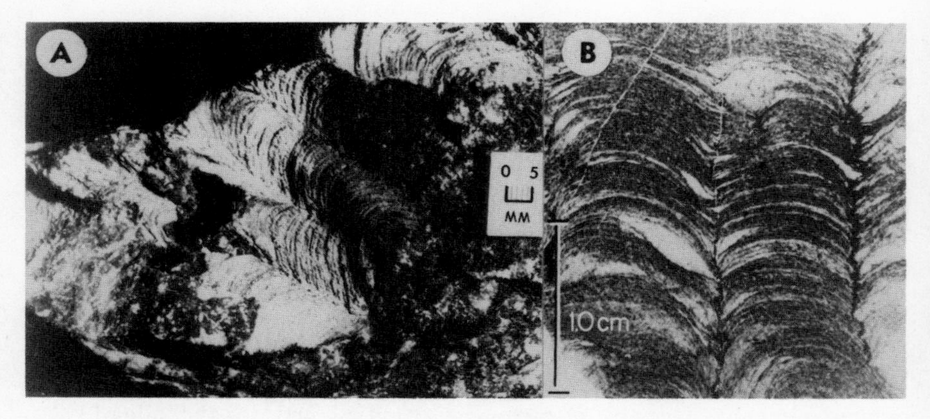

FIG. 6. Stromatolites from the Early Precambrian Bulawayan Group of Rhodesia. A: Weathered surface showing columnar structures (Schopf *et al.*, 1971). B: Optical photomicrograph of petrographic thin section showing the laterally contiguous organic mats that comprise the stromatolitic columns (Schopf *et al.*, 1971).

insight into the physiological characteristics of the primitive biota. Present-day stromatolites are formed in shallow-water, near-shore environments (within the photic zone) through the activities of microbiological communities dominated by blue-green algae; their distinctive laminated organization results from deposition of alternating layers of mineralic material (principally carbonates) and mucilagenous algal mats. The successive stacking of the organic mats is a consequence of the phototactic habit of the algal components of the community; as detritus and precipitated carbonate accumulate on an algal mat, the microorganisms grow or move (by gliding motility) upward, toward available light, and form a new mucilagenous sheet which binds the underlying sediment. It seems evident that fossil stromatolites are the result of a similar light-induced response; they therefore constitute firm evidence of the existence of phototrophic life. Further, laminated stromatolitic communities appear to be microbiologically distinctive; in all cases studied, both modern (Monty, 1967; Sharp, 1969; Gebelein, 1969) and fossil (Schopf, 1968; Schopf and Blacic, 1971; Cloud *et al.*, 1969; Gutstadt and Schopf, 1969; Licari, 1961; Barghoorn and Tyler, 1965; Cloud, 1965), they are composed predominantly of mat-building, filamentous (commonly oscillatoriacean) cyanophytes. Although it is conceivable, of course, that during the essentially anoxic Early Precambrian, structures of this type *might* have been produced by anaerobic photobacteria—a possibility that deserves serious consideration in view of the recent discovery of bacteriochlorophyllous mat-building "flexibacteria" (Pierson and Castenholz, 1971) that seem morphologically somewhat similar to the Middle Precambrian microorganism *Gunflintia minuta* (Barghoorn and Tyler, 1965)—there is no compelling evidence at present to support this contention, and the Bulawayan structures do not differ significantly from younger stromatolites of unquestionable algal origin. Thus, at present, the Bulawayan stromatolites appear to constitute strong presumptive evidence for the existence of oxygen-producing, photoautotrophic, blue-green algae. And, since the Bulawayan and Fig Tree deposits appear to be of similar age, it is tempting to suggest (and is consistent with the geochemical data) that the "alga-like" prokaryotes preserved in the Fig Tree cherts could be of cyanophytic rather than bacterial affinities.

In summary, the reported occurrence in Early Precambrian sediments of metalloporphyrin complexes and isoprenoid hydrocarbons, presumed to be geochemical derivatives of chlorophyll, is consistent with and seems suggestive of the early evolution of chlorophyllous microorganisms. Although the earliest phototrophs may have been photoassimilators rather than photosynthesizers, carbon isotopic data suggest that the CO_2-fixing pathways of photoautotrophy had become established as early as 3.2 billion years ago. By the time of deposition of the Bulawayan limestones, possibly prior to 3 billion years ago, anaerobic photoautotrophs had apparently given rise to prokaryotic, oxygen-producing photosynthesizers—the several stages in biochemical evolution leading to the origin of the Cyanophyta had been completed, the environment was becoming increasingly oxygenic near sites of photosynthetic activity, and the way was paved for diversification and "radial evolution" of the group.

PHYLOGENY OF THE CYANOPHYTA

Coccoid Line ("Coccogoneae")

Differences of opinion have long existed among phycologists as to the probable phylogenetic relationships among extant blue-green algal families (Desikachary and Padmaja, 1970). Nevertheless, most authorities are in agreement that (1) the most primitive members of the group are simple, unicellular chroococcaceans, and thus that that (2) the filamentous habit is a derived condition (although there is rather vigorous debate as to the identity of the ancestral filamentous stock). These views, of course, are based almost entirely on morphological data, the underlying assumption being that, in general, evolution can be expected to proceed from relatively simple forms to forms relatively more complex and structurally differentiated. Recent biochemical studies of living blue-green algae have provided evidence in support of this contention. For example, the fatty acid composition of chroococcaceans tends to be substantially less complex than

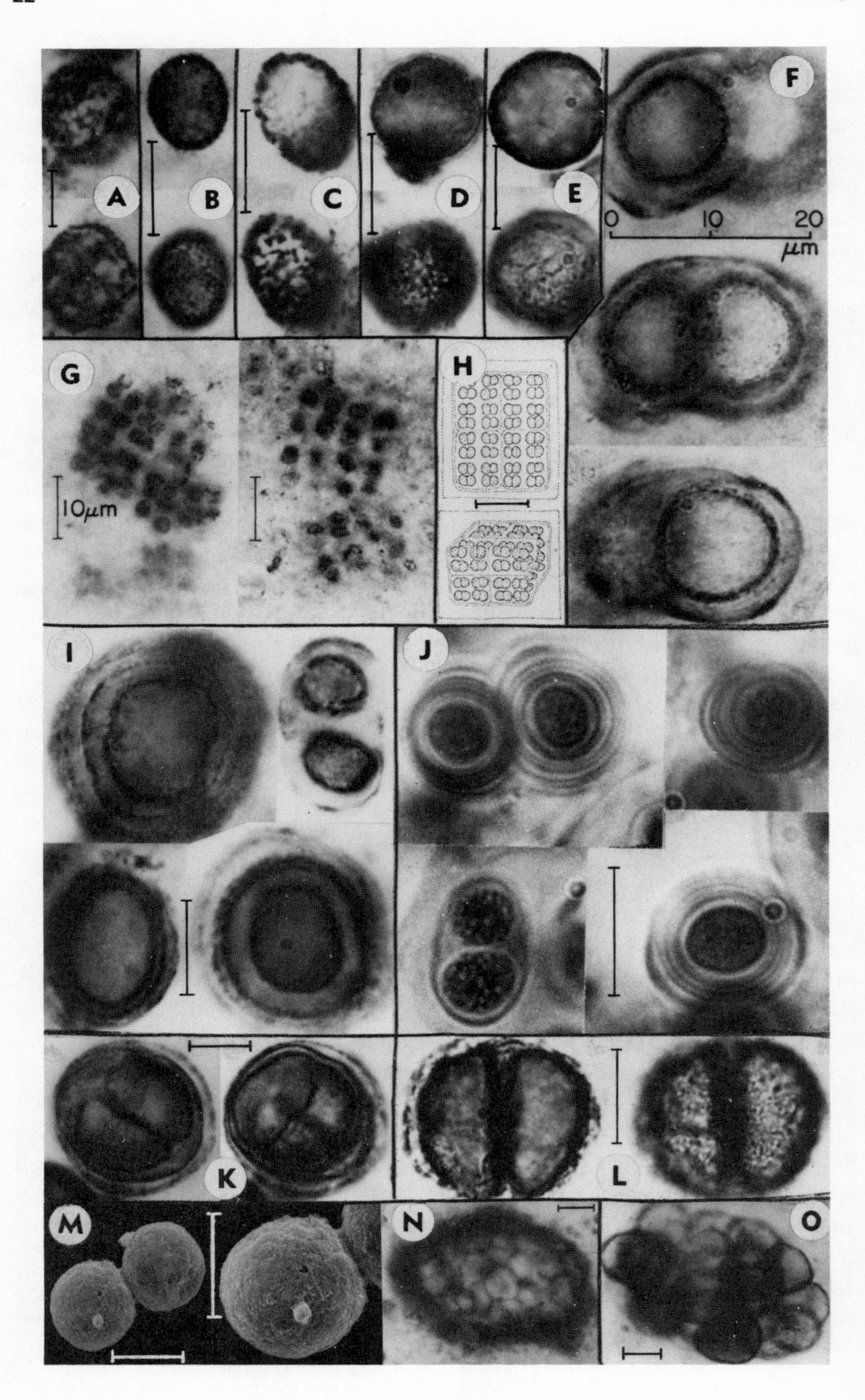
A
B
C
D
E
F
0
10
20
μm
G
10μm
H
I
J
K
L
M
N
O

that of other cyanophytes (Holton *et al.*, 1968; Holton, 1969; Holton and Blecker, 1970); indeed, certain of these coccoid unicells have very simple fatty acid patterns, the most simple known among all oxygen-producing photosynthesizers (Holton and Blecker, 1970). In contrast, the capability to fix atmospheric nitrogen under aerobic conditions, a biochemically complex process, is restricted among cyanophytes to morphologically differentiated, heterocystous varieties (Stewart *et al.*, 1969). As Holton and Blecker have indicated, it appears likely that "the development of biosynthetic abilities [in the Cyanophyta] is directly correlated with the evolution of morphological complexities" (Holton and Blecker, 1970, p. 126).

The foregoing considerations suggest that the earliest blue-green alga would have been a coccoid, sheathless, noncolonial chroococcacean of a rather generalized type, perhaps roughly similar in organization to certain members of the modern genus *Synechocystis*. It is interesting, therefore, and seems more than coincidental, that the oldest "alga-like" microfossils now known (Fig. 7A; *Archaeosphaeroides barbertonensis* from the Fig Tree cherts of South Africa) fit this general description. Moreover, although the physiological characteristics of these approximately 3.1-billion-year-old microorganisms are unknown, their morphology seems markedly similar to that of planktonic algae preserved in younger Precambrian and Phanerozoic deposits (compare Fig. 7A with Fig. 7B–E). Further, spheroidal, cyanophyte-like microfossils have been reported (Laberge, 1967; Cloud and Licari, 1968) from several other sediments of Early Precambrian age (older than 2.5×10^9 years); these forms could provide a connecting "link," approximately intermediate in age, between the Fig Tree organisms and the relatively rich chroococcacean assemblage of the Middle Precambrian. Thus it seems reasonable to regard the Fig Tree spheroids (and comparable "alga-like" microfossils of the Early Precambrian) as quite probably being planktonic chroococcaceans, representing a very early stage in an evolutionary continuum that extends from the Early Precambrian to the present.

FIG. 7. Coccoid cyanophytes ("Coccogoneae") and "alga-like" microfossils. With the exception of M (which shows scanning electron micrographs of cells in a hydrofluoric acid–resistant residue), all microfossils are shown in optical photomicrographs of petrographic thin sections; lines for scale represent 10 μm. A–E: Unicellular microfossils from cherts of the Early Precambrian (A, Fig Tree Group, Schopf and Barghoorn, 1967), Middle Precambrian (B, Gunflint Iron Formation, Barghoorn and Tyler, 1965), Late Precambrian (C, Skillogalee dolomite, Schopf and Barghoorn, 1969; D, Bitter Springs Formation, Schopf, 1968), and Eocene (E, Clarno chert, Oregon). F,K–O: Chroococcacean cell pairs (F,L,M), decussate quartet (K), and globular colonies (N,O) from the Bitter Springs cherts (Schopf, 1968; Schopf and Blacic, 1971). G,H: Late Precambrian *Eucapsis*-like colonies (G, Paradise Creek Formation, Licari *et al.*, 1969) for comparison with modern *Eucapsis* (H). I,J: Late Precambrian *Gloeocapsa*-like microfossils (I, Bitter Springs cherts, Schopf and Blacic, 1971) for comparison with modern *Gloeocapsa* (J).

A diverse assemblage of coccoid microfossils—more than a score of forms, almost all of which appear to be of chroococcacean affinities—has been reported (Barghoorn and Tyler, 1965; Cloud, 1965; LaBerge, 1967; Cloud and Licari, 1968; Hofmann and Jackson, 1969; Moore, 1918; Schidlowski, 1966; Bondensen *et al.*, 1967; Vologdin and Kordé, 1965) from sediments of the Middle Precambrian (between 2.5 and 1.7 $\times$ 10^9 years old). The most fossiliferous and best known of these units is the cherty stromatolitic horizon of the approximately 1.9-billion-year-old Gunflint Iron Formation of Ontario, Canada (Barghoorn and Tyler, 1965; Cloud, 1965). Interspersed among the filamentous cyanophytes that dominate the Gunflint microbiota are a variety of chroococcaceans including sheathless unicells (Fig. 7B), sheath-enclosed spheroids, and irregular colony-like aggregations. Thus by Gunflint time the Chroococcaceae had become moderately diversified, and forms apparently leading toward the most advanced level of organization exhibited by the family, that of an ordered, three-dimensional colony, were already in evidence. The majority of these cyanophytes, particularly the unicellular varieties, were partially or predominantly planktonic in habit and were therefore widely dispersed within "local" basins (e.g., throughout much of the 2500 km^2 of the Gunflint basin). Wind currents, hurricanes, and similar agents would have provided effective means for long-distance dispersal, such as that required between disconnected basins; Middle Precambrian chroococcaceans were no doubt cosmopolitan in distribution and presumably occupied terrestrial as well as aquatic habitats (terrestrial forms are abundant among modern Chroococcales).

Chroococcaceans exhibiting highly ordered colonies are first known from the Late Precambrian (extending from 1.7 to about 0.6 $\times$ 10^9 years ago). The oldest examples of such organization, produced by sequential cell division in three mutually perpendicular planes, are the *Eucapsis*-like colonies shown in Fig. 7G, detected in cherts about 1.6 billion years old from northeastern Australia (Licardi *et al.*, 1969); for comparison, cuboidal colonies of modern *Eucapsis* are shown in Fig. 7H.

Among all known Late Precambrian cyanophytic assemblages, by far the best preserved and most diverse is that contained in laminated black cherts of the Bitter Springs Formation of central Australia, having an age of approximately 1 billion years (Wells *et al.*, 1970). Of the 50 species of microorganisms recognized in the Bitter Springs cherts, 11 have been referred to the Chroococcaceae (Schopf and Blacic, 1971). By Bitter Springs time, evolutionary diversification of the family was essentially completed; sheathless spheroids (Fig. 7M) and sheath-enclosed unicells (Fig. 7I), cell-pairs (Fig. 7F,I,L), decussate quartets (Fig. 7K), and colonies (Fig. 7N,O) are prominent members of this Late Precambrian community. Certain of these fossil chroococcaceans bear marked resemblance to well-known living cyanophytes (compare to the Bitter Springs unicells shown in Fig. 7I with modern *Gloeocapsa* in Fig. 7J), a resemblance extending from characteristics of gross morphology (e.g., cell size and shape) to such detailed features as the degree of lamellation of encompassing sheaths (Fig. 7I,K) and the ordered arrangement of daughter cells produced by "binary fission" (Fig. 7G,K). Many Phanerozoic chroococcaceans exhibit a similar degree of comparability with modern taxa (Baschnagel, 1942, 1966; Zalessky, 1917; Rao, 1957; Mehta, 1954; Bradley, 1931; Kordé, 1958). Thus it is apparent that by the Late Precambrian the Chroococcaceae was essentially "modern" in character and that subsequent evolution within the family has been typified by extreme morphological conservatism, a rather remarkable feature that appears to be characteristic of the Cyanophyta in general (Schopf, 1968; Schopf and Blacic, 1971).

Although the fossil record of nonchroococcacean "coccogoneans" is very poorly known, available data suggest that most and perhaps all of the principal coccoid groups originated early in evolutionary history. For example, although members of the Entophysalidaceae (a chroococcalean family closely related to and apparently derived from the Chroococcaceae) are known from only a few deposits, the fossil record of the family extends well into the Precambrian (Vologdin and Kordé, 1965; Laristschev, 1952). Similarly, although few fossils have been referred to the Pleurocapsales (Johnson, 1937), the recent discovery of well-preserved *Myxosarcina*-like (Pleurocapsaceae) colonies in cherty stromatolites about 1.1 billion years old from south Australia* establishes that this order also dates from the Precambrian. In fact, only one relatively minor "coccogonean" group, the Chamaesiphonales (an order represented in the modern flora by but two principal genera), is as yet apparently unreported from geological strata; the relationship of this group to other cyanophytic orders is uncertain (Desikachary, 1959; Desikachary and Padmaja, 1970).

* These colonies are components of a diverse, well-preserved microbiological community, composed of more than a dozen distinct types of microorganisms and dominated by filamentous blue-green algae, which I discovered in 1971 in black cherts occurring in as yet unnamed dolomitic formations in the Peak and Denison ranges about 30 miles northwest of William Creek ("Boorthanna North" section; approximate coordinates 135°53′E, 28°40′S). Mr. Thomas R. Fairchild and I are currently preparing a manuscript which will describe in detail this new assemblage. The samples were collected and kindly forwarded for investigation by Dr. H. Wopfner (South Australia Department of Mines), who regards these fossil-bearing units as being stratigraphic equivalents of the Skillogalee dolomite of south Australia, shown recently to contain filamentous and spheroidal cyanophytes (Schopf and Barghoorn (1969).

Filamentous Line ("Hormogoneae")

The mode of origin of the filamentous condition in blue-green algae and the probable identity of the ancestral filamentous stock have long been matters of controversy and speculation among "cyanophycologists." As summarized in an article by Desikachary and Padmaja (1970), authorities can be characterized as advocating either a "progressive" or a "retrogressive" concept of cyanophytic phylogeny. Proponents of the former concept view evolutionary diversification of "hormogoneans" as having resulted from a progressive increase in morphological and organizational complexity. Thus the Oscillatoriaceae (forms having structurally undifferentiated trichomes) are regarded as primitive; the heterocystous Nostocaceae, Rivulariaceae, and Scytonemataceae are considered moderately advanced (and derived, directly or via intermediates, from oscillatoriacean progenitors); and stigonemataleans (relatively elaborate, commonly multiseriate, heterocystous forms which exhibit true branching and heterotrichous organization) are thought to represent the most highly evolved condition. In contrast, advocates of the "retrogressive" concept hold that stigonemataleans comprised the ancestral plexus, that the remaining filamentous groups were therefore derived by *reduction*, and that the Oscillatoriaceae are thus the most recently evolved and *least* primitive "hormogonean" family.

Some proponents of the "retrogressive" concept, noting that certain "primitive" stigonemataleans (e.g., *Mastigocladus laminosus*) are abundant in hot spring environments, have sought to bolster their position by associating it with the Relict Hypothesis, a postulate asserting that cyanophytes originated in Precambrian thermal areas and that modern thermophiles represent "relicts" that have survived to the present by virtue of their adaptation to this relatively inhospitable environment. The Relict Hypothesis, however, is essentially a "red herring," having little, if any, bearing on the central issue of the debate (e.g., *Phormidium laminosum* and other oscillatoriaceans, regarded by "retrogressivists" as being highly evolved, are abundant in modern hot springs). In fact, although this hypothesis may have seemed attractive in earlier years when relatively little was known of either the duration or the environment of the Precambrian (and when the landscape of this "primordial" era was popularly romanticized as being replete with hot bubbling pools and towering, smoldering volcanoes), support of it now appears unwarranted and, indeed, has been contraindicated by geological and paleobiological data. All of the major Precambrian cyanophytic assemblages now known (e.g., Schopf, 1968; Schopf and Blacic, 1971; Cloud *et al.*, 1969; Licari, 1971; Barghoorn and Tyler, 1965) are distributed over very extensive areas (thus markedly differing from a local hot spring setting) and appear to have been preserved near the sediment–water interface in "warm" (temperate or tropical), shallow, shelflike, marine environments.

The central issue separating the "progressive" and "retrogressive" concepts of cyanophytic phylogeny is the position of the Stigonematales relative to the remaining "hormogonean" families. Although paleobiological data can be expected to shed little light on the precise *mechanisms* of evolution in the Cyanophyceae (i.e., the complex series of genetic, phenotypic, and environmental changes resulting in diversification), the fossil record can provide crucial evidence regarding the relative timing of various evolutionary events—in this case, the probable times of origin of the major filamentous families. These data suggest that the earliest filamentous cyanophytes were probably oscillatoriaceans, that moderately complex heterocystous forms (e.g., nostocaceans and rivulariaceans) were also of early appearance, and that the Stigonematales appeared relatively late in the evolutionary sequence; the evidence strongly supports the "progressive" phylogenetic concept.

With regard to the mode of origin of the filamentous condition, the principal step appears to have been the transition from chroococcoid to true trichomatous organization. As Desikachary (1959) has indicated, extant chrooroccaceans such as *Johannesbaptistia* suggest the general organization of such transitional forms; the pseudotrichomatous condition of "coccogoneans" of this type (having "filaments" composed of a uniseriate row of discoidal cells encompassed by a cylindrical sheathlike matrix) results from the occurrence of successive cell divisions perpendicular to the axis of the "filament." Although *Johannesbaptistia* exhibits the most highly developed tendency toward filamentous organization known among extant chroococcaleans, a similar tendency occurs in other genera; indeed, recent studies have shown that short filament-like forms occur during the ontogeny of many coccoid blue-green algae (Desikachary and Padmaja, 1970). Thus the earliest "Hormogoneae" were probably nonheterocystous oscillatoriaceans derived from pseudotrichomatous progenitors in which the discoidal daughter cells remained together, rather than separating, following cell division.

The known fossil record contains limited evidence of the earliest stages in "hormogonean" diversification. By analogy with modern mat-building blue-green algal communities, the occurrence of stromatolites in Early Precambrian strata, such as the limestones of the Bulawayan Group, seems indicative of the existence of filamentous cyanophytes and, in particular, of the Oscillatoriaceae since members of this family are dominant in modern stromatolitic assemblages. However, like most calcareous stromatolites, the Bulawayan structures are devoid of cellular microfossils (Schopf *et al.*,

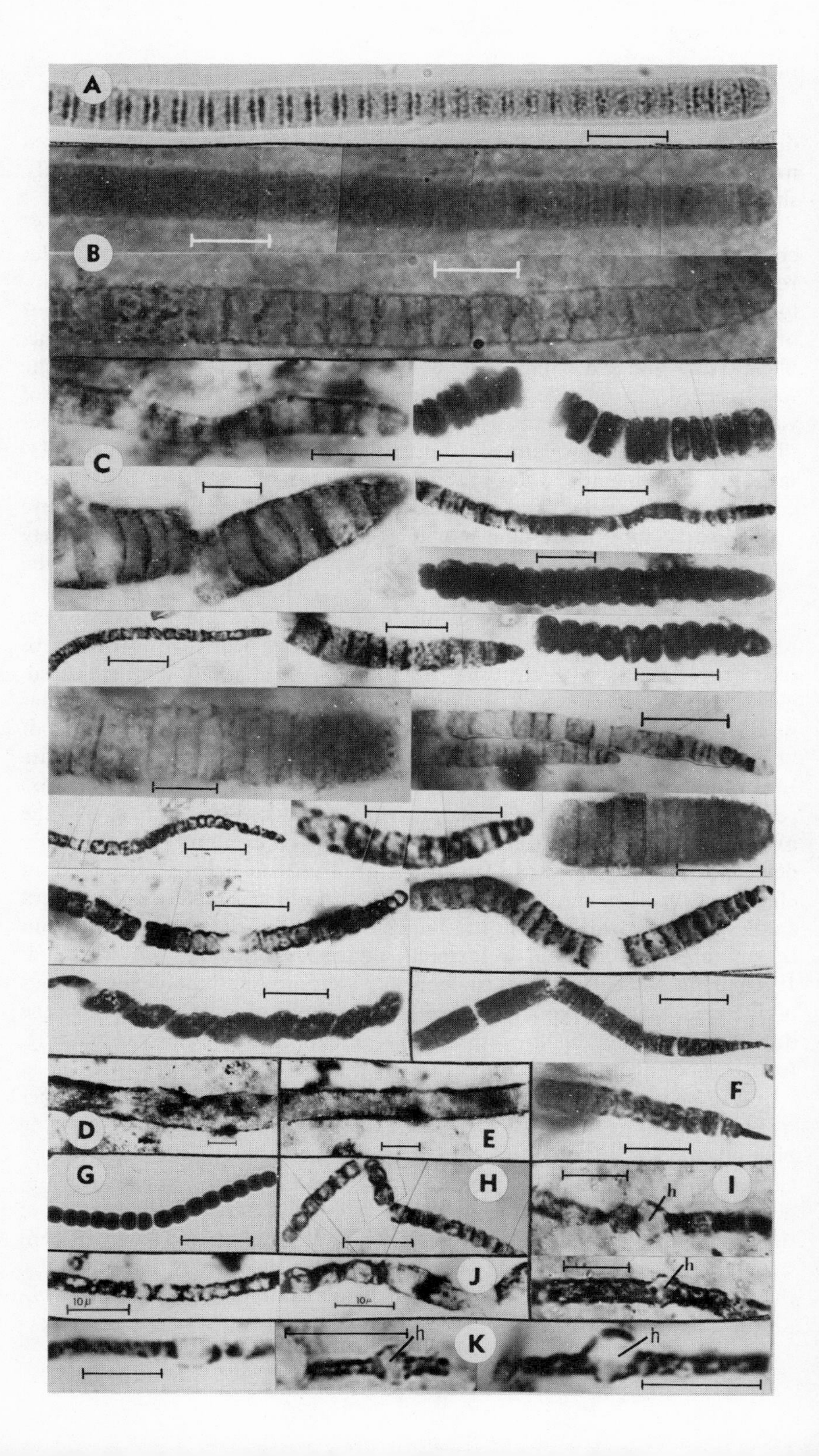
A
B
C
D
E
F
G
H
I
h
J
10μ
10μ
h
K
h
h

1971). The known fossil record of well-preserved "hormogoneans" dates back "only" into the Middle Precambrian, by which time several filamentous families had evolved.

As is true of coccoid cyanophytes, the most impressive Middle Precambrian evidence of the "Hormogoneae" has come from studies of the Gunflint cherts (Barghoorn and Tyler, 1965; Cloud, 1965); and, like the "Coccogoneae," the filamentous "tribe" was already relatively diversified by the time of Gunflint sedimentation (approximately 1.9×10^9 years ago). Represented in the Gunflint assemblage are both oscillatoriaceans (e.g., Fig. 8E, the sheath of an *Oscillatoria*-like cyanophyte) and nostocaceans (e.g., Fig. 8J), some containing enlarged cells (Fig. 8K) that appear to represent heterocysts and akinetes (Licari and Cloud, 1968). Although evidence for the occurrence of rivulariaceans in the Gunflint biota is marginal, the family may have been extant during the Middle Precambrian (Butin, 1959).

The apparently early appearance of heterocyst-bearing nostocaceans is of considerable interest since heterocysts are the principal site of nitrogen fixation in modern cyanophytes. Unlike other reactions of the nitrogen cycle (e.g., denitrification, ammonification, and nitrification), which are energy-yielding, biological fixation of atmospheric nitrogen is an expensive energy-requiring process; it is therefore resorted to by living blue-green algae only if ammonia and nitrate are essentially unavailable in the immediate environment. For such an "expensive" process to have become established, it originally must have been of substantial selective advantage, presumably during a time when or in localities where usable nitrogen was in short supply. Significantly, among extant organisms, the capability to fix atmospheric nitrogen is restricted entirely to prokaryotic microbes. Thus it seems evident that nitrogen fixation is a "primitive" process extending well back into the geological past and that its development was probably interrelated with atmospheric evolution and the availability of combined nitrogen.

Several lines of evidence suggest that between about 2.5 and 1.8 billion years ago the earth's atmosphere evolved through a "transitional phase"

FIG. 8. Filamentous cyanophytes ("Hormogoneae"). All microfossils are shown in optical photomicrographs of petrographic thin sections; lines for scale represent 10 μm. A–C: Modern *Oscillatoria* (A) for comparison with Devonian *Oscillatoria*-like filaments (B, Rhynie chert, *cf.* Kidston and Lang, 1921) and with Late Precambrian oscillatoriaceans (C, Schopf, 1968; Schopf and Blacic, 1971). D–E: Oscillatoriacean sheaths from the Late Precambrian (D, Bitter Springs cherts, Schopf, 1968) and Middle Precambrian (E, Gunflint cherts, Barghoorn and Tyler, 1965). F: Rivulariaceans from the Late Precambrian Bitter Springs cherts (Schopf, 1968; Schopf and Blacic, 1971). G,H: Late Precambrian *Nostoc*-like filament (H, Bitter Springs cherts, Schopf and Blacic, 1971) for comparison with modern *Nostoc* (G). I–K: Nostocacean filaments from the Late Precambrian (I, Skillogalee dolomite, Schopf and Barghoorn, 1969) and Middle Precambrian (J,K, Gunflint cherts, Barghoorn and Tyler, 1965; Licari and Cloud, 1968), some with enlarged heterocyst-like cells ("h").

from a relatively anoxic to a relatively oxygenic composition (Rutten, 1962; Cloud, 1968). Prior to and during the early stages of this transition, sources of fixed nitrogen were probably quite limited; in the absence of an oxygenic atmosphere and a resulting ultraviolet-absorbing ozone layer, concentrations of ammonia would have been maintained generally at very low levels (Abelson, 1966), and direct oxidation of N_2 by lightning during thunderstorms and photooxidation of NH_3 at the air–water interface, important sources of combined nitrogen in the modern environment (Hutchinson, 1954), would have produced only minimal amounts of nitrate. In modern cyanophytes, the nitrogenase enzyme complex is known to be oxygen sensitive; nitrogen fixation is thus an anaerobic or microaerophilic process (Stewart and Pearson, 1970). Prior to the onset of aerobic conditions, it is therefore possible that various types of blue-green algae fixed atmospheric N_2 (e.g., at night). However, as the environment became increasingly oxygenic, especially near sites of photosynthetic activity, the nitrogen-fixing apparatus would have been "poisoned" by free oxygen, and cyanophytes may have experienced a "nitrogen crisis" (somewhat analogous to the "carbon crisis" thought to have preceded the origin of autotrophy). Possibly, blue-green algae would have become restricted to rather specialized niches, tied by metabolic requirements to local areas in which concentrations of nitrogenous compounds could have been sustained at moderate levels by bacterial activity or in which oxygen concentrations would have been depleted by reaction with inorganic substrates (*cf.* Cloud, 1968). In such a "crisis," cyanophytes possessing heterocysts, which serve to separate physically the oxygen-sensitive pathways of nitrogen fixation from the oxygen-evolving centers of photosynthesis (Thomas, 1970) and in which oxygen concentrations are kept at low levels by active aerobic respiration, would have had considerable selective advantage. In fact, the apparent occurrence of heterocysts in Middle Precambrian cyanophytes provides strong biological evidence, in some ways perhaps even more compelling than available geological data (Rutten, 1962; Cloud, 1968), for the existence of a relatively oxygenic environment and aerobiosis.

The origin of heterocysts, and thus of the Nostocaceae, would have enabled those biological communities containing such nitrogen fixers to spread rapidly into previously sterile environments, a penchant for colonization that has survived to the present, as evidenced in areas newly created by volcanic activity (Raven and Curtis, 1970, p. 377) or denuded by atomic blasts (Shields and Drouet, 1962). It thus seems significant that nostocaceans appear to have formed the dominant component of Middle Precambrian stromatolitic communities (Barghoorn and Tyler, 1965; Licari and Cloud, 1968) and that stromatolites appear to be prevalent only in sediments having an age less than about 2.5 or 2.3 billion years. Possibly, therefore, the first widespread appearance of stromatolites in the geological record, near the beginning of the Middle Precambrian, may mark the time of origin of heterocystous cyanophytes. As the atmosphere continued to become increasingly oxygenic (spurred now by the increased photosynthetic biomass and a concomitant increase in amount and rate of burial of reduced carbon), microbes completing the aerobic portion of the present-day nitrogen cycle would have become established. With this development, the "nitrogen-crisis" would have been surmounted and nostocaceans would have become relatively less important members of the biota, a decline in prominence that appears to be reflected in the subsequently deposited fossil record.

By the Late Precambrian, the Oscillatoriaceae had largely supplanted the Nostocaceae as the major "hormogonean" family; the billion-year-old Bitter Springs assemblage (Schopf and Blacic, 1971), for example, includes 22 species of oscillatoriaceans (Fig. 8C,D) as compared with only three species of nostocaceans (Fig. 8H) and two species of rivulariaceans (Fig. 8F). As is true of coccoid cyanophytes, filamentous forms known from the Middle Precambrian have morphological counterparts in the Late Precambrian biota (e.g., compare the Middle Precambrian heterocystous filaments shown in Fig. 8K with those shown in Fig. 8I, from the Late Precambrian; or compare *Animikiea* from the Gunflint cherts in Fig. 8E with the Bitter Springs oscillatoriacean sheath shown in Fig. 8D). And, in further similarity to the "Coccogoneae," many Late Precambrian "hormogoneans" bear marked resemblance to Phanerozoic and modern blue-green algae (e.g., compare modern *Nostoc*, Fig. 8G, with the Bitter Springs equivalent shown in Fig. 8H; or compare modern *Oscillatoria*, Fig. 8A, with the *Oscillatoria*-like Devonian filaments in Fig. 8B and with the Bitter Springs oscillatoriaceans shown in Fig. 8C).

The known fossil records of the remaining two major filamentous families, the Scytonemataceae and the Stigonemataceae, are relatively restricted. Although several members of each family have been reported from sediments of Devonian and younger age (Croft and George, 1959; Goswami, 1955; Bradley, 1970), neither is known from the Precambrian. Nevertheless, the Scytonemataceae seem rather closely related to the Oscillatoriaceae—probably representing derivatives of *Lyngbya*-like forms that acquired heterocysts and false branching (Fritsch, 1965, p. 842)—and it would not at all be surprising if scytonemataceans were to be detected in deposits of Late or perhaps Middle Precambrian age; certainly, the available data do not preclude a relatively early origin for the group. In contrast, the Stigonemataceae, although also possibly dating from the Precambrian, appear to represent a rather marked divergence from other types of "hormogonean" organization; the origin of the group, presumably from heterocystous progenitors and apparently a relatively recent event in cyanophytic evolution, remains obscure (Desikachary and Padmaja, 1970).

Thus, in general, the evolutionary history of the "Hormogoneae" parallels that of the "Coccogoneae" (Fig. 9); both "tribes" originated during the Early Precambrian, had become moderately diversified by the Middle Precambrian, and were highly diversified and apparently quite "modern" in character by the Late Precambrian, and both include members that subsequently have exhibited a marked degree of morphological evolutionary conservatism.

Evolutionary Conservatism in the Cyanophyta

The detailed morphological resemblance between blue-green algae of the modern biota and taxa of the Late Precambrian and, indeed, the simi-

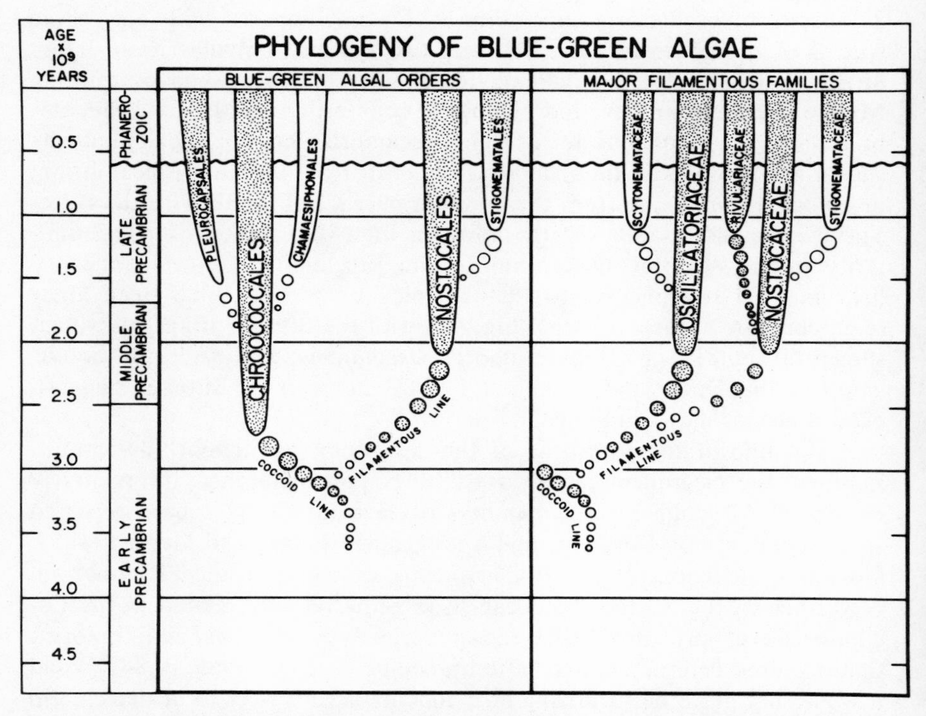

FIG. 9. Inferred phylogenetic relationships. "Phylogenetic trees" showing inferred geological distributions, times of phyletic divergences, and phylogenetic relationships of blue-green algal orders (left) and major "hormogonean" families (right). Stippled areas show distributions of reported fossil evidence; stippled circular symbols indicate a degree of uncertainty regarding the interpretation of available data; clear areas indicate relationships or distributions inferred in the absence of known fossil evidence.

larities generally between living cyanophytic species and fossils of diverse ages have become increasingly appreciated in recent years. Although many such examples are now known, the most compelling evidence has come from studies of the billion-year-old Bitter Springs microbiota (Schopf, 1968; Schopf and Blacic, 1971), which includes "hormogoneans" comparable to members of at least ten extant genera (*Lyngbya, Oscillatoria, Phormidium, Porphyrosiphon, Schizothrix, Microcoleus, Spirulina, Nostoc, Anabaenopsis,* and *Homoeothrix*) and "coccogoneans" closely resembling members of at least three modern genera (*Anacystis, Chroococcus*, and *Gloeocapsa*). It is apparent, therefore, that as the Cyanophyta diversified, various growth forms became well established and have been subject to little subsequent modification. This evolutionary conservatism is striking, both because of the high degree of comparability between fossil and modern taxa and because of the extreme duration of the various lineages, several of which approach or exceed half the presently accepted age of the earth. Although the Cyanophyta are thus truly "archaic," and although the prominence of the group markedly declined with the appearance of sexual, eukaryotic organisms late in the Precambrian (Schopf and Blacic, 1971; Schopf, 1972), cyanophytes are by no means "antiquated." In fact, if biological "success" is measured by longevity of existence, blue-green algae constitute the most successful form of aerobic, photoautotrophic organization to have appeared during the long course of biological history. This success and the bradytelic evolution of the group are no doubt closely intertwined.

As has been discussed in considerable detail by Schopf and Blacic (1971), the morphological conservatism of the Cyanophyta, presumably paralleled by a comparable degree of biochemical conservatism, seems a result of the wide ecological tolerance, versatile physiology, and unusually stable genetic system characteristic of the group. The environmental limits for survival, growth, and reproduction of blue-green algae are notably broad in comparison with all other photosynthetic organisms; cyanophytes survive and often thrive at or near environmental extremes (e.g., of temperature, Eh, pH, total salinity, altitude, aridity, and exposure to ionizing and ultraviolet radiation) inimical to "usual" biological activity. In physiological characteristics, blue-green algae are among the most versatile of living systems (growing aerobically, as photoautotrophs; anaerobically, in the presence of hydrogen sulfide; heterotrophically under either aerobic or anaerobic conditions, assimilating exogenous organic compounds either as photoassimilators or without the aid of light energy; and, in many heterocystous varieties, as nitrogen-fixing aerobic photoautotrophs which require only light, water, N_2, CO_2, and a few trace elements), a versatility that frees cyanophytes from dependence on factors that limit the distribution of most other photosynthetic organisms. And finally, whether "polygenomic," as seems possible (Schopf, 1968), or essentially haploid, like many bacterial prokaryotes, cyanophytes have a genetic system that is markedly resistant to mutagens, and, perhaps even more importantly, they appear to be exclusively asexual (for possible exceptions, see Bazin, 1968; Shestakov and Khyen, 1970; Singh, 1967) and hence apparently lack the potential for genetic recombination, the major source of variability in populations of higher organisms. Cyanophytes are thus genetically stable, highly successful "generalists"—a suitable ecological niche, relatively free from competitors and thus from pressure for evolutionary change, has been available to these adaptive, primitive microorganisms since very early geological time.

SUMMARY

1. During the past two decades, and especially since 1965, impressive progress has been made toward documenting the geological distribution and diversity of fossil blue-green algae. Much of this progress has accrued from studies of fine-grained, Precambrian chert deposits in which cyanophytes are preserved as three-dimensional, structurally intact, organic residues. In such deposits and in similar chemical sediments of the Phanerozoic, evidence of ecological setting, growth habit, general morphology, detailed cellular anatomy, and mode of reproduction commonly can be discerned, features that provide a sound basis for comparison of extant and fossil taxa.

2. The known cyanophytic fossil record is, of course, far from complete. Nevertheless, the most active phase of blue-green algal evolution, that occurring during the Middle and Late Precambrian, is relatively well documented; overall, the quantity, quality, and geological distribution of available fossil evidence seem sufficient for tentative identification both of major events in blue-green algal evolution and of phylogenetic relationships among certain of the principal cyanophytic families.

3. Consideration of the biochemistry, physiology, and cellular organization of extant microorganisms suggests that the evolutionary sequence leading from the earliest forms of life to the origin of the Cyanophyta was principally characterized by modification and refinement of porphyrin-based metabolism and that the immediate precursors of the group were anaerobic bacterial photoautotrophs.

4. Paleobiological data, including the occurrence of both morphological evidence (*viz.*, cellular microfossils and stromatolitic sediments) and "chemical fossils" (e.g., geochemical derivatives of chlorophyll and C^{13}:C^{12} values) in very ancient sediments, seem to indicate that (presumably

bacterial) photoautotrophy had become established as early as 3.2 billion years ago and that oxygen-producing, prokaryotic blue-green algae probably were extant earlier than 3 billion years ago.

5. The earliest blue-green algae apparently were unicellular, noncolonial coccoid forms ("Coccogoneae") of the Chroococcales (Fig. 9). By Middle Precambrian time (2.5–1.7 $\times$ 10^9 years ago), "coccogoneans" had become moderately diversified; by the Late Precambrian (extending from 1.7 to about 0.6 $\times$ 10^9 years ago), the "tribe" was highly diverse and quite "modern" in character, represented by members of at least three extant families (Chroococcaceae, Entophysalidaceae, and Pleurocapsaceae) and a large variety of growth forms (unicells, sheath-enclosed pairs, decussate quartets, irregular colonies of a few to many cells, and highly ordered cuboidal and spheroidal colonies).

6. The occurrence of algal stromatolites in Early Precambrian sediments suggests that the filamentous "tribe" ("Hormogoneae"), derived from chroococcacean progenitors, was extant prior to 2.8 billion years ago and that the earliest "hormogoneans" were relatively simple forms of the oscillatoriacean type (Fig. 9). At least two, and probably three, "hormogonean" families (Oscillatoriaceae, Nostocaceae, and apparently, Rivulariaceae) evolved earlier than 2 billion years ago; the Middle Precambrian biota appears to have been dominated by heterocystous, presumably nitrogen-fixing, nostocaceans. By Late Precambrian time, the Nostocaceae had been supplanted by the Oscillatoriaceae as the dominant cyanophytic family and most, and possibly all, major "hormogonean" families had become established. The fossil evidence, therefore, strongly supports the "progressive" rather than the "retrogressive" concept of cyanophytic phylogeny.

7. The course of cyanophytic evolution has been influenced markedly by the evolution of the earth's atmosphere from an essentially anoxic to a relatively oxygenic state; this atmospheric evolution, in turn, has been strongly influenced by the evolution of the Cyanophyta. Thus the development of an oxygenic environment was almost entirely the result of cyanophytic photosynthesis, and, as oxygen concentrations increased, cyanophytes evolved mechanisms for both coping with (e.g., heterocysts to "protect" the oxygen-sensitive nitrogenase enzyme complex) and utilizing (e.g., aerobic respiration) the newly available free oxygen. The reported occurrence of heterocysts in Middle Precambrian cyanophytes provides strong evidence for the existence of relatively oxygenic conditions; it seems possible that the earliest widespread occurrence in the geological record of algal stromatolites, near the beginning of the Middle Precambrian, may reflect the origin of heterocystous nostocaceans capable of fixing atmospheric nitrogen in an aerobic environment.

8. The evolutionary histories of the "Coccogoneae" and "Hormo-

goneae" seem notably similar; both cyanophycean "tribes" originated during the Early Precambrian, had become moderately diversified by the Middle Precambrian, and were highly diversified and quite "modern" in character by the late Precambrian. Subsequently, members of both "tribes" have exhibited an extreme degree of evolutionary conservatism—as evidenced by detailed similarities in morphology, ecology, growth habit, and mode of reproduction among Precambrian, Phanerozoic, and extant taxa—a feature attributable to the wide ecological tolerance, versatile physiology, and unusually stable genetic system characteristic of the class. The Cyanophyceae represent perhaps the best example of bradytelic evolution now known in any biological group.

9. Thus the Precambrian was truly "the age of blue-green algae," with the group originating early in earth history, dominating the primitive biota, and reaching its zenith about 1 billion years ago. Cyanophytic dominance may have begun to decrease somewhat with the origin of eukaryotic algae, perhaps 1.6 billion years ago (Schopf, 1970*a*); this effect was not marked, however, since the earliest eukaryotes were planktonic, apparently asexual forms that were of subsidiary importance only in benthic communities where cyanophytes were prevalent. As is suggested in Fig. 10, the im-

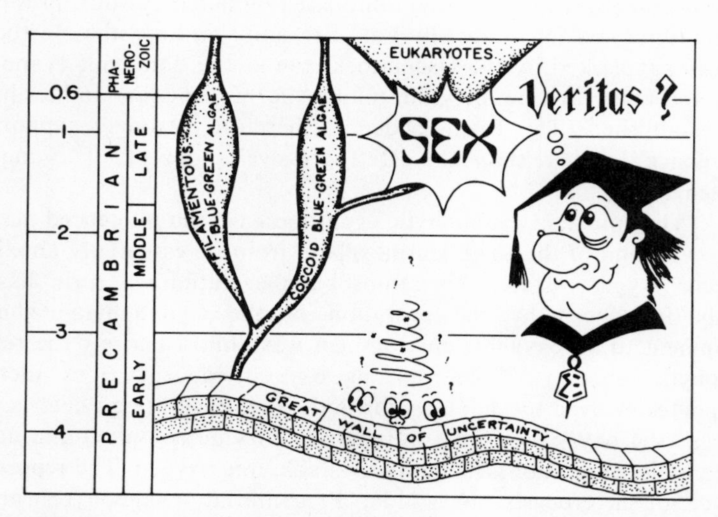

FIG. 10. "Truncated tree of life." Cartoon view of early biological evolution suggesting an inverse correlation between the Late Precambrian emergence of sexual eukaryotes (Schopf and Blacic, 1971; Schopf, 1972) and the decline from dominance of the Cyanophyta (Garrett, 1970). Prior to the past two decades, knowledge of blue-green algal evolution was effectively limited by the Cambrian-Precambrian "boundary"; in recent years, this "Great Wall of Uncertainty" has been pushed back through time until it now approaches the age of the oldest known sedimentary rocks.

mediate "trigger" leading to the rapid decline in the dominance of the Cyanophyta was probably the origin of meiosis and sexuality in eukaryotes (Schopf and Blacic, 1971; Schopf, 1972), which resulted in greatly increased evolutionary rates (due to the almost limitless variability produced by genetic recombination); this event led to the origin of mobile eukaryotic heterotrophs and subsequently to the development of grazing metazoans and a concomitant decrease in abundance and distribution of cyanophytic communities early in the Paleozoic (Garrett, 1970; Awramik, 1971).

10. It has long been assumed by phycologists that fossil evidence can make no useful contribution to understanding cyanophytic evolution. For example, Fritsch (1965, p. 859) has asserted that "even should some of the fossil types referred to Blue-green Algae actually belong to this class, as well they may, they afford no morphological data that might help in the elucidation of structural features or of the evolutionary sequence"; this theme has been often repeated (e.g., Desikachary, 1959; Desikachary and Padmaja, 1970). Indeed, it is believed by some, and seems widely accepted, that a fossil record of thallophytic evolution is "lacking" and that this is "a state of affairs which is unlikely to be remedied" (Klein and Cronquist, 1967, p. 256). Such views are no longer tenable. It is now evident that the fossil record represents, both actually and potentially, a rich source of significant data that can make a major contribution toward deciphering the course of thallophytic phylogeny.

Acknowledgments

This chapter has evolved from material which I first prepared for an invited lecture at the Eastern Canada Biostratigraphy Seminar on "Precambrian Fossils" held in Sudbury, Ontario, in October 1970. A greatly condensed version of many of these concepts was presented at the Joint Meeting of the Canadian Botanical Association and the American Institute of Biological Sciences (Botanical Society of America) in Edmonton, Alberta, in June 1971. And, most recently, in December 1971 I presented this material in essentially its present form as an invited "Special Lecture" at the Silver Jubilee Celebration of the Birbal Sahni Institute of Palaeobotany in Lucknow, India. Abstracts of these three papers have been published (Schopf, 1970*b*, 1971*a,b*).

The significance of heterocysts as possibly being indicative of aerobic conditions during the Middle Precambrian occurred to me following a discussion in 1969 with Dr. W. D. P. Stewart at the Eleventh International

Botanical Congress in Seattle and was reinforced by our subsequent correspondence. Discussions with Dr. J. P. Thornber have contributed significantly to my understanding of the events in biochemical evolution that may have led from the earliest heterotrophs to the origin of the Cyanophyta. Dr. R. Kozlowski generously provided samples of Ordovician chert containing *Schizothrichites ordoviciensis* (Fig. 5). Photomicrographs of Precambrian cyanophytes were kindly provided by Dr. E. S. Barghoorn (Figs. 7A and 8E,J) and by Drs. P. Cloud and G. R. Licari (Figs. 7G and 8K). Mrs. C. Lewis helped in compiling the data summarized in Figs. 2 and 3; Fig. 10 was drafted by Miss J. Guenther. I very much appreciate the cooperation and assistance of all these individuals.

Research leading to this publication has been supported by the Earth Sciences Section, National Science Foundation, NSF Grant GA-23741, and, in part, by NASA Grant NGR 05-007-292.

REFERENCES

Abelson, P. H., 1966, Chemical events on the primitive earth, *Proc. Natl. Acad. Sci.* **55**:1365–1372.

Awramik, S. M., 1971, Precambrian columnar stromatolite diversity: Reflection of metazoan appearance. *Science* **174**:825–827.

Banks, H. P., 1968, The early history of land plants, in: *Evolution and Environment* (E. T. Drake, ed.), pp. 73–107, Yale University Press, New Haven.

Barghoorn, E. S., and Schopf, J. W., 1965, Microorganisms from the Late Precambrian of central Australia, *Science* **150**:337–339.

Barghoorn, E. S., and Tyler, S. A., 1965, Microorganisms from the Gunflint chert, *Science* **147**:563–577.

Barghoorn, E. S., and Schopf, J. W., 1966, Microorganisms three billion years old from the Precambrian of South Africa, *Science* **152**:758–763.

Baschnagel, R. A., 1942, Some microfossils from the Onondaga chert of central New York, *Bull. Buffalo Soc. Nat. Sci.* **17**(**3**):1–8.

Baschnagel, R. A., 1966, New fossil algae from the Middle Devonian of New York, *Trans. Am. Microsc. Soc.* **85**:297–302.

Bazin, M. J., 1968, Sexuality in a blue-gree alga: Genetic recombination in *Anacystis nidulans, Nature* **218**:282–283.

Becker, R. H., 1971, Carbon and oxygen isotope ratios in iron-formation and associated rocks from the Hamersley Range of western Australia and their implications, Ph.D. thesis, Department of Chemistry, University of Chicago.

Bendoraitis, J. G., Brown, G. L., and Hepner, L. S., 1962, Isoprenoid hydrocarbons in petroleum, *Anal. Chem.* **34**:49–53.

Bondesen, E., Pedersen, K. R., and Jorgensen, L., 1967, Precambrian organisms and the isotopic composition of organic remains in the Ketilidian of south-west Greenland, *Medd. Grønland* **164**(**4**):41 pp.

Bradley, W. H., 1931, Origin and Microfossils of the Oil Shale of the Green River Formation of Colorado and Utah, U.S. Geological Survey Professional Paper No. 168, 58 pp.

Bradley, W. H., 1970, Eocene algae and plant hairs from the Green River Formation of Wyoming, *Am. J. Bot.* **57**:782–785.

Brock, T. D., 1970, *Biology of Microorganisms*, 737 pp., Prentice-Hall, Englewood Cliffs, N.J.

Brooks, J., 1971, Some chemical and geochemical studies on sporopollenin, in *Sporopollenin* (J. Brooks, P. R. Grant, M. Muri, P. van Gijzel, and G. Shaw, eds.), pp. 351–407, Academic Press, New York.

Brooks, J., and Muir, M. D., 1971, Morphology and chemistry of the organic insoluble matter from the Onverwacht Series Precambrian chert and the Orgueil and Murray carbonaceous meteorites, *Grana* **11**:9–14.

Butin, R. V., 1959, Iskopaemye Cyanophyceae v Proterozoyskikh karbonatnykh otlozheniyakh yuzhnoy Karelli [Fossil Cyanophyceae in Proterozoic carbonate oncolites from southern Karelia], *Karelskii filial Petrozavsdsk Izvestia Karalskogo i Kol'skogo filialov, Akad. Nauk SSSR* **2**:47–51.

Calvin, M., 1969, *Chemical Evolution*, 278 pp., Oxford University Press, New York.

Carr, N. G., and Hallaway, M., 1965, The presence of α-tocopherolquinone in blue-green algae, *Biochem. J.* **97**:9c–10c.

Case, G. D., Parson, W. W., and Thornber, J. P., 1970, Photooxidation of cytochromes in reaction center preparations from *Chromatium* and *Rhodopseudomonas viridis*, *Biochim. Biophys. Acta* **223**:122–128.

Chaloner, W. G., 1970, The rise of the first land plants, *Biol. Rev.* (*Camb.*) **45**:353–377.

Cloud, P. E., Jr., 1965, Significance of the Gunflint (Precambrian) microflora, *Science* **148**:27–35.

Cloud, P. E., Jr., 1968, Pre-Metazoan evolution and the origins of the Metazoa, in *Evolution and Environment* (E. T. Drake, ed.), pp. 1–72, Yale University Press, New Haven.

Cloud, P. E., Jr., and Licari, G. R., 1968, Microbiotas of the banded iron formations, *Proc. Natl. Acad. Sci.* **61**:779–786.

Cloud, P. E., Jr., Licari, G. R., Wright, L. A., and Troxel, B. W., 1969, Proterozoic eucaryotes from eastern California, *Proc. Natl. Acad. Sci.* **62**:623–630.

Croft, W. N., and George, E. A., 1959, Blue-green algae from the Middle Devonian of Rhynie, Aberdeenshire, *Brit. Mus.* (*Nat. Hist.*) *Bull. Geol.* **3**:339–353.

Degens, E. T., 1969, Biogeochemistry of stable carbon isotopes, in: *Organic Geochemistry* (G. Eglinton and M. T. J. Murphy, eds.), pp. 304–330, Springer-Verlag, New York.

Desikachary, T. V., 1959, *Cyanophyta*, 686 pp., Indian Council of AgriculturaL Research, New Delhi.

Desikachary, T. V., and Padmaja, T. D., 1970, Origin of filamentous condition and phylogeny in the blue green algae. *Rev. Algol. (Paris)* **10**:8–17.

Engel, A. E. J., Nagy, B., Nagy, L. A., Engel, C. G., Kremp, G. O. W., and Drew, C. M., 1968, Alga-like forms in Onverwacht Series, South Africa: Oldest recognized lifelike forms on earth, *Science* **161**:1005–1008.

Fay, P., 1965, Heterotrophy and nitrogen fixation in *Chlorogloea fritschii*, *J. Gen. Microbiol.* **39**:11–20.

Fritsch, F. E., 1965, *The Structure and Reproduction of the Algae*, Vol. 2, 939 pp., Cambridge University Press, London.

Garrett, P., 1970, Phanerozoic stromatolites: Non-competitive ecologic restriction by grazing and burrowing animals, *Science* **169**:171–173.

Gebelein, C. D., 1969, Distribution, morphology, and accretion rate of Recent subtidal algal stromatolites, *J. Sed. Petrol.* **39**:49–69.

Gerloff, G. C., Fitzgerald, G. P., and Skoog, F., 1950, The isolation, purification, and culture of blue-green algae, *Am. J. Bot.* **37**:216–218.

Goswami, S. K., 1955, Occurrence of *Scytonema* sp. in the lignite of Kashmir Valley, *Curr. Sci.* **24**(**2**):56.

Gruner, J. W., 1925, Discovery of life in the Archaean, *J. Geol.* **33:**151–152.

Gutstadt, A. M., and Schopf, J. W., 1969, Possible algal microfossils from the Late Pre-Cambrian of California, *Nature* **223:**165–167.

Han, J., and Calvin, M., 1969, Occurrence of fatty acids and aliphatic hydrocarbons in a 3.4 billion-year-old sediment, *Nature* **224:**576–577.

Hoering, T. C., 1967, The organic geochemistry of Precambrian rocks, in: *Researches in Geochemistry*, Vol. 2 (P. H. Abelson, ed.), pp. 87–111, Wiley, New York.

Hofmann, H. J., and Jackson, G. D., 1969, Precambrian (Aphebian) microfossils from Belcher Islands, Hudson Bay, *Canad. J. Earth Sci.* **6:**1137–1144.

Holmes, A., 1954, The oldest dated minerals of the Rhodesian Shield, *Nature* **173:**612.

Holm-Hansen, O., 1968, Ecology, physiology, and biochemistry of blue-green algae, *Ann. Rev. Microbiol.* **22:**47–70.

Holton, R. W., 1969, Blue-green algae, primitive cells, and comparative biochemistry, in: *Abstracts of the XI International Botanical Congress, Seattle, Wash.*, p. 93.

Holton, R. W., and Blecker, H. H., 1970, Fatty acids of blue-green algae, in: *Properties and Products of Algae* (J. E. Zajic, ed.), pp. 115–127, Plenum Press, New York.

Holton, R. W., Blecker, H. H., and Stevens, T. S., 1968, Fatty acids in blue-green algae: Possible relation to phylogenetic position, *Science* **160:**545–547.

Hutchinson, G. E., 1954, The biochemistry of the terrestrial atmosphere, in *The Earth as a Planet* (G. P. Kuiper, ed.), pp. 371–433, University of Chicago Press, Chicago.

Ikan, R., and Bortinger, A., 1971, Normal and isoprenoid alkanes from an Israeli shale, *Geochim. Cosmochim. Acta* **35:**1059–1065.

Johnson, J. H., 1937, Algae and algal limestones from the Oligocene of South Park Colorado, *Geol. Soc. Am. Bull.* **48:**1227–1235.

Kidston, R., and Lang, W. H., 1921, On Old Red Sandstone plants showing structure, from the Rhynie Chert Bed, Aberdeenshire. V. The Thallophyta occurring in the peat-bed, the succession of the plants throughout a vertical section of the bed, and the conditions of accumulation and preservation of the deposit, *Roy. Soc. Edinburgh Trans.* **52:**855–902.

Klein, R. M., and Cronquist, A., 1967, A consideration of the evolutionary and taxonomic significance of some biochemical, micromorphological, and physiological characters in the thallophytes, *Quart. Rev. Biol.* **42:**105–296.

Kordé, K. B., 1958, K sistematie iskopaemykh *Cyanophyceae* [The systematics of fossil Cyanophyceae], *Materialy k Osnovam Paleontologii, Akad. Nauk SSSR Paleont. Inst.* **2:**99–111.

Kvenvolden, K. A., and Hodgson, G. W., 1969, Evidence for porphyrins in Early Precambrian Swaziland System sediments, *Geochim. Cosmochim. Acta* **33:**1195–1202.

Kvenvolden, K. A., Hodgson, G. W., Peterson, E., and Pollock, G. E., 1968, Organic geochemistry of the Swaziland System, South Africa, in: *Program of the Annual Meeting of the Geological Society of America, Mexico City*, pp. 167–168 (abst.).

LaBerge, G. L., 1967, Microfossils and Precambiran iron-formations, *Geol. Soc. Am. Bull.* **78:**331–342.

Laristschev, A. A., 1952, O novoy iskopaemoy vodorsli yuvskogo vozrasta [On new fossil cyanophycean algae of Jurassic age], *Otdel Sporovykh Rastenii, Botanicheskie Materialy, Akad. Nauk SSSR Bot. Inst.* **8:**47–50.

Licari, G. R., 1971, Paleontogy and paleoecology of the Proterozoic Beck Spring Dolomite of eastern California, Ph.D. thesis, Department of Geology, University of California, Los Angeles.

Licari, G. R., and Cloud, P. E., Jr., 1968, Reproductive structures and taxonomic affinities of some nannofossils from the Gunflint Iron Formation, *Proc. Natl. Acad. Sci.* **59:**1053–1060.

Licari, G. R., Cloud, P. E., Jr., and Smith, W. D., 1969, A new chroococcacean alga from the Proterozoic of Queensland, *Proc. Natl. Acad. Sci.* **62**:56–62.

Macgregor, A. M. (1940), A pre-Cambrian algal limestone in Southern Rhodesia, *Trans. Geol. Soc. S. Afr.* **43**:9–16.

MacLeod, W. D., Jr., 1968, Combined gas chromatography–mass spectrometry of complex hydrocarbon trace residues in sediments, *J. Gas Chromatogr.* **6**:591–594.

Margulis, L., 1970, *Origin of Eukaryotic Cells,* 349 pp., Yale University Press, New Haven.

McCarthy, E. D., and Calvin, M., 1967, The isolation and identification of the C_{17} saturated isoprenoid hydrocarbon, 2,6,10-trimethyltetradecane, from a Devonian shale: The role of squalene as a possible precursor, *Tetrahedron* **23**:2609–2619.

Mehta, K. R., 1954, A fossil member of the Chroococcales from the lower Gondwans of India, *Palaeobotanist* (*Lucknow*) **3**:38–59.

Meinschein, W. G., 1967, Paleobiochemistry, in: *1967 McGraw-Hill Yearbook of Science and Technology,* pp. 283–285, McGraw-Hill, New York.

Monty, C. L. V., 1967, Distribution and structure of recent stromatolitic algal mats, eastern Andros Island, Bahamas, *Soc. Geol. Belg. Ann.* **90**(**1–3**):55–102.

Moore, E. S., 1918, The iron formation on Belcher Islands, Hudson Bay, with special reference to its origin and its associated algal limestones, *J. Geol.* **26**:412–438.

Nagy, B., and Nagy, L. A., 1969, Early Pre-Cambrian Onverwacht microstructures: Possibly the oldest fossils on earth? *Nature* **223**:1226–1229.

Oehler, J. H., and Schopf, J. W., 1971, Artificial microfossils: Experimental studies of permineralization of blue-green algae in silica, *Science* **174**:1229–1231.

Oehler, D. Z., Schopf, J. W., and Kvenvolden, K. A., 1972, Carbon isotopic studies of organic matter in Precambrian rocks, *Science* **175**:1246–1248.

Olson, J. M., 1970, The evolution of photosynthesis, *Science* **168**:438–446.

Oró, J., and Nooner, D. W., 1967, Aliphatic hydrocarbons in Pre-Cambrian rocks, *Nature* **213**:1082–1085.

Park, R., and Epstein, S., 1960, Carbon isotopic fractionation during photosynthesis, *Geochim. Cosmochim. Acta* **21**:110–126.

Park, R., and Epstein, S., 1961, Metabolic fractionation of C^{13} and C^{12} in plants, *Plant Physiol.* **36**(**2**):133–138.

Peach, B. N., Horne, J., Gunn, W., Clough, C. T., and Hinxman, L. W., 1907, The geologic structure of the northwest highlands of Scotland, *Gr. Brit. Geol. Survey Mem.,* 668 pp.

Pflug, H. D., 1967, Structured organic remains from the Fig Tree Series (Precambrian) of the Barberton Mountain Land (South Africa), *Rev. Palaeobot. Palynol.* **5**:9–29.

Pflug, H. D., Meinel, W., Neumann, K. H., and Meinel, M., 1969, Entwicklungstendenzen des frühen Lebens auf der Erde, *Naturwissenschaften* **56**:10–14.

Pierson, B. K., and Castenholz, R. W., 1971, Bacteriochlorophylls in gliding filamentous prokaryotes from hot springs, *Nature* **233**:25–27.

Rao, A. R., 1957, Algal remains of the Tertiary lignites of Palana (Eocene), Bikaner, *Curr. Sci.* **26**:177–178.

Raven, P. H., 1970, A multiple origin for plastids and mitochondria, *Science* **169**:641–646.

Raven, P. H., and Curtis, H., 1970, *Biology of Plants,* 706 pp., Worth, New York.

Rutten, M. G., 1962, *The Geological Aspects of the Origin of Life on Earth,* 146 pp., Elsevier, New York.

Sagan, L. (Margulis, L.), 1967, On the origin of mitosing cells, *J. Theoret. Biol.* **14**:225–275.

Schidlowski, M., 1966, Zellular strukturierte Elemente aus de Präkambrium des Witwatersrand-Systems (Sudafrica), *Z. Deutsch. Geol. Ges.* **115**:783–786.

Schopf, J. M., 1971, Notes on plant tissue preservation and mineralization in a Permian deposit of peat from Antarctica, *Am. J. Sci.* **271**:522–543.

Schopf, J. W., 1968, Microflora of the Bitter Springs Formation, Late Precambrian, central Australia, *J. Paleontol.* **42**:651–688.

Schopf, J. W., 1970*a*, Precambrian micro-organisms and evolutionary events prior to the origin of vascular plants, *Biol. Rev.* (*Camb.*) **45**:319–352.

Schopf, J. W., 1970*b*, Evolution and geologic distribution of blue-green algae, in: *Abstracts of the Eastern Canada Biostratigraphy Seminar on Precambrian Fossils, Sudbury, Ont.*, pp. 14–15.

Schopf, J. W., 1971*a*, Evolution and geologic distribution of blue-green algae, *Am. J. Bot.* **58**:472 (abst.).

Schopf, J. W., 1971*b*, Antiquity and evolution of the earliest plants, in: *Abstracts of the Palaeobotanical Conference, Birbal Sahni Institute of Palaeobotany, Silver Jubilee Celebration, Lucknow, India*, p. 56.

Schopf, J. W., 1972, Evolutionary significance of the Bitter Springs (Late Precambrian) microflora, in: *Proceedings of the 24th International Geological Congress, Section 1, Precambrian Geology, Montreal*:68–77.

Schopf, J. W., and Barghoorn, E. S., 1967, Alga-like fossils from the Early Precambrian of South Africa, *Science* **156**:508–512.

Schopf, J. W., and Barghoorn, E. S., 1969, Microorganisms from the Late Precambrian of south Australia, *J. Paleontol.* **43**:111–118.

Schopf, J. W., and Blacic, J. M., 1971, New microorganisms from the Bitter Springs Formation (Late Precambrian) of the north-central Amadeus Basin, Australia, *J. Paleontol.* **45**:925–960.

Schopf, J. W., Oehler, D. Z., Horodyski, R. J., and Kvenvolden, K. A., 1971, Biogenicity and significance of the oldest known stromatolites, *J. Paleontol.* **45**:477–485.

Scott, W. M., Modzeleski, V. E., and Nagy, B., 1970, Pyrolysis of Early Pre-Cambrian Onverwacht organic matter ($> 3 \times 10^9$ yr old), *Nature* **225**:1129–1130.

Sharp, J. H., 1969, Blue-green algae and carbonates—*Schizothrix calcicola* and algal stromatolites from Bermuda, *Limnol. Oceanog.* **14**:568–578.

Shestakov, S. V., and Khyen, N. T., 1970, Evidence for genetic transformation in blue-green alga *Anacystis nidulans*, *Mol. Gen. Genet.* **107**:372–375.

Shields, L. M., and Drouet, F., 1962, Distribution of terrestrial algae within the Nevada Test Site, *Am. J. Bot.* **49**:547–554.

Singh, H. N., 1967, Genetic control of sporulation in the blue-green alga *Anabaena doliolum* Bharadwaja, *Planta* (*Berl.*) **75**:33–38.

Stanier, R. Y., Doudoroff, M., and Adelberg, E. A., 1970, *The Microbial World*, 3rd ed., 873 pp., Prentice-Hall, Englewood Cliffs, N.J.

Starmach, K., 1963, Blue-green algae from the Tremadocian of the Holy Cross Mountains (Poland), *Acta Palaeontol. Pol.* **8**:451–462.

Stewart, W. D. P., and Pearson, H. W., 1970, Effects of aerobic and anaerobic conditions of growth and metabolism of blue-green algae, *Proc. Roy. Soc. Lond. Ser. B* **175**:293–311.

Stewart, W. D. P., Haystead, A., and Pearson, H. W., 1969, Nitrogenase activity in heterocysts of blue-green algae, *Nature* **224**:226–228.

Thomas, J., 1970, Absence of the pigments of Photosystem II of photosynthesis in heterocysts of a blue-green alga, *Nature* **228**:181–183.

Tyler, S. A., and Barghoorn, E. S., 1954, Occurrence of structurally preserved plants in pre-Cambrian rocks of the Canadian shield, *Science* **119**:606–608.

Vail, J. R., and Dodson, M. H., 1969, Geochronology of Rhodesia, *Trans. Geol. Soc. S. Afr.* **72**:79–113.

Vologdin, A. G., and Kordé, K. B., 1965, Neskol'ko vidov drevnikh Cyanophyta i ikh tsenozy [Several species of ancient Cyanophyta and their coenoses], *Dokl. Akad. Nauk SSSR* **164**(**2**):429–432.

Walcott, C. D., 1914, Pre-Cambrian Algonkian algal flora, *Smithson. Misc. Coll.* **64(2):**77–156.

Wald, G., 1964, The origins of life. *Proc. Natl. Acad. Sci.* **52:**595–611.

Wells, A. T., Forman, D. J., Ranford, L. C., and Cook, P. J., 1970, Geology of the Amadeus Basin, central Austrialia, *Bureau Mineral. Resources Austral. Bull.* **100:**222 pp.

Whittaker, R. H., 1969, On the broad classification of organisms, *Science* **163:**150–159.

Witkin, E. M., 1966, Radiation-induced mutations and their repair, *Science* **152:**1345–1353.

Zalessky, M. P., 1917, Sur le sapropélite marin de l'âge Silurien formé par une algue Cyanophycée, *Ann. Soc. Paleontol. Russ.* **1:**25–42.

2

Five-Kingdom Classification and the Origin and Evolution of Cells

LYNN MARGULIS

Department of Biology
Boston University
Boston, Massachusetts

PLANTS AND ANIMALS: BOTANISTS AND ZOOLOGISTS

This chapter will argue that modern biologists, in spite of social pressures and historical precedents, need to replace the traditional two-kingdom animal–plant distinction, which has outlived its usefulness, with a multikingdom classification of living organisms. For reasons discussed below, based on recent discoveries from a variety of disciplines, it seems that Whittaker's five-kingdom system (Whittaker, 1969) is the most logical and consistent yet devised. Whittaker's system is expanded below the phylum level and slightly modified on the basis of cell evolutionary considerations; suggestions for its adoption by zoologists, botanists, and microbiologists are made.*

Since antiquity and in many different cultures (e.g., see Dibble and Anderson, 1963, for Pre-Columbian Mexican natural history), the world's organisms have been divided into the all-inclusive two great groups: the animals and the plants. This distinction, so appropriate for the large organisms, persists today in publications as general and recent as the second edition of the *Biology Data Book* (Altman and Dittmer, 1972) and

* References listed but not cited in the text were used in devising the classification (see p. 75). A popularized version of this classification has appeared in the *Handbook of Genetics*, Chapter 1 (see Margulis, 1974*a*).

that critically important encyclopedia of bacteriological information, ***Bergey's Manual*** (Breed *et al.*, 1957). Because the application of the concepts and terms "animal" and "plant" to the microbial world has been an uncomfortable extrapolation from the metazoans and the tracheophytes, several attempts to replace this traditional grouping have been made (Copeland, 1956; Dodson, 1971; Whittaker, 1969). Table I presents (in a simplified manner modeled after Keeton, 1972) the various published kingdom-level systems in recent publications. In fact, none but the traditional has wide practical usage among taxonomists. The plant–animal dichotomy also permeates the organization and teaching of biological science; this is clearly reflected in the social structure of professional biologists in liberal arts institutions who are often grouped into botany and zoology departments. Microbiologists have traditionally either been excluded or allied with botanists. Microbiology has emerged from the last century as a branch of various practical arts: medicine, food processing, sewage treatment, soil science, and the like. The development of microbial genetics and molecular biology has effected the integration of microbiology into the rest of biology only very recently. The understanding obtained from these new disciplines in many cases has still not permeated the scientific community enough to alter the organization of professionals. Many professional societies (e.g., International Association of Plant Taxonomy, Society of Systematic Zoology, International Botanical Congress, Society of Protozoologists, Phycological Society of America) are still formed around the plant–animal distinction and thus they are still serving to selectively exclude information or transmit it to members in a way that reinforces this ill-defined categorization.

Considerable thought and work have established internationally acceptable botanical and zoological "codes": *International Code of Botanical Nomenclature* (F. A. Staffleu, ed., 1969) and *International Code of Zoological Nomenclature* (Stoll *et al.*, eds., 1964). These provide rules and recommendations for naming and grouping the living organisms of the world, presumably on a foundation of evolutionary relationships. Unfortunately (for the higher taxa especially, e.g., phyla and divisions, classes, orders, families) the rules and name endings are not uniformly applied to animals, plants, and microbes. In fact, in the case of many forms neither clearly multicellular animals nor clearly plants, such as the slime molds or the colonial photosynthetic flagellates (that is, all organisms traditionally considered thallophytes or protozoans), all of the taxa above the genus level are different in botanical vs. zoological classification schemes. This being the case, it is not possible to construct an overall classification system of higher taxa that will be acceptable to both taxonomic botanists and zoologists without some serious deviation from tradition and established recommendations of the *Codes*. This general confusion and nomenclatural inconsistency has led to an attitude of disdain and skepticism on the part of physiologically and biochemically oriented scientists even if they have deep interests in the evolution of biological diversity. The tendency among many bacteriologists has been to explicitly ignore or argue about any attempts at taxonomy above the level of genus (see Cowan, 1962, and other articles in that volume). In spite of their sound intellectual base, proposals to include the blue-green algae in *Bergey's Manual* have been defeated.

Thus in the task I undertake it is intrinsically impossible to satisfy the traditionalists of one camp without alienating those of the other. My attempt to present an expansion of the Whittaker five-kingdom system below the level of phyla will be based on the following considerations:

1. Comparative genetic and molecular biological criteria of relatedness will be utlized (see Margulis, 1968, 1970, 1971*b*, for discussion).
2. Familiar and widely used names and endings will be retained with a meaning as close as possible to existing usage.
3. The three morphologically complex eukaryotic kingdoms (plants, animals, and fungi) will be treated as consistently as possible, optimizing both tradition and logic. The approval of specialists is sacrificed for comprehension by geneticists, molecular biologists, and physiologists. Definitions and brief explanations will be included wherever possible.
4. In the more well-known groups, such as flowering plants and mammals, classification will extend down to classes; in the more obscure groups (e.g., priapulids and tardigrades), no attempt will be made to subdivide phyla.
5. The deviations from Whittaker's (1969) original treatment are based on a certain evolutionary model for the origin and evolution of eukaryotic cells: cell symbiosis theory. This deviation and the reasons for it will be explained (Margulis, 1971*a*).

Whittaker (1969), Dodson (1971), and others who recently have tried to resolve some of these problems raised by the arbitrary assignment of bacteria, algae, protozoans, and fungi to the animal and plant kingdoms owe a great debt to the magnificent work of Copeland (1956). Copeland failed to include common names and explain the principles on which he was reclassifying these organisms; he failed thus to communicate. This work lay sadly unappreciated until Whittaker integrated it into his own system (Whittaker, 1969), accepting and expanding its strengths and replacing its

TABLE I. Kingdom Classifications

System 1	System 2	System 3	System 4	System 5	System 6	System 7
Traditional, see Altman and Dittmer (1972)	Dodson (1971)	Curtis (1968)	Stanier *et al.* (1970)	Copeland (1956)	Whittaker (1969)	Whittaker (modified by Margulis, 1974*a*, 1971*a*, and in this chapter)
PLANTAE	MONERA (MYCHOTA)	PROTISTA	PROTISTA	MONERA	MONERA	MONERA
Bacteria	Bacteria	Bacteria	Bacteria	Bacteria	Bacteria	Bacteria
Blue-green algae	Blue-green algae	Blue-green algae	Blue-green algae	Blue-green algae	Blue-green algae	Blue-green algae
Green algae		Protozoa	Protozoa			
Chrysophytes	PLANTAE	Slime molds	Green algae	PROTOCTISTA	PROTISTA	PROTISTA
Brown algae			Chrysophytes			
Red algae	Green algae	PLANTAE	Brown algae	Protozoa	Protozoa	Protozoa
Slime molds	Chrysophytes		Red algae	Green algae	Chrysophytes	Green algae

Fungi	Brown algae	Green algae	Slime molds	Chrysophytes		Chrysophytes
Bryophytes	Red algae	Chrysophytes	Fungi	Brown algae	PLANTAE	Brown algae
Tracheophytes	Slime molds	Brown algae		Red algae		Red algae
	Fungi	Red algae	PLANTAE	Slime molds	Green algae	Slime molds
ANIMALIA	Bryophytes	Fungi		Fungi	Brown algae	Flagellated fungi
	Tracheophytes	Bryophytes	Bryophytes		Red algae	
Protozoa		Tracheophytes	Tracheophytes	PLANTAE	Bryophytes	FUNGI
Multicellular animals	ANIMALIA				Tracheophytes	
		ANIMALIA	ANIMALIA	Bryophytes		Amastigote fungi
	Protozoa			Tracheophytes	FUNGI	
	Multicellular animals	Multicellular animals	Multicellular animals			PLANTAE
				ANIMALIA	Slime molds	
					Fungi	Bryophytes
				Multicellular animals		Tracheophytes
					ANIMALIA	
						ANIMALIA
					Multicellular animals	
						Multicellular animals

weaknesses with sounder and more defensible groupings. In this chapter, I have attempted to expand Whittaker's unified classification system so that it is detailed enough that each of us working with certain genera and species can easily place our organisms into the appropriate higher taxa. As Whittaker emphasized, the traditional animal–plant dichotomy is appropriate for the morphologically and nutritionally distinct organisms of these two great groups; it should thus be retained. For the microorganisms, the Chatton–Kluyver–Van Niel–Stanier distinction between the prokaryotes and the eukaryotes is far more meaningful, useful, and unambiguous (see Stanier *et al.,* 1970; and Brock, 1970, for extensive discussion). Recognition of this microbial dichotomy results in the assignment of all prokaryotes to the monera kingdom (in agreement with Copeland, 1956, and Dodson, 1971) and all nucleated organisms to eukaryote kingdoms. Whittaker's establishment of four such eukaryote kingdoms (protists, animals, fungi, and plants) is recognized and expanded herein. I believe his arguments concerning the vast differences between the fungi and the protists relative to the green plants and invertebrate animals are valid. His five-kingdom scheme, with my minor alterations, is summarized in Table II.

By this publication I hope to encourage at least the dialogue that must precede the development of a modern, logical phylogenetic classification system. At the First International Congress of Systematic and Evolutionary Biology (Boulder, Colorado, 1973), in which for the first time the International Association of Plant Taxonomists met with the Society for Systematic Zoology, some of these issues were formally discussed (Taylor, 1974; Leedale, 1974, in a symposium, Margulis 1974*c*). Tentative plans for the formation of a society for evolutionary protistology and plans of the protozoologists to meet with the phycologists are evidence of the contemporary reunification of the science of biology. This sort of formal communication and cooperation is required before an overall classification of organisms will ever be accepted and utilized.

PROKARYOTES AND EUKARYOTES

Results of genetic, ultrastructural, and biochemical studies of microorganisms have entirely supported the concept that the prokaryote–eukaryote dichotomy is far more profound than the traditional animal–plant distinction. Prokaryotic organisms, which are generally composed of small cells, include a diverse assortment of bacteria and, of the algae, only the blue-green algae. The organization of prokaryotic cells is distinctive. Although the genetic material within them is composed of

DNA, the DNA is not complexed with RNA and proteins that typically make up the chromosomes of eukaryotes. Unlike eukaryotes, the prokaryotes do not contain this class of arginine- and lysine-rich chromosomal proteins, the histones. Prokaryotic organisms reproduce asexually by fission or budding. They contain only one genophore ("bacterial chromosome," located in the electron-less-dense area of the cell, the nucleoid), except in cases where DNA replication is relatively more rapid than cell division. When this occurs, several duplicate genomes may be present in each cell. In bacterial genomes the genes are represented, one copy each for most of them, usually as a single covalently closed circle of DNA with a circumferential length greater than some tens of microns (i.e., mol wt DNA $> 10^8$). Newly synthesized DNA is probably segregated to daughter genophores via a direct attachment of DNA to membrane; never is a "mitotic apparatus" formed (see Luykx, 1970, for review of chromosome distribution). A very small amount of RNA has been implicated in the maintenance of the highly compacted state of the DNA of the prokaryote nucleoid (Worcel and Burgi, 1972). Sexual systems are unidirectional in those prokaryote organisms in which sexual recombination is present. That is, in the sexual process genes are transferred in one direction from the donor cell to the partner (the recipient cell). Only genes and generally only a small fraction of the total genes are transferred. The new recombinant organism consists thus of the recipient cell itself with some genes replaced (those derived from the donor). All prokaryotic cells lack certain organelles routinely found in eukaryotes such as mitochondria, chloroplasts, basal bodies, and centrioles.

Familiar organisms (nucleated algae, protozoans, fungi, invertebrate and vertebrate animals, and green plants) are composed of eukaryotic cells. These cells have membrane-bounded nuclei containing chromosomes complexed with RNA and histone and other proteins. The number of chromosomes per nucleus varies greatly in eukaryotes from two (in haploid *Haplopappus*, a composite plant) to greater than 1000 (in some radiolarians). Each chromosome contains only part of the entire genome and acts as an independent linkage group in meiosis. Never in eukaryotes is the entire genome one single uncomplexed DNA molecule making up a single chromosome as in prokaryotes. The amounts of DNA in eukaryote nuclei are in general 10^3 times or so more than the amounts in prokaryote nucleoids. Most animals and plants seem to have large amounts of DNA composed of repetitive nucleotide sequences in their nuclei in addition to unique-sequence DNA. The reason for this is unknown (see Thomas, 1971, for discussion). The number and morphology of chromsomes (i.e., the karyotype) and indeed even the amount of DNA per nucleus may vary even among

TABLE II. Summary of the Five Kingdoms

Kingdom	Examples of organisms	Genetic organization	Approximate time of diversification and documented first appearance (millions of years ago)	Major traits that environmental selection pressures acted on to produce	Major significant selective factor in the environment
Monera	All prokaryotes: bacteria, blue-green algae, mycelial bacteria, gliding bacteria, etc.	Prokaryote chromoneme merozygotes only; sex unidirectional	Early-middle Precambrian (3000–1000): Fig Tree microfossils (>3000); Bulawayan stromatolites (>3000) (Barghoorn, private communication)	Ultraviolet photoprotection, photosynthesis, motility, and aerobiosis	Solar radiation, increasing atmospheric oxygen concentration, depletion of nutrients
Protista	All eukaryotic algae: green, yellow-green, red and brown, and golden-yellow; all protozoa; flagel-	Eukaryote chromosomes, ploidy levels, meiotic sexual systems vary, gametes and zygotes	Late Precambrian, early Paleozoic (1500–500): possibly Bitter Springs microflora (≈900) (Schopf,	Mendelian genetic systems, mitosis and meiosis: obligate recombination each generation; phagocytosis,	Depletion of nutrients

	lated fungi; slime molds; and slime net molds		1972); Ediacaran animals (750) (Glaessner, 1968; see Margulis, 1974*b*, for discussion)	pinocytosis, intracellular motility	
Animalia	Metazoa: all animals developing from blastulas	Diploids, meiosis precedes gametogenesis	Phanerozoic (600 on); Ediacaran fauna (750) (Glaessner, 1968)	Tissue development for heterotrophic specializations; ingestive nutrition	Transitions from aquatic to terrestrial and aerial environments
Plantae	Metaphyta: all green plants developing from embryos	Alternation between haplo- and diplophases	Phanerozoic (600 on); rhyniophytes, Downtonian, Wales, Czechoslovakia, New York State (405) (Banks, 1972)	Tissue development for autotrophic specializations; photosynthetic nutrition	Transitions from aquatic to terrestrial environments
Fungi	Amastigomycota: conjugation fungi, sac fungi (molds), club fungi (mushrooms), yeasts	Haploid and dikaryotic; zygote formation followed by meiosis and haploid spore formation	Phanerozoic (600 on); Rhynie chert, *Paleomyces asteroxyli* (400) (Arnold, 1947)	Advanced mycelial development; absorptive nutrition	Transitions from aquatic to terrestrial environments; nature of nutrient source, nature of host

closely related organisms. The chromosomes of most eukaryotes uncoil during the time in the life cycle that RNA is transcribed (the interphase) and then coil again prior to mitosis. Mitotic cell division, resulting in an equipartition of the chromosomal complement of the cell, is typical of eukaryotes. The biparental inheritance of characteristics in the "mendelian" pattern is present only in eukaryotes. In sexually reproducing eukaryotes, each parent in fertilization makes an equal contribution of nuclear genes to each offspring. Table III summarizes the differences between these two cell types.

The classification system presented here is based on the evolutionary views of the symbiotic theory (see p. 63); that is, Whittaker (1969; Table 1, system 6) is accepted with minor modification (Margulis, 1971*a*; Table 1, system 7). Whittaker bases his three multicellular eukaryotic kingdoms on concepts of nutrition and ecological role: the plants are photoautotrophic primary producers, the animals are heterotrophic consumers, and the fungi are absorptive parasites and saprophytes. However, Whittaker, like Copeland (1956) and Dodson (1971) among others, has wrestled with the problem of "blurred boundaries" between these three kingdoms and the protists from which they presumably arose. Filamentous, colonial, and mycelial habits have evolved convergently in many groups, including prokaryotes. All these are fundamentally multicellular. In no modern classification can all protists be considered single cells (*cf.* the labyrinthulids, the cellular slime molds, or the colonial euglenids). It thus is not feasible to draw the line between protists and the other three eukaryote kingdoms on the basis of the "unicellular–multicellular" distinction. A solution to this problem involves a closer restriction of the three multicellular eukaryote (fungi, animal, and plant) kingdoms and a positive definition of their boundaries. Thus the fungi kingdom here excludes all of the flagellated fungi and the amoeboid forms with a variety of life cycles, and it therefore contains only the presumed homologous series of amastigote (flagella-less) haploid or dikaryotic forms, e.g., the zygo-, asco-, and basidiomycetes (Fig. 1). The animal kingdom is limited to those multicellular organisms which develop from a blastula (metazoans), and the plant kingdom comprises only embryophytes, multicellular archegoniate green plants (that is, the primarily terrestrial bryophytes and tracheophytes). The protists become that remaining miscellaneous group of phyla in which profound "evolutionary experimentation" occurred. They can be comprehended as a natural group if it is realized that, concomitant with their diversification, mitosis and mendelian genetic patterns of inheritance originated and evolved and meiosis eventually stabilized (Margulis, 1974*b*). This resulted in the appearance of a diverse assemblage of

TABLE III. Major Differences Between Prokaryotes and Eukaryotes

Prokaryotes	Eukaryotes
Mostly small cells (1–10 μm). All are microbes. The most morphologically complex are filamentous or cycelial with fruiting bodies (e.g., actinomycetes, myxobacteria, blue-green algae)	Mostly large cells (10–100 μm). Some are microbes, most are large organisms. The most morphologically complex are the vertebrates and the flowering plants
Nucleoid, not membrane-bounded	Membrane-bounded nucleus
Cell division direct, mostly by "binary fission." Chromatin body which contains DNA, but no protein. Does not stain with the Feulgen technique. No centrioles, mitotic spindle, or microtubules	Cell division by various forms of mitosis. Many chromosomes containing DNA, RNA, and proteins; they stain bright red with Feulgen technique. Centrioles present in many; mitotic spindle (or at least some arrangement of microtubules) occurs
Sexual systems absent in most forms; when present, unidirectional transfer of genetic material from donor to host takes place	Sexual systems present in most forms; equal participation of both partners (male and female) in fertilization
Multicellular organisms never develop from diploid zygotes. No tissue differentiation	Meiosis produces haploid forms, diploids develop from zygotes. Multicellular organisms show extensive development of tissues
Includes strict anaerobes (these are killed by O_2) and facultatively anaerobic, microaerophilic, and aerobic forms	All forms are aerobic (these need O_2 to live); exceptions are clearly secondary modifications
Enormous variations in the metabolic patterns of the group as a whole. Mitochondria absent; enzymes for oxidation of organic molecules bound to cell membranes, i.e., not packaged separately	Same metabolic patterns of oxidation within the group (i.e., Embden-Meyerhof glucose metabolism, Krebs cycle oxidations, molecular oxygen combines with hydrogens from foodstuffs, and catalyzed by cytochrome, water is produced). Enzymes for oxidation of 3-carbon organic acids are packaged within membrane-bounded sacs (mitochondria)
Simple bacterial flagella, if flagellated, flagellin proteins	Complex (9+2) flagella or cilia if flagellated or ciliated, tubulin proteins
If photosynthetic, enzymes for photosynthesis bound to cell membrane (chromatophores), not packed separately. Anaerobic and aerobic photosynthesis, sulfur deposition, and O_2 elimination	If photosynthetic, enzymes for photosynthesis packaged in membrane-bounded chloroplasts. O_2-eliminating photosynthesis
No intracellular movement	Intracellular movement: phagocytosis, pinocytosis, streaming, etc.

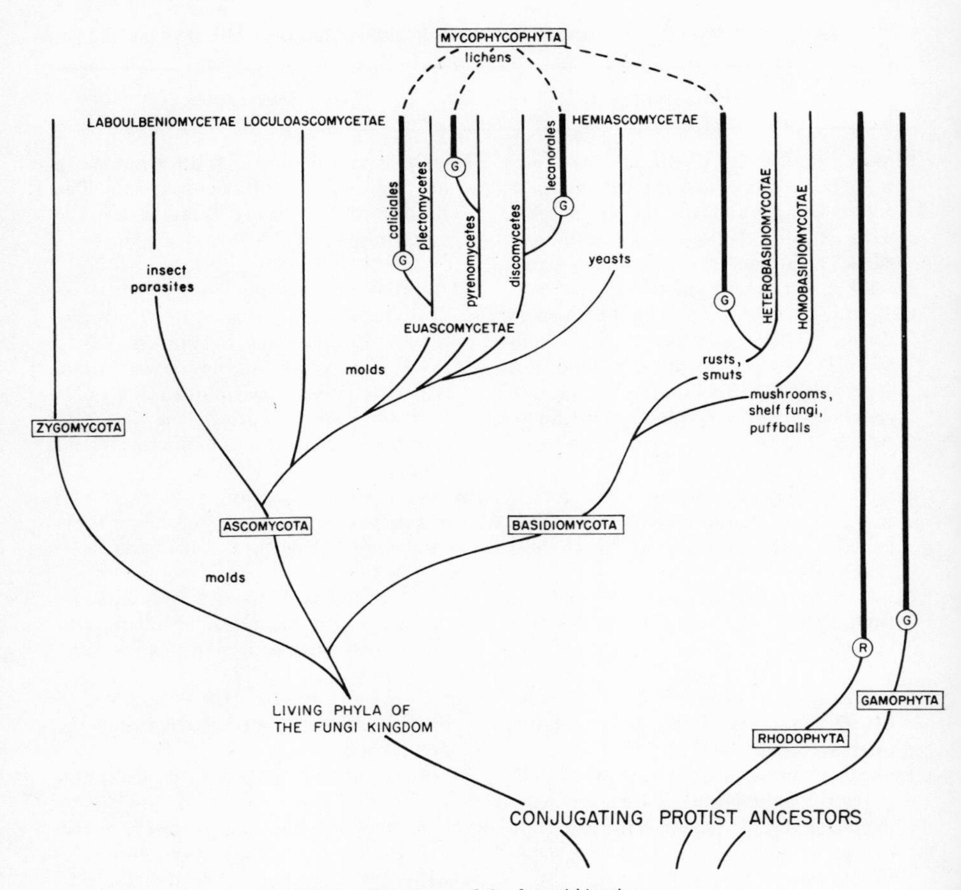

FIG. 1. Phylogeny of the fungal kingdom.

heterotrophic organisms many of which subsequently became hosts to photosynthetic prokaryotes that eventually evolved into plastids (Table IV and Fig. 4). The name "protist" is retained (in favor of, for example, Copeland's "protoctist") simply for its familiarity as a designation for eukaryotes, primarily microorganisms, which are neither metazoan animals, embryophyte green plants, nor amastigote fungi.

The viruses, which are not cells, are incapable of autonomous replication. Since they are best thought of as parts of organisms, each virus group might logically be classified in the appropriate taxa of its host. Since this would be unwieldy, a listing of viruses has been omitted, but see the classification of Lwoff and Tournier (1966), and Whitehouse, (1969).

TABLE IV. The Five Kingdoms of Whittaker Together with a Listing of the Prokaryote Components of Eukaryotic Cells Based on the Cell Symbiosis Theory of Margulis

Kingdom		Description of major groups in kingdoms	Number of original genomes hypothesized
Monera		In order of evolution, from early to middle Precambrian	
		1. Anaerobic heterotrophs (Embden-Meyerhof fermentation = proteoukaryotes) A	Monogenomic
		2. Motile anaerobes = spirochete-like protoflagella F	Monogenomic
		3. Photoautotrophs (CO_2 fixation, O_2 elimination) = photosynthetic protoplastids P	Monogenomic
		4. Aerobic heterotrophs (respiration via the Krebs cycle) = protomitochondria M	Monogenomic
Protista		In order of acquisition of organelles	
	A+M	Some amoebae = mitochondria-containing protists, lacking flagellar homologues	Digenomic
	A+M+F = AMF	The (9+2) amoeboflagellate ancestor, leading to the origin of mitosis and origin of protozoans, slime molds, flagellated fungi, etc.	Trigenomic
	AMF + P = AMFP	Ancestral flagellate with photosynthetic plastids which evolved into chloroplasts, rhodoplasts, etc., leading to the evolution of nucleated algae: brown and red seaweeds, green algae, euglenids, diatoms, etc.	Tetragenomic
Fungi	AMF	Haploid and dikaryotic true fungi: zygomycetes, ascomycetes, and basidiomycetes, from ancestral amoeboflagellate; sacrificed (9+2) flagella during the evolution of mitosis and meiosis	Trigenomic
Animalia	AMF	Metazoa, derived from the ancestral amoeboflagellates	Trigenomic
Plantae	AMFP	Green plants; bryophytes to angiosperms, derived from ancestral green algae	Tetragenomic

PHYLOGENIES AND THE FOSSIL RECORD

Phylogenies, often represented in the form of family trees, should be eclectic summaries of evolutionary information, from many sources, which relate groups of organisms to each other on plots against time. Organisms at a branching fork in an evolutionary line represent the most recent common ancestor of organisms out on a branch. The sequencing of amino acids in protein and nucleotide bases in the nucleic acids is leading to the establishment of quantitative methods of phylogeny generation. For example, estimations of the number of mutations that have occurred since two organisms diverged from a common ancestor can be made and partial phylogenies can be drawn on the basis of primary amino acid sequence data from cytochrome *c* or hemoglobin. Computers have been employed to make the simplest tree that is most consistent with the data (Dayhoff, 1972). These partial phylogenies can be compared with those based primarily on studies of skeletal structures of living organisms, comparative anatomy of soft tissue, and the relationships of these to the fossil forms. The newer comparative biochemical techniques have tended in general to confirm and extend the phylogenetic information concerning metazoans and higher plants gathered over the past century. Preliminary phylogenies generated by comparisons of cytochrome *c* amino acid sequences from 45 species of organisms as diverse as ***Rhodospirillum*** and man have tended to confirm the five-kingdom concept (McLaughlin and Dayhoff, 1973).

In the case of well-recorded animal groups, "the number of phyla is the same as it was about 470 million years ago in the mid-Ordovician as it is now and the number of classes slightly smaller" (Simpson, 1960). The animal kingdom and its living phyla are the same as those of the Ordovician, and at this high taxonomic level the subsequent equilibrium has been static. Within the phyla, several classes and an increasing proportion of the less inclusive groups have become extinct and replaced by groups of separate origin. The proportion of Ordovician species that have living descendants is minute. If Simpson's analysis of the animal kingdom is, in principle, extrapolatable to the other kingdoms, we might expect great stability of the higher taxa (kingdoms and phyla) once they arose, relative to great turnover at the species level. Although paleontology alone can inform us if this has been the case, clearly the notion of "kingdom" and "phylum" influences the data we seek from the record. Phylogenies for the higher taxa of the five kingdoms (as deduced from cell symbiosis theory and based on the Whittaker model) are drawn to the phylum level for the reader's scrutiny in this context (Figs. 1–5). Many of these microbial forms have been raised to higher taxonomic status, and it must be realized that the phylogenies for the monera and protist kingdoms are in a state of flux.

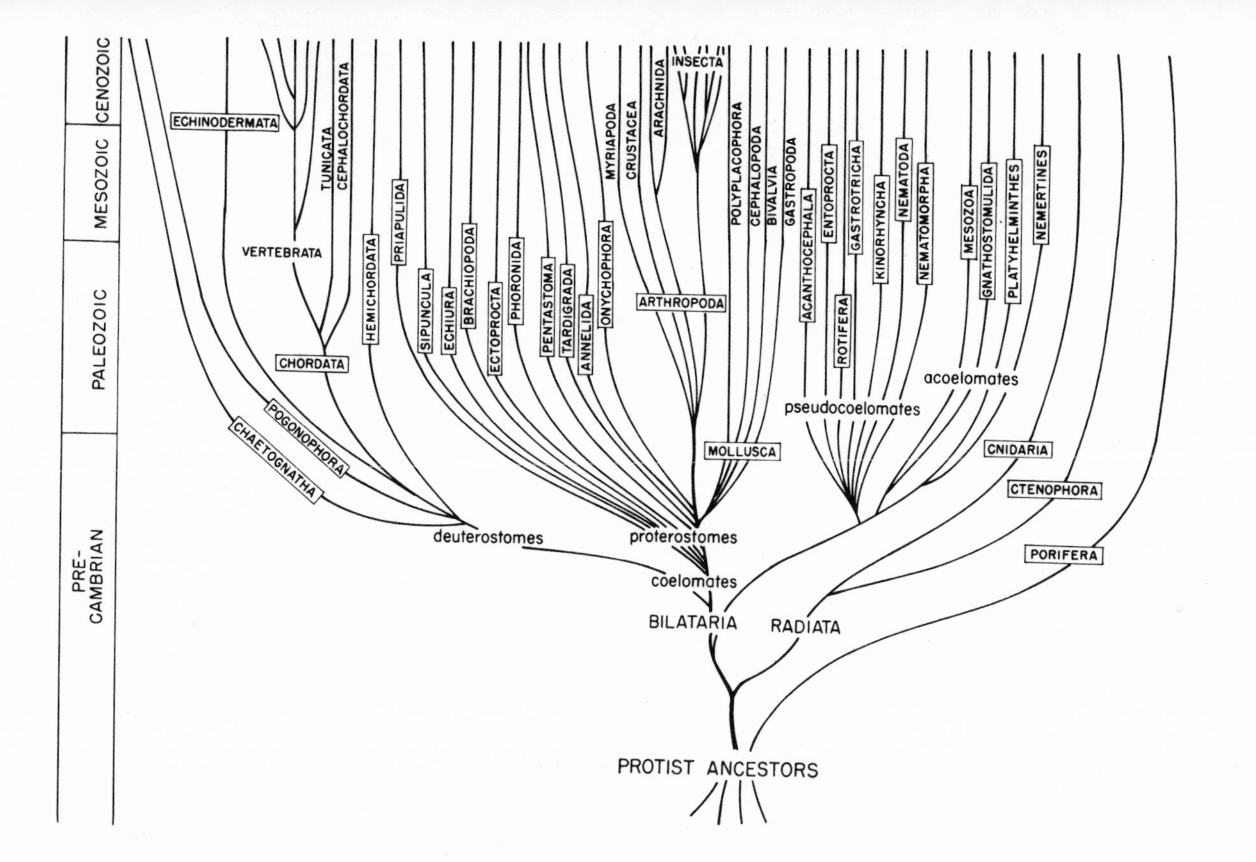

FIG. 2. Phylogeny of the animal kingdom (boxes, phyla, capitals, classes)

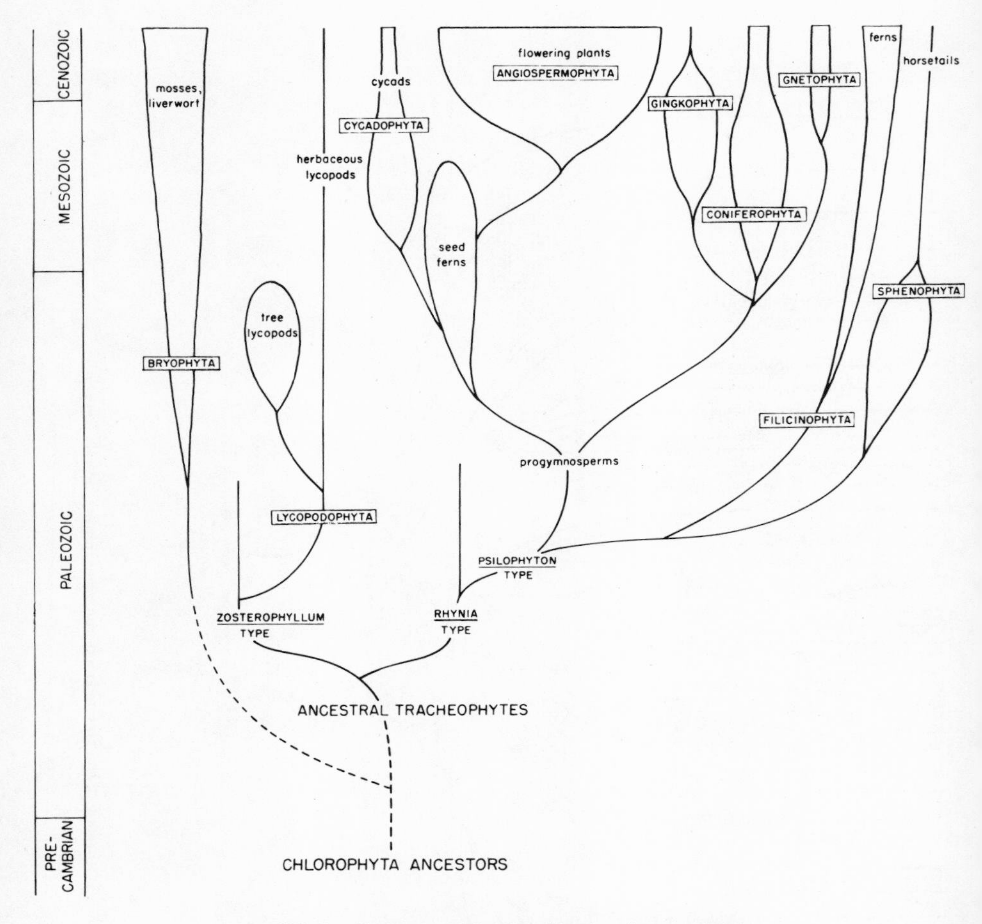

FIG. 3. Phylogeny of the plant kingdom (boxes, phyla).

Because of the fragmentary nature of the fossil record of fungi, that of the fungi kingdom is only slightly more reliable.

However, the phylogenies of the metaphytes (green plants) and metazoans (animals) are based on distinctly more substantial information. Although they have required continual revision as new evidence has accumulated, the classification system in general use summarizes a large body of information concerning the evolutionary history of these larger organisms. The broad outlines of their history during the last 600 million years of the Phanerozoic Eon are known (e.g., see Banks, 1970*a,b,* 1972, for plants and Simpson, 1954, 1961, and Romer, 1968, 1970, for animals). This information is presented in the form of phylogenies for the animal kingdom in Fig. 2 and for the plant kingdom in Fig. 3. The evolutionary process that resulted in the diversity of organisms both extant and preserved in the fossil record can be understood in broad outline on the basis of the neodarwinian synthetic theory of evolution (Lerner, 1968; Simpson, 1954; Mayr, 1970). Data from the fossil record have been a critical source of input for animals and plants. As we have seen, the systematics of monerans, protists, fungi, and so forth has been in general confusion. For a long time, it was thought that microorganisms left no fossil record. Yet it is now realized that the evolutionary history of microbial forms dates back over 3 billion years (Schopf, 1972). Although blue-green algae, radiolarians, foraminiferans, and a few other groups of shelled or sediment-trapping microorganisms are exceptions, in general the microbial fossil record has been so deficient that the inadequate "taxonomy and systematics" of such groups have been based primarily on comparisons of morphology and growth requirements of living forms. Even though the relatively simple structure of these forms provides much less information about relationships than the complex struc-

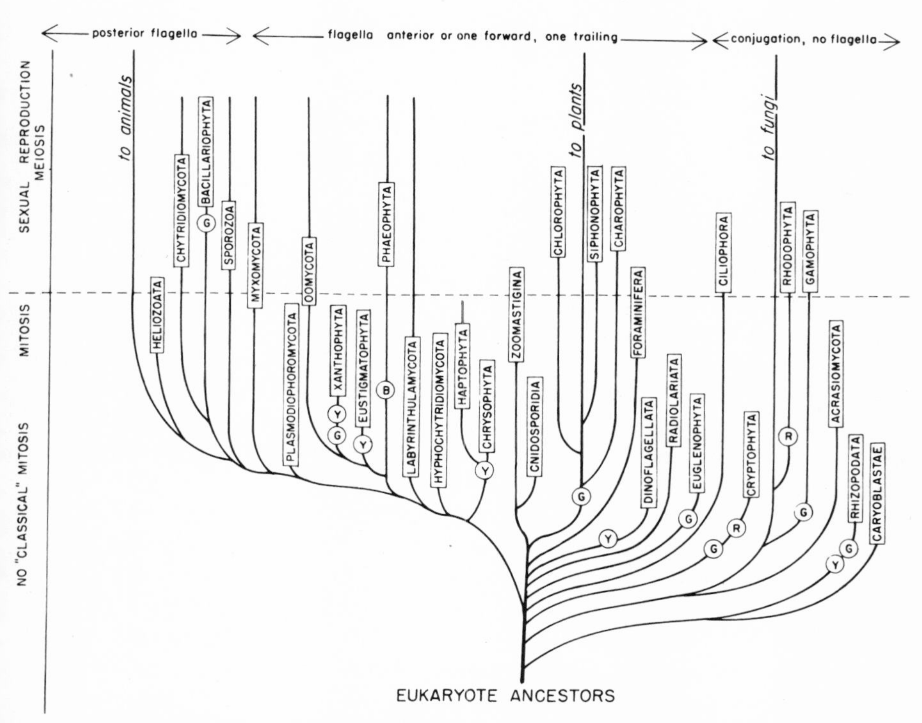

FIG. 4. Phylogeny of the protist kingdom (boxes, phyla; G-green, R-red, Y-yellow plastid colors).

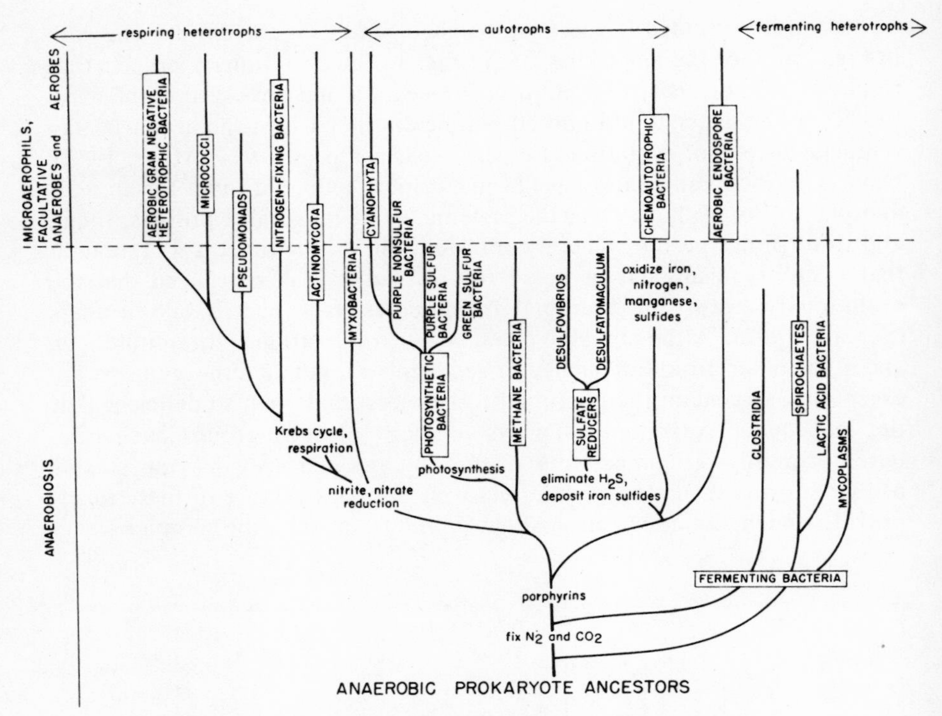

FIG. 5. Phylogeny of the monera kingdom.

ture of multicellular plants and animals, microscopic appearance has been of paramount importance in traditional microbial systematics. Since many of these smaller organisms have no obvious sexual stages, the definition of "species" in practice has depended solely on morphological and physiological frequency distributions in extant microbes. Both the lack of a basis for systematics in the extant forms and the inadequacy of the fossil record have led to an extremely limited knowledge of microbial evolution. Recent developments in molecular biology and micropaleontology are slowly clarifying this picture.

Only since S. Tyler and E. S. Barghoorn developed the techniques to observe ancient microbial fossils embedded in cherts has it been possible to study the micropaleontology of soft-bodied microbial forms. The adaptation of these techniques has already proved immensely fruitful in the interpretation of ancient environments (see Schopf, 1972, for review). It is becoming steadily more apparent that Precambrian calcium carbonate structures such as stromatolites and oncolites are biogenic, sedimentary structures produced in the main by microbial communities dominated by blue-green algae (Golubic, 1973). Slowly, the extent of the Precambrian record, which comprises the first seven-eights of earth history, is being realized. This work in progress, especially combined with organic geochemistry (Eglinton and Murphy, 1969) and comparative biochemistry of modern microbes (DeLey, 1968; Mandel, 1969; Margulis, 1971*b*), presages a much more brilliant future for the reconstruction of the evolutionary history of microbial forms than previously could be anticipated. Although the fossil record is likely never to be as important a tool in the reconstruction of microbial history as in metazoan and metaphytan history, there is little doubt that information from the Precambrian record will alter our notions of the origin and early evolution of life as more becomes known.

SYNOPSIS OF THE SYMBIOTIC THEORY OF THE ORIGIN OF EUKARYOTIC CELLS

The classification presented here (system 7 of Table I) is summarized in Table II. Unlike the classical two- or three-kingdom scheme (systems 1–4 in Table I), systems 5, 6, and 7 are consistent with the cell symbiosis theory (Margulis, 1970). This theory proposes that prokaryotic organisms originated and diversified during the early and middle Precambrian Eon (from 3 to 1.5×10^9 years ago) and that the eukaryotes arose as products of a series of specific symbioses during the middle to late Precambrian (from 2 to 0.6×10^9 years ago). Hereditary endosymbiosis, the process by which one kind of cell becomes an intracellular self-reproducing inhabitant of another kind of cell, is considered to be a major evolutionary mechanism in the origin of certain eukaryotic organelles. A brief résumé of the cell symbiosis theory follows, yet it must be realized that certain hypotheses of the theory are currently being tested and that utility is the reason it is used here as a basis for the classification of the morphologically less complex eukaryotes. For partial phylogenies and a classification of these same organisms based on the nonsymbiotic view, see, for example, Klein and Cronquist (1967). For a critical assessment of the present status of the symbiotic theory of the origin of eukaryotic organelles, see Taylor, (1974).

All terrestrial living systems contain certain organic constituents and a minimal amount of DNA-coded hereditary information. The minimal number of biochemical components (enzymes, nucleic acids, etc.) necessary for free-living replication has been estimated to be greater than 50 (Morowitz, 1967; Margulis, 1971*b*). At least certain organelles of eukaryotic cells, for example, chloroplasts and mitochondria (Cohen, 1970), fulfill "minimal self-replicating criteria," that is, contain most of the items essential for self-replication. I have postulated that these items are present inside chloroplasts and mitochondria precisely because these organelles originated as free-living self-replicating cells (Margulis, 1970). That is, certain free-living prokaryotes are hypothesized to have later become obligate symbionts in the population of complex cellular entities ancestral to all of the organisms in the four eukaryote kingdoms: Protista, Plantae, Animalia, and Fungi. These concepts are outlined in Fig. 6 and Table IV.

The earliest Precambrian microorganisms were, by inference, anaerobic heterotrophs, and probably fermentative bacteria, and it is assumed that their source of food was abiotically produced organic matter (Sylvester-Bradley, 1971). These prokaryotes eventually gave rise to an enormous range of bacterial cells, among them spirochaetes, small motile heterotrophs, and photosynthesizers. Blue-green algae are essentially highly evolved, complex bacterial photosynthesizers (Echlin and Morris, 1965). Mutations of photosynthetic bacteria which permitted the use of water (instead of organic hydrogen, molecular hydrogen, or hydrogen sulfide) as

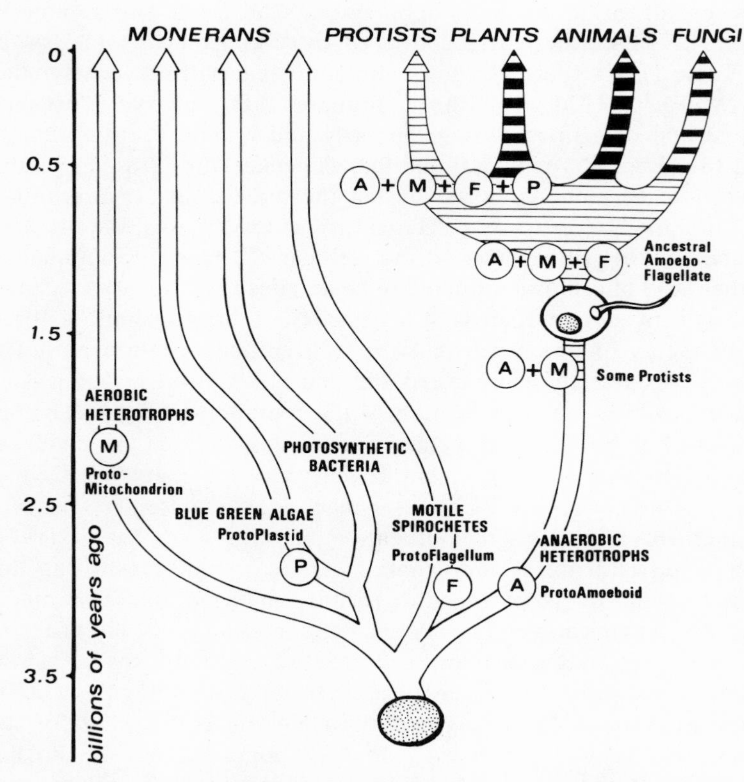

FIG. 6. Phylogenies of the five kingdoms based on the cell symbiosis theory.

hydrogen donors in CO_2 reduction during photosynthesis led eventually to cells that eliminated molecular oxygen into the atmosphere. The blue-green algae arose from these, and they intensively speciated and flourished during the Precambrian. A highly diversified moneran biota then existed, setting the stage for the evolution of eukaryotes.

The production of oxygen forced the development of oxygen tolerance in many of the prokarytotic microbes. Oxygen toleration was followed by the evolution of systems for oxygen utilization. Microaerophilic organisms (those requiring O_2 in quantities less than in the ambient atmosphere) and aerobic bacteria evolved, including a population of bacteria thought to be "protomitochondria." Such protomitochondria are thought to have invaded an anaerobic host, perhaps something like the invasion of ***Pseudomonas*** by ***Bdellovibrio*** (Starr and Seidler, 1971). Such an invasion eventually resulted in the formation of a stable symbiotic complex leading to the primitive mitochondria-containing amoebae. Some of these primitive amoebae became amoeboflagellates when they acquired still another set of symbionts which evolved into the eukaryote flagellum (this aspect of the cell symbiosis theory is the least defensible; see Younger *et al.*, 1972, Margulis, 1974*b*).

The common ancestors of all mitotic eukaryotes were thus amoeboflagellate heterotrophs, cells that are thought to be trigenomic, that is, products of three endocellular symbioses: the protoameboid host Ⓐ plus the protomitochondrial Ⓜ and protoflagellar Ⓕ symbionts (see Fig. 6). Like eukaryotes today, the ancestral eukaryotes were fundamentally aerobic; they depended on many oxygen-mediated metabolic steps including the biosynthesis of steroids. It was in the ancestral eukaryote population that mitosis and meiosis evolved, giving rise to the higher nucleated algae and plants, the fungi, and the metazoans. During the evolution of mitosis and eventually meiosis, some eukaryotes acquired photosynthesis by further hereditary endocellular symbioses, incorporating into their cells prokaryotic algae (such as blue-greens) which, with time, became the plastids Ⓟ. This scheme implies that all nucleate algae and multicellular plants are tetragenomic; that is, they contain a total of the four original genomes abbreviated in Fig. 6 by the letters Ⓐ (amoeboid host), Ⓜ (mitochondria), Ⓕ (flagellum), and Ⓟ (plastid).

This symbiosis theory is consistent with the fact that over the vast stretches of Precambrian time the earth was populated with microbial rather than macroscopic life (Schopf, 1972; Schopf and Blacic, 1971). It utilizes the concept that the metabolism of blue-green algae during the middle Precambrian was the major cause of the transition to an oxidizing atmosphere. The prokaryotic blue-green algae differ radically from all other algae and higher plants, except in their photosynthetic metabolism. Not

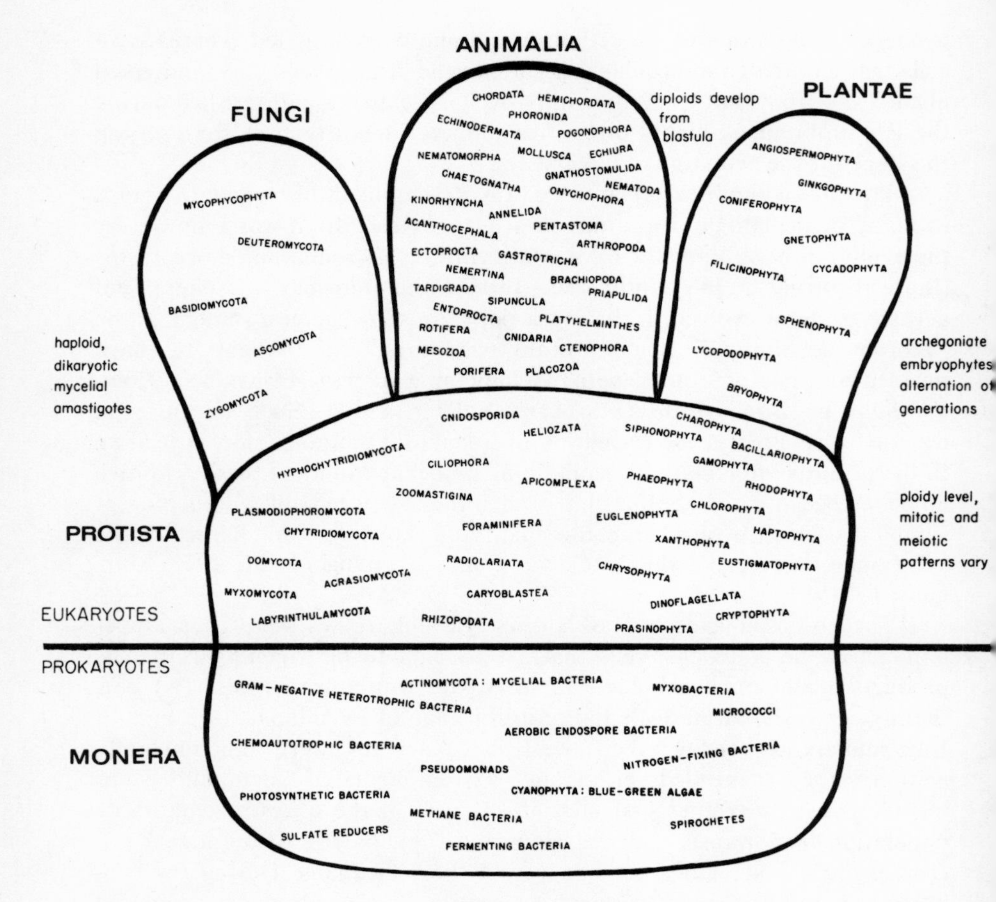

FIG. 7. Phyla of the five kingdoms.

until oxygen was widespread in the terrestrial atmosphere did eukaryotic cells evolve. The elegant mitotic mechanism of eukaryotic chromosome segregation is thought to have thus evolved in a series of protists, resulting in the origin of miscellaneous groups such as dinoflagellates, radiolarians, euglenids, cryptomonads, and others (see Margulis, 1974*b*, and Pickett-Heaps, 1974, for discussion). The stabilization of spindle-mediated mitosis and eventually meiosis was a prerequisite to the subsequent diversification of the fungi, plants, and animals (Margulis, 1970, 1974*b*). The presumed history and the characteristics of the five kingdoms thus defined are summarized in Table II.

The classification of living forms which follows has been devised on the basis of the discussion above. The suggested phyla are shown in Fig. 7. Like all phylogenies and classification systems, it can only be tentative. It should be considered a "working paper." All criticism and comment are actively solicited, for these are problems whose solutions are entirely a function of cooperation between knowledgeable and concerned individuals.

SUPERKINGDOM (CHROMONEMAL ORGANIZATION = PROKARYOTA)

KINGDOM MONERA

A, F, P, M, and others†

Prokaryotic cells, nutrition absorptive, chemosynthetic, photoheterotrophic, or photoautotrophic. Anaerobic, facultative, or aerobic metabolism. Reproduction asexual; chromonemal; recombination unidirectional or viral mediated. Nonmotile or motile by bacterial flagella composed of flagellin proteins or by gliding. Solitary unicellular, filamentous, colonial, or mycelial.

SUBKINGDOM 1. Bacteria

GRADE EUBACTERIA Eubacteria, true bacteria (include A, M, P, and F)

PHYLUM 1. Fermenting bacteria unable to synthesize porphyrins
- Class 1. Mycoplasmas (*Mycoplasma, Thermoplasma*)
- Class 2. Chlamydias (psittacosis group, bedsonias)
- Class 3. Lactic acid bacteria (*Streptococcus, Diplococcus, Leuconostoc*)
- Class 4. Clostridia (anaerobic endospore formers, *Clostridium*)

PHYLUM 2. Spirochaetae, spirochetes (*Cristispira, Treponema, Spirochaeta, Leptospira*) F

PHYLUM 3. Anaerobic sulfate reducers (synthesis of iron hemes limited)
- Class 1. Desulfovibrios (spore-forming sulfate reducers, *Desulfovibrio*)
- Class 2. Desulfatomaculum (non-spore-forming sulfate-reducing bacteria, *Desulfatomaculum*)

PHYLUM 4. Methane bacteria (anaerobic chemoautotrophs: use CO_2 as electron acceptor for anaerobic respiration reducing it to CH_4, *Methanococcus, Methanosarcina, Methanobacterium*)

* Living forms only.

† These letters refer to genomes identified in Fig. 6, explained in Table IV.

PHYLUM 5. Photosynthetic bacteria (synthesis of iron- and magnesium-chelated hemes; carotenoids)
- Class 1. Purple nonsulfur bacteria (*Rhodopseudomonas, Rhodomicrobium, Rhodospirillum*)
- Class 2. Green sulfur bacteria (*Chlorobium*)
- Class 3. Purple sulfur bacteria (*Chromatium*)

PHYLUM 6. Nitrogen-fixing aerobic bacteria (*Azotobacter, Beijerinckia, Rhizobium*)

PHYLUM 7. Pseudomonads (*Pseudomonas, Photobacterium, Hydrogenomonas, Halobacterium*)

PHYLUM 8. Aerobic endospore bacteria, gram positive (*Bacillus*)

PHYLUM 9. Micrococci, gram-positive aerobes, entire Krebs cycle present (*Micrococcus, Sarcina, Gaffyka*)

PHYLUM 10. Chemoautotrophic bacteria
- Class 1. Sulfur-oxidizing bacteria (*Thiobacillus*)
- Class 2. Ammonia-oxidizing bacteria (*Nitrobacter, Nitrosomonas, Nitrosocystis*)
- Class 3. Iron-oxidizing bacteria (*Ferrobacillus*)

PHYLUM 11. Aerobic gram-negative heterotrophic bacteria
- Class 1. Enterobacteria (coliforms), facultative anaerobes (*Escherichia coli, Salmonella, Shigella, Serratia*)
- Class 2. Prosthecate bacteria (caulobacters, hyphomicrobial budding bacteria)
- Class 3. Sphaerotilus group (form distinctive cell aggregates)
- Class 4. Acetic acid bacteria (rectangular sheaths, *Gluconobacter, Acetobacter*)
- Class 5. Moraxella–Neisseria nonflagellated group (*Neisseria, Moraxella, Acinetobacter*)
- Class 6. Predatory bacteria (reproduce inside host, polar flagellated: *Bdellovibrio*)
- Class 7. Microaerophils (polar flagellated helical cells, *Spirillum;* single thick polar flagellum, *Campylobacter*)
- Class 8. Rickettsias, glutamate oxidizers (*Coxiella*)

GRADE GREATER BACTERIA (heterotrophs and chemoautotrophs with distinct morphology at light microscopic level)

PHYLUM 12. Myxobacteria, heterotrophic gliding bacteria
- Class 1. Flexibacteria (*Flexibacter, Saprospira*)
- Class 2. Filamentous gliding bacteria (*Beggiatoa, Leucothrix, Vitreoscilla*)
- Class 3. Fruiting myxobacteria (*Myxococcus, Chondromyces, Podangium*)

PHYLUM 13. Actinomycota, gram-positive coryneform and mycelial bacteria
Class 1. Coryneform bacteria (*Arthrobacter*, *Propionibacterium*)
Class 2. Proactinomycetes (*Mycobacterium*, *Actinomyces*, *Nocardia*)

SUBKINGDOM 2. Aerobic photosynthesizers
PHYLUM 14. Cyanophyta, blue-green algae, aerobic photosynthesis, photosystem II P
Class 1. Coccogoneae, coccoid blue-greens (*sensu lato*)
Class 2. Hormogoneae, filamentous blue-greens (*sensu stricto*)

SUPERKINGDOM (CHROMOSOMAL ORGANIZATION = EUKARYOTA)

KINGDOM PROTISTA (or PROTOCTISTA)

AMF and AMFP

Eukaryotic cells, nutrition ingestive, absorptive or, if photoautotrophic, photosynthesis in photosynthetic plastids. Premitotic and eumitotic asexual reproduction. In eumitotic forms meiosis and fertilization present but life cycle and ploidy levels vary from group to group. Solitary unicellular, colonial unicellular, or multicellular. Complex flagella or cilia composed of microtubules in the (9 + 2) pattern.

GRADE NONMITOTICA (AM)
PHYLUM 1. Caryoblastea, amitotic amoebae (*Pelomyxa palustris*)

GRADE MITOTICA (AMF, AMFP) Microtubular (9 + 2) homologue phyla. Mitotic spindle or equivalent composed of 250 Å microtubules
PHYLUM 2. Dinoflagellata, mesokaryota, dinoflagellates (*Gymnodinium*, *Peridinium*) AMF, AMFP
PHYLUM 3. Chrysophyta, chrysophyte golden yellow algae (*Ochromonas*, *Echinochrysis*, *Sarcinochrysis*) AMFP
PHYLUM 4. Haptophyta, haptophytes or coccolithophorids (*Hymenomonas*, *Pontosphaera*) AMFP
PHYLUM 5. Euglenophyta, euglenids (*Euglena*, *Peranema*, *Astasia*) AMF, AMFP
PHYLUM 6. Cryptophyta, cryptomonads (*Cryptomonas*, *Cyanomonas*, *Cyathomonas*) AMF, AMFP
PHYLUM 7. Zoomastigina, animal flagellates AMF
Class 1. Opalinida (opalinids) (*Opalina*)
Class 2. Choanoflagellida (collared flagellates) or Craspedomonadaceae (*Monosiga*)

Class 3. Bicoecidea (shelled biflagellates, one flagellum attached) (*Bicoeca*)
Class 4. Diplomondida (diplomonads) (*Diplomonas*, *Giardia*)
Class 5. Kinetoplastida (bodos and typanosomes) (*Bodo*, *Crithidia*, *Trypanosoma*)
Class 6. Oxymonadida (oxymonads) (*Oxymonas*)
Class 7. Trichomonadida (trichomonads) (*Trichomonas*)
Class 8. Hypermastigida (hypermastigotes) (*Saccinobaculus*, *Barbulanympha*, *Trichonympha*)

PHYLUM 8. Rhizopodata, rhizopod amoebas (*Difflugia*, *Rhizochrysis*, *Chrysarachnion*) AMF, AMFP
PHYLUM 9. Xanthophyta, yellow-green algae (*Tribonema*, *Botryococcus*) AMFP
PHYLUM 10. Eustigmatophyta, eustigmatophytes (*Pleurochloris*, *Vischeria*) AMFP
PHYLUM 11. Prasinophyta, flagellate algae (*Pyramimonas*, *Platymonas*, *Prasinocladus*) AMFP
PHYLUM 12. Bacillariophyta, diatoms (*Surirella*, *Nitzschia*, *Planktoniella*) AMFP
PHYLUM 13. Phaeophyta, brown algae (*Fucus*, *Dictyota*) AMFP
PHYLUM 14. Rhodophyta, red algae (*Porphyra*, *Nemalion*) AMFP
PHYLUM 15. Gamophyta, conjugating green algae (*Spirogyra*, *Zygnema*, desmids) AMFP
PHYLUM 16. Chlorophyta, grass-green algae (*Chlamydomonas*, *Tetraspora*, *Volvox*) AMFP
PHYLUM 17. Siphonophyta, siphonaceous syncitial green algae (*Acetabularia*, *Caulerpa*, *Codium*) AMFP
PHYLUM 18. Charophyta, stoneworts or charophyte green algae (*Nitella*, *Chara*) AMFP
PHYLUM 19. Heliozoata, sun animalicules (*Echinosphaerium*, *Acintophrys*) AMF
PHYLUM 20. Radiolariata, radiolarians (*Acantharia*, *Thalassicola*) AMF
PHYLUM 21. Foraminifera, foraminiferans (*Globigerina*, *Nodosaria*) AMF
PHYLUM 22. Ciliophora, ciliates AMF
Class 1. Kinetofragmophora (*Tokophyra*, *Entodiniomorpha*, *Trachelocerca*)
Class 2. Oligohymenophora (*Tetrahymena*, *Paramecium*)
Class 3. Polyhymenophora, spirotrichs (*Stentor*, *Stylonichia*, *Euplotes*, tintinnids)

PHYLUM 23. Apicomplexa (Sporozoa, or Telosporidea), sporozoan parasites AMF
- Class 1. Sporozoasida
  - Subclass 1. Gregarinasina (*Gregarina*)
  - Subclass 2. Coccidiasina (*Eimeria, Isospora*) and hemosporidians, (*Plasmodium, Haemoproteus*)
- Class 2. Piroplasmasida (*Babesia*)

PHYLUM 24. Cnidosporidia, cnidosporidian parasites, polar capsules
- Class 1. Microsporida (*Nosema*)
- Class 2. Myxosporida (*Myxobolus, Henneguya*)
- Class 3. Actinomyxida (*Sphaeractinomyxon*)

PHYLUM 25. Labyrinthulamycota, slime net amoebas (*Labyrinthula, Labyrinthorhiza*) AMF

PHYLUM 26. Acrasiomycota, cellular slime molds AMF
- Class 1. Dictyostelia (*Dictyostelium, Polysphondylium*)
- Class 2. Acrasia (*Guttulinopsis, Acrasis*)

PHYLUM 27. Myxomycota, plasmodial slime molds AMF
- Class 1. Protostelida, protostelids (*Protostelium, Ceratiomyxa*)
- Class 2. Myxogastria, myxomycetous plasmodial slime molds (*Physarum, Echinostelium*)

PHYLUM 28. Plasmodiophoromycota, plasmodiophores (*Plasmodiophora, Spongospora, Woronina, Polymyxa*) AMF

PHYLUM 29. Hyphochytridiomycota, hyphochytrids, anterior mastigonemate uniflagellum (*Rhizidiomyces, Anisolpidium*) AMF

PHYLUM 30. Oomycota, oomycetous water molds (*Saprolegnia, Albugo, Achlya, Pythium, Phytophthora*) AMF

PHYLUM 31. Chytridiomycota, chytrids, posteriorly uniflagellated aquatic fungi AMF
- Class 1. Chytridia (*Olpidium, Synchytrium*)
- Class 2. Blastocladia (*Allomyces, Blastocladiella*)
- Class 3. Monoblepharida (*Monoblepharis, Gonapodya*)
- Class 4. Harpochytria (*Harpochytrium*)

KINGDOM FUNGI

AMF 6

Haploid or dikaryotic, mycelial or secondarily unicellular, chitinous walls, absorptive nutrition, lack cells with (9 + 2) homologue motile organelles (i.e., cilia or flagella), body plan branched coenocytic filament which may be divided by perforate septa. Zygotic meiosis, propagate by haploid spores. (Eumycophytes, Denison and Carroll, 1966, Alexopoulos, 1962; amastigote fungi, Copeland, 1956; Whittaker, 1969; Hale, 1967.)

PHYLUM 1. Zygomycota, zygomycetous molds (*Phycomyces*, *Rhizopus*, *Mucor*)

PHYLUM 2. Ascomycota, sac fungi or ascomycetes, ascomycetous molds
- Class 1. Hemiascomycetae, yeasts, leaf curl fungi (*Endomyces*, *Saccharomyces*, *Schizosaccharomyces*)
- Class 2. Euascomycetae, mycelial ascomycetes
- Class 3. Loculoascomycetae, ascostromatic fungi (*Elsinoe*, *Ophiobus*, *Curcurbitana*)
- Class 4. Laboulbeniomycetae, insect parasites (*Laboulbenia*, *Herpomyces*)

PHYLUM 3. Basidiomycota, club fungi
- Class 1. Heterobasidiomycetae
- Class 2. Homobasidiomycetae

PHYLUM 4. Deuteromycota, fungi imperfecti (*Cryptococcus*, *Candida*, *Monilia*, *Histoplasma*)

PHYLUM 5. Mycophycophyta, lichens AMF (fungal component) + AMF (blue-green algal component) or AMF (fungal component + AMFP (green algal component)
- Class 1. Ascolichenes (all species lichenized with ascomycetes)
- Class 2. Basidiolichenes (lichenized with basidiomycetes including species of Herpothallaceae, Coraceae, Clavariaceae, etc.)
- Class 3. Imperfectilichenes (lichenized with fungi imperfecti, fruiting bodies unknown) (e.g., *Cystocoleus*, *Lepraria*, *Lichenothrix*)

KINGDOM ANIMALIA

AMF 6

Multicellular animals, develop from diploid blastula, gametic meiosis, gastrulation occurs. Nutrition heterotrophic, ingestive, phagocytosis, extensive cellular and tissue differentiation.

SUBKINGDOM PARAZOA

PHYLUM 1. Placozoa (diploblastic, dorsoventral organization, no polarity or bilateral symmetry, e.g., *Trichoplax*)

PHYLUM 2. Porifera (sponges)
- Class 1. Calcarea, calcitic spicules
- Class 2. Demospongiae, spongin network with or without siliceous spicules

Class 3. Sclerospongiae, spongin network with or without siliceous spicules and basal skeleton of aragonite
Class 4. Hexactinellida, siliceous spicules with three axes

SUBKINGDOM EUMETAZOA

BRANCH RADIATA (radially symmetrical organisms)

PHYLUM 1. Cnidaria (coelenterates)
Class 1. Hydrozoa (hydroids)
Class 2. Scyphozoa (true jellyfish)
Class 3. Anthozoa (corals and sea anemones)
PHYLUM 2. Ctenophora (comb jellies)

BRANCH BILATERIA (bilaterally symmetrical organisms)

GRADE ACOELOMATA

PHYLUM 3. Mesozoa (mesozoans)
PHYLUM 4. Platyhelminthes (flatworms)
Class 1. Turbellaria (planarians)
Class 2. Trematoda (flukes)
Class 3. Cestoda (tapeworms)
PHYLUM 5. Nemertina (nemertine worms)
PHYLUM 6. Gnathostomulida (gnathostome worms)

GRADE PSEUDOCOELOMATA

PHYLUM 7. Acanthocephala (spiny-headed worms)
PHYLUM 8. Entoprocta (entoprocts or kamptozoa)
PHYLUM 9. Rotifera (rotifers)
PHYLUM 10. Gastrotricha (gastrotrichs)
PHYLUM 11. Kinorhyncha (kinorhynchs)
PHYLUM 12. Nematoda (round worms, *Ascaris*, *Caenorhabditis*)
PHYLUM 13. Nematomorpha (Gordiaceae) (horsehair worms)

GRADE COELOMATA

DIVISION PROTEROSTOMA

PHYLUM 14. Priapulida (priapulids)
PHYLUM 15. Sipuncula (sipunculids or peanut worms)
PHYLUM 16. Mollusca (mollusks)
Class 1. Monoplacophora (monoplacophorans)
Class 2. Aplacophora (solenogasters)
Class 3. Polyplacophora (chitons)
Class 4. Scaphopoda (toothshells)
Class 5. Gastropoda (snails)
Class 6. Bivalvia (bivalves or pelecypods: clams and oysters)
Class 7. Cephalopoda (squids, octopuses)
PHYLUM 17. Echiura (echiuroids)
PHYLUM 18. Annelida (segmented worms)

Class 1. Polychaeta (polychaete worms)
Class 2. Clitellata (leeches, earthworms)
PHYLUM 19. Tardigrada (water bears)
PHYLUM 20. Pentastoma (tongue worms, pentastomes)
PHYLUM 21. Phoronida (phoronids)
PHYLUM 22. Ectoprocta (bryozoa = moss animals)
PHYLUM 23. Brachiopoda (brachiopods)
PHYLUM 24. Onychophora (*Peripatus*)
PHYLUM 25. Arthropoda (jointed-foot animals)
Superclass Chelicerata
Class 1. Merostomata (horseshoe crabs)
Class 2. Pycnogonida (sea spiders)
Class 3. Arachnida (scorpions, harvestmen, ticks, mites, spiders)
Superclass Mandibulata
Class 4. Crustacea
Class 5. Myriapoda
Class 6. Insecta

DIVISION DEUTEROSTOMA

PHYLUM 26. Echinodermata (echinoderms)
Class 1. Crinoidea (sea lilies)
Class 2. Asteroidea (starfish)
Class 3. Ophiuroidea (brittle stars)
Class 4. Echinoidea (sea urchins)
Class 5. Holothuroidea (sea cucumbers)
PHYLUM 27. Chaetognatha (arrow worms)
PHYLUM 28. Pogonophora (beard worms)
PHYLUM 29. Hemichordata (acorn worms)
PHYLUM 30. Chordata (notochord-bearing animals)
SUBPHYLUM 1. Tunicata (urochordates, tunicates, ascidians)
SUBPHYLUM 2. Cephalochordata (lancelets, *Amphioxus*, cephalochordates)
SUBPHYLUM 3. Vertebrata
Class 1. Agnatha (jawless fish, lampreys, hagfishes)
Class 2. Chondrichthyes (cartilagenous fish)
Class 3. Osteichthyes (bony fish)
Class 4. Amphibia
Class 5. Reptilia
Class 6. Aves (birds)
Class 7. Mammalia
Subclass 1. Protheria (egg-laying mammals, *Echidna*, *Platypus*)

Subclass 2. Metatheria (marsupials, kangaroos, opossums)
Subclass 3. Eutheria (placental mammals)

KINGDOM PLANTAE

Autotrophic green plants, advanced tissue differentiation, diploid phase develops from embryos, haplophase develops gamete nuclei by mitosis, primarily terrestrial. Archegoniate.

GRADE BRYOPHYTA (bryophytes, haplophase conspicuous)
- PHYLUM 1. Bryophyta
  - Class 1. Anthocerotae (hornworts)
  - Class 2. Hepaticeae (liverworts)
  - Class 3. Musci (mosses)

GRADE TRACHEOPHYTA (vascular plants, diplophase conspicuous)
- PHYLUM 2. Lycopodophyta (club mosses and quillworts, *Lycopodium, Selaginella, Isoetes*)
- PHYLUM 3. Sphenophyta (Equisetophyta, horsetails, *Equisetum*)
- PHYLUM 4. Filicinophyta (pteridophytes, ferns)
  - Class 1. Filicinae
- PHYLUM 5. Cycadophyta (cycads, *Zamia, Cycas*)
- PHYLUM 6. Ginkgophyta (ginkgos)
- PHYLUM 7. Coniferophyta (conifers)
- PHYLUM 8. Gnetophyta (*Gnetum, Ephedra, Welwitschia*)
- PHYLUM 9. Angiospermophyta (Anthophyta, Magnoliophyta, flowering plants)
  - Class 1. Monocotyledoneae (Liliatae) (monocots)
  - Class 2. Dicotyledoneae (Magnoliatae) (dicots)

ACKNOWLEDGMENTS

I am indebted to many systematic biologists whose work has been referred to and utilized in devising this classification. The referenced published schemes have been modified based on suggestions and comments by my colleagues. I am particularly grateful to S. Banerjee, E. S. Barghoorn, H. P. Banks, H. Bold, H. Booke, J. Bulaffi, G. Carroll, J. Corliss, J. A. Doyle, S. Duncan, A. Echternacht, S. Golubic, K. Grell, W. Hartman, A. Humes, G. E. Hutchinson, G. Leedale, N. D. Levine, L. Olive, G. G. Simpson, N. Todd, T. Varghese, and R. H. Whittaker. I thank NASA (NGR 22-004-025) and the Boston University Graduate School for support.

REFERENCES

Alexopoulos, C. J., 1962, *Introductory Mycology,* 2nd ed., Wiley, New York.

Altman, P. L., and Dittmer, D. S. (eds.), 1972, *Biology Data Book,* Federation of American Societies for Experimental Biology, Bethesda, Md.

Arnold, C. A., 1947, *Introduction to Paleobotany,* McGraw-Hill, New York.

Banks, H. P., 1970*a*, Major evolutionary events and the geological record of plants, *Biol. Rev.* **45**:451–454.

Banks, H. P., 1970*b*, *Evolution and Plants of the Past,* 170 pp., Wadsworth, Belmont, Calif.

Banks, H. P., 1972, The stratigraphic occurrence of early land plants, *Paleontology* **15**:365–397.

Bold, H. C., 1967, *Morphology of Plants,* 2nd ed., Harper and Row, New York and Evanston.

Breed, R. S., Murray, E. G. D., and Smith, N. R., 1957, *Bergey's Manual of Determinative Bacteria,* 7th ed., Balliere, Trudall and Cox, London.

Brock, T. D., 1970, *Biology of Microorganisms,* Prentice-Hall, Englewood Cliffs, N.J.

Campbell, L. L., and Postgate, J. R., 1965, Classification of the spore forming sulfate reducing bacteria, *Bacteriol. Rev.* **29**:359–363.

Cohen, S. S., 1970, Are/were mitochondria and chloroplasts microorganisms? *Am. Scientist* **58**:281–289.

Copeland, H. F., 1956, *Classification of the Lower Organisms,* Pacific Books, Palo Alto, Calif.

Cowan, S. T., 1962, The microbial species—A macromyth? in: *Microbial Classification* (12th Symposium of the Society for General Microbiology) (G. C. Ainsworth and P. H. A. Sneath, eds.), pp. 433–455, Cambridge University Press, London.

Cronquist, A., 1968, *Evolution and Classification of Flowering Plants,* Houghton-Mifflin, Boston.

Cronquist, A., 1971, *Introductory Botany.* 2nd ed., pp. 365–374. Harper & Row, Publ., New York.

Curtis, H., 1968, *Biology,* Worth, New York.

Dayhoff, M. O., 1972, *Atlas of Protein Sequence and Structure,* National Biomedical Research Organization, Bethesda, Md.

Dibble, C. E., and Anderson, A. J. O., 1963, *Florentine Codex, Earthly Things, 11th Book Which Telleth of the Different Animals, the Birds, the Fishes: and the Trees and the Herbs; the Metals Resting in the Earth—Tin, Lead, and Still Others; and the Different Stones,* Published by School of American Research and the University of Utah, Santa Fe, N.M.

Dodson, E. O., 1971, The kingdoms of organisms, *Syst. Zool.* **20**:265–281.

DeLey, J., 1968, Molecular biology and bacterial phylogeny, *Evol. Biol.* **2**:104–154.

Echlin, P., and Morris, I., 1965, The relationship between blue-green algae and bacteria, *Biol. Rev.* **40**:143.

Eglinton, G., and Murphy, M. T., 1969, *Organic Geochemistry,* Springer-Verlag, New York.

Fritsch, F. E., 1935, *The Structure and Reproduction of the Algae,* Vol. 1, Cambridge University Press, London.

Glaessner, M. F., 1968, Biological events and the Precambrian time scale, *Canad. J. Earth Sci.* **5**:585–590.

Golubic, S., 1973, The relationship between blue-green algae and carbonate deposition, in: *The Biology of Blue-Green Algae* (N. G. Carr and B. A. Whitton, Eds.) University of California Press, p. 439–472.

Grant, V., 1971, *Plant Speciation,* Columbia University Press, New York.

Greenwood, P. H., Rosen, D. E., Weitzman, S. H., and Myers, G. S., 1966, Phyletic studies of teleostean fishes, with a provisional classification of living forms, *Bull. Am. Mus. Nat. Hist.* **131**:339–456.

Hale, M. E., Jr., 1967, *The Biology of Lichens*, Edward Arnold, London.

Honigberg, B. M., Balamuth, W., Bovee, E. C., Corliss, J. O., Godjics, M., Hall, R. D., Kudo, R. R., Levine, N. D., Leoblich, A. R., Jr., Weiser, J., and Wenrich, D. H., 1964, A revised classification of the phylum Protozoa, *J. Protozool.* **11**:7–20.

Hutchinson, G. E., 1967, *Treatise on Limnology*, Vol. 2, Wiley, New York.

Hutchinson, J., 1959, *The Families of Flowering Plants*, 2nd ed., Vol. 1: *Dicotyledons*, Clarendon Press, Oxford.

Hutchinson, J., 1969, *Evolution and Phylogeny of Flowering Plants. Dictoyledons: Facts and Theory*, Academic Press, London and New York.

International Code of Botanical Nomenclature (F. A. Stafleu, ed.), 1969, Eleventh International Botanical Congress, Seattle.

International Code of Zoological Nomenclature, (N. R. Stoll, R. P. Dollfus, J. Forest, N. D. Riley, C. W. Sabrosky, C. W. Wright, and R. V. Melville, eds.), 1964, XV International Congress of Zoology, London.

Leedale, G., 1974, How many are the kingdoms of organisms? *Taxon* **23**:37–47.

Lerner, I. M., 1963, *Heredity, Evolution and Society*, Freeman, San Francisco.

Luykx, P., 1970, *Cellular Mechanisms of Chromosome Distribution*, Academic Press, New York.

Lwoff, A., and Tournier, M., 1966, Classification of viruses, *Ann. Rev. Microbiol.* **20**:45–74.

Keeton, W., 1972, *Biological Science*, 2nd ed., 888 pp., Norton, New York.

Klein, R. M., and Cronquist, A., 1967, A consideration of the evolutionary and taxonomic significance of some biochemical, micromorphological and physiological characters in the Thallophyta, *Quart. Rev. Biol.* **42**:105–296.

Mandel, M., 1969, New approaches of bacterial taxonomy: Perspective and prospects, *Ann. Rev. Microbiol.* **23**:239–274.

Margulis, L., 1968, Evolutionary criteria in thallophytes: A radical alternative, *Science* **161**:1020–1022.

Margulis, L., 1970, *Origin of Eukaryotic Cells*, Yale University Press, New Haven.

Margulis, L., 1971*a*, Whittaker's five kingdoms: Minor modifications based on considerations of the origins of mitosis, *Evolution* **25**:242–245.

Margulis, L., 1971*b*, Early cell evolution, in: *Exobiology* (C. Ponnamperuma, ed.), pp. 342–368, North-Holland, Amsterdam.

Margulis, L. 1974*a*, The classification of prokaryotes and eukaryotes, in: *Handbook of Genetics* (R. C. King, ed.), Chap. 1, Plenum Press, New York.

Margulis, L. 1974*b*, On the origin and possible mechanism of colchicine-sensitive mitotic movements, *BioSystems* **6**:16–36.

Margulis, L., 1974*c*, Origin and evolution of the eukaryotic cell, *Taxon* **23**:225–226.

Mayr, E., 1970, *Populations, Species and Evolution*, Harvard University Press, Cambridge, Mass.

McLaughlin, P., and Dayhoff, M. O., 1973, Eukaryote evolution: A view based on cytochrome *c* sequence data, *J. Mol. Evol.* **2**:99–116.

Morowitz, H. J., 1967, Biological self-replicating systems, *Progr. Theoret. Biol.* **1**:35–58.

Olive, L. S., 1970, The Mycetozoa: A revised classification, *Bot. Rev.* **36**:59–89.

Pickett-Heaps, J., 1974, Evolution of mitosis and the eukaryote condition, *BioSystems*, **6**:37–45.

Romer, A. S., 1968, *The Procession of Life* (1972 Anchor Books edition), 384 pp., World, Cleveland.

Romer, A. S., 1970, *The Vertebrate Body,* 4th ed., 452 pp., Saunders, Philadelphia.

Schopf, J. W., 1972, Precambrian Paleobiology, in: *Exobiology* (C. Ponnamperuma and R. Buvet, eds.), pp. 16–61, North-Holland, Amsterdam.

Schopf, J. W., and Blacic, J. M., 1971, New microorganisms from the Bitter Springs Formation (Late Precambrian) of the north-central Amadeus Basin, Australia, *J. Paleontol.* **45**:925–961.

Schulthorpe, C. D., 1967, *The Biology of Aquatic Vascular Plants,* Edward Arnold, London.

Simpson, G. G., 1954, *The Meaning of Evolution,* Harper and Row, New York.

Simpson, G. G., 1960, The history of life, in: *Evolution After Darwin* (S. Tax, ed.), pp. 117–180, University of Chicago Press, Chicago.

Simpson, G. G., 1961, *Principles of Animal Taxonomy,* 247 pp., Columbia University Press, New York.

Simpson, G. G., 1963, *Major Features of Evolution,* Columbia University Press, New York.

Stafleu *et al.* (see International Code).

Stanier, R., Douderoff, M., and Adelberg, E., 1970, *The Microbial World,* 3rd ed., Prentice-Hall, Englewood Cliffs, N.J.

Starr, M. P., and Seidler, R. J., 1971, The Bdellovibrios, *Ann. Rev. Microbiol.* **25**:649–678.

Stoll *et al.* (see International Code).

Sylvester-Bradley, P., 1971, Carbonaceous chondrites and the prebiological origin of food, in: *Molecular Evolution* (L. Buvet and C. Ponnamperuma, eds.), pp. 499–504, North-Holland, Amsterdam.

Taylor, F. J. R., 1974, Implications and extensions of the serial endosymbiosis theory of the origin of eukaryotes, *Taxon* **23**:229–258.

Thomas, C. A., Jr., 1971, The genetic organization of chromosomes, *Ann. Rev. Genet.* **5**:237–256.

Worcel, A., and Burgi, E., 1972, On the structure of the folded chromosome of *E. coli, J. Mol. Biol.* **71**:127–138.

Whitehouse, H. K. L., 1969, *Towards an Understanding of the Mechanism of Heredity,* 2nd ed., St. Martin's Press, New York.

Whittaker, E. H., 1969, New concepts of the kingdoms of organisms, *Science* **163**:150–160.

Younger, K. B., Banerjee, S., Kelleher, J. K., Winston, M., and Margulis, L., 1972, Evidence that the synchronized production of new basal bodies is not associated with DNA synthesis in *Stentor coeruleus, J. Cell Sci.* **11**:621–637.

3

Biochemical Parameters of Fungal Phylogenetics

H. B. LÉJOHN

Department of Microbiology
University of Manitoba
Winnipeg, Manitoba, Canada

INTRODUCTION

The phylogeny of the rather diverse and varied forms of organisms called the Fungi has been a hotbed of debate among mycologists for several decades. Until the late 1950s, almost all phylogenetic schemes of the fungi were based on morphological criteria. But, in the absence of reliable fossil records, all phylogenetic trees proposed had to remain hypothetical since it could not be proven unequivocally whether a particular morphological homology between taxa developed through convergent or parallel evolution.

It is now a universally accepted fact that the basic biochemistry of most living organisms is the same and variations on this theme may be profitably used to determine relatedness of organisms. Biochemistry has its roots in physiology, and Cantino (1955) attempted to devise a phylogenetic network for the Phycomycetes based on the dietary habits of these organisms. In reflection, Cantino's efforts were probably premature, because as the techniques and materials used in growth studies have become improved we find that several organisms previously considered incapable of growing in the absence of certain inorganic and organic compounds are now known to be able to. Furthermore, until chemically defined media are developed that can support the complete growth and reproduction of such organisms, the information obtained from such studies will remain inconclusive.

In 1929, Mez started what can be considered as a pioneering effort to incorporate biochemistry into fungal taxonomy. His approach was based on serological comparisons of fungal species, but his conclusions have since been disputed by other data obtained in recent years by Vogel (1964) and Hutter and DeMoss (1967). These investigators used different biochemical techniques; they analyzed metabolic pathways and the properties of enzymes involved in a pathway in special groups of fungi.

Prior to the mid 1950s, knowledge of the biochemical characteristics of fungi were fragmentary and, in many ways, even rudimentary. With the advent of sensitive and penetrating molecular biology techniques such as nucleic acid hybridization, characterization of proteins by determination of their primary sequences, characterization of ribosomal RNAs by their molecular weights, determination of DNA and RNA base compositions, and detection of the allosteric and regulatory properties of enzymes that act at key loci to regulate the flow of metabolites through the intertwined grid of biochemical pathways, it now appears possible that biologists can develop a more comprehensive system of molecular taxonomy to indicate phylogenetic relationships.

This review will attempt to integrate primarily biochemical and genetic data on the fungi that have gradually been accumulating during the past decade. Some of the material will of necessity be repetitious since it has been considered splendidly by others (Bartnicki-Garcia, 1968, 1970; Vogel, 1965; Vogel *et al.,* 1970; Klein and Cronquist, 1967; Storck and Alexopoulos, 1970; Raper and Flexer, 1970; LéJohn, 1971***a***). Where pertinent, we will attempt to reinterpret data that may not complement the conclusions reached by these authors. Hopefully, this will improve our biochemical reasoning about the fungi.

GENOTYPE: PHENOTYPE–PROTEIN BRIDGE

Proteins occupy a strategic position among all macromolecules in living organisms. Their amino acid sequences reflect the genotypic information encoded in the genome, and, at the same time, proteins act as the primary mediators of cellular metabolism. Consequently, the phenotypic expression of the organism is merely the largely irreversible end product of the diverse protein activities. In this manner, a protein acts as a mobile and plastic bridge linking the genotype with the phenotype, and consequently it should always be endowed with important specialized properties (Watts, 1970). In the case of enzymes, this special property is catalysis. Since the biochemistry of life is unified, it means that similar biochemical reactions would be catalyzed by enzymes of similar though not necessarily identical composition. If an enzyme is to retain its special catalytic attribute, its "active" center(s) would have to be conserved without undergoing radical change during evolution. The rules governing mutational drifts of any *functional* region of the enzyme which does not participate directly as the active site may not be very different from those of the catalytic site. Other than the role of maintaining a specific and favored conformation of the protein, such secondary "active" sites may interact with small molecules (sometimes with the substrates themselves, and at other times with stereochemically unrelated molecules) to facilitate or negate the catalytic process of the enzyme. These "second sites" or allosteric regions of the protein would not experience frequent evolutional changes, although, for adaptive reasons, they might be more susceptible to change than the catalytic regions. A schematic representation of this hypothesis is presented in Fig. 1. It is clear that the environmental forces of natural selection would select the molecular form of an enzyme that befits the physiological and biochemical requirements of metabolism of a particular organism. It is to be expected that differentiation and the course of development would have roots at this molecular level of selection and in time this would be reflected in the morphological expression of the organism. This notion is a reasonable one considering what we now know about gene–enzyme relationships and the properties of enzymes with respect to their allosteric effects, substrate affinities, and amino acid sequences at active sites. These ideas are by no means new. They have their roots in the theory of retrograde evolution of biochemical pathways developed by Horowitz (1945, 1965), various facets of which have been examined by Bonner *et al.* (1965), Watts (1970, 1971), Rutter (1965), Kaplan (1965), Margoliash and Smith (1965), Smith (1970), Fitch and Margoliash (1970), and Joshi *et al.* (1965).

To our way of thinking, the "phenotypic" expression of a regulatory enzyme should act as a dimly-lit beacon of the genotype when homologous enzymes from different taxa are compared. If a particular enzyme has a sufficiently large number of regulatory parameters, then by comparing homologous enzymes that catalyze the same reaction from different species for their regulatory traits, the biological evolutionist would have an easy and simple diagnostic tool. This technique, besides its inherent simplicity and rapidity, can be employed even when very small amounts of material are available, and it has very powerful predictive potentialities. The only disadvantage is that for statistical reliability the enzyme should show more than one regulatory property so as to minimize the risk of complete loss of the property through molecular selection, as indicated in Fig. 1. It is also of

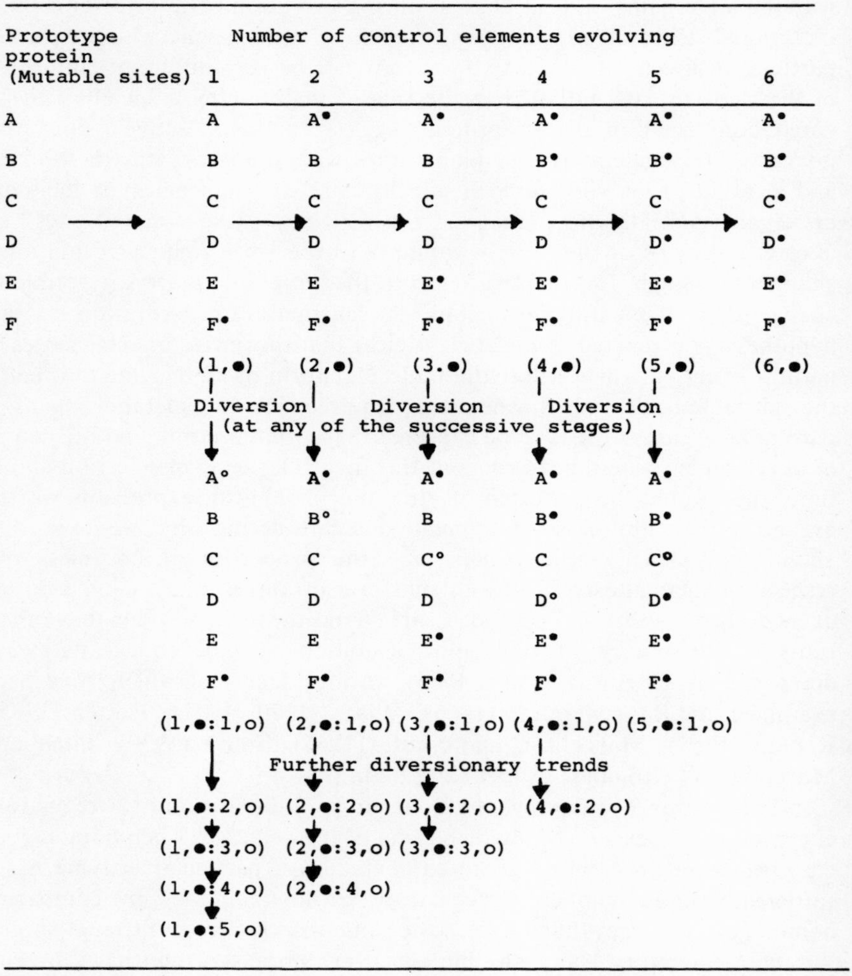

Prototype protein (Mutable sites)	Number of control elements evolving: 1	2	3	4	5	6
A	A	A•	A•	A•	A•	A•
B	B	B	B	B•	B•	B•
C	C	C	C	C	C	C•
D	D	D	D	D	D•	D•
E	E	E	E•	E•	E•	E•
F	F•	F•	F•	F•	F•	F•
	(1,●)	(2,●)	(3.●)	(4,●)	(5,●)	(6,●)

Diversion Diversion Diversion
(at any of the successive stages)

1	2	3	4	5
A°	A•	A•	A•	A•
B	B°	B	B•	B•
C	C	C°	C	C°
D	D	D	D°	D•
E	E	E•	E•	E•
F•	F•	F•	F•	F•
(1,●:1,o)	(2,●:1,o)	(3,●:1,o)	(4,●:1,o)	(5,●:1,o)

Further diversionary trends

1	2	3	4
(1,●:2,o)	(2,●:2,o)	(3,●:2,o)	(4,●:2,o)
(1,●:3,o)	(2,●:3,o)	(3,●:3,o)	
(1,●:4,o)	(2,●:4,o)		
(1,●:5,o)			

FIG. 1. A theoretical treatment of possible progress in evolution of regulatory sites in proteins. ● Mutation to activation; ○, mutation to inhibition. A–F, Sites other than catalytic that affect the function of the protein.

value to have more than one regulatory parameter because a single comparable control characteristic could develop independently through parallel evolution. A unique case of multiple controls on a single enzyme type has been uncovered among the fungi and will be the subject of the latter part of this chapter.

BIOCHEMICAL PARAMETERS

Cell Wall Chemistry

Bartnicki-Garcia (1970) delineated three distinct categories of biochemical criteria that could be used in the phylogenetic organization of fungi. These are (1) relatedness based on chemical similarities of structural components of cells within and between taxa; (2) relatedness based on the distribution of uniquely different biochemical sequences that lead to the same end product; and (3) relatedness based on the fine properties of enzymes such as catalysis, solubility, precipitability, thermolability, and molecular size. Considering the first parameter of chemical similarity, the gross chemical constitution of fungal cell walls can now be appraised by a diversity of techniques such as X-ray diffraction, electron microscopy, enzymatic digestion, infrared spectroscopy, and direct chemical analyses. Each method has been used with varying degrees of success. X-ray studies resolved the conflicting questions about the coexistence of "cellulose" and chitin in the walls of some members of the Phycomycetes—a class that has now disappeared in the taxonomic organization of some modern mycologists (see Alexopoulos, 1962). Fuller and Barshad (1960) used X-ray methods to detect both chitin and cellulose in the walls of the phylogenetically enigmatic mold *Rhizidiomyces,* considered as a member of the class Hyphochytridiomycetes. Lin and Aronson (1970) employed similar techniques to show that *Apodachlya* of the Leptomitales, and a member of the "cellulose fungi," Oomycetes, contains chitin in addition to cellulose. Earlier, Rosinski and Campana (1964) had shown by X-ray studies that cellulose is not confined to the Oomycetes because *Ceratocystis ulmi* of the Sphaeriales, and a member of the Ascomycetes, contains both cellulose and chitin. A report by Cook (1966), which needs X-ray confirmation, is that the desmid parasite *Entophlyctis reticulospora*, a member of the marine Chytridiales, contains cellulose and chitin. The only major group of fungi in which cellulose is yet to be detected is the Basidiomycetes. It would not be surprising if cellulose is found in the Basidiomycetes at some future date for reasons discussed below. On the basis of the spurious distribution of cellulose in all taxonomic groups of fungi, one is forced to seriously consider the adaptive responses of organsms that elaborate this polymer. One thing immediately clear is that those fungal groups in which cellulose is found have an established history of association with higher plants. Many of the Oomycetes, particularly members of the Peronosporales, are parasitic on plants; *Ceratocystis* is the agent of dutch elm disease and *Entophlyctis reticulospora* parasitizes desmids. This is not to imply that other plant parasites such as rusts, vascular wilt disease–producing fungi (*Fusarium*), downy mildews, and a host of others contain cellulose. It is, by the same token, not unlikely that some of the species do. The correlation we suggest should serve, hopefully, as a springboard for new excursions into the subject of cell-wall chemistry of the fungi.

The astute reader might wonder how the capacity to elaborate cellulose could develop in some plant pathogens and not in others. Those fungi with this capacity could have acquired it in ancestral times, first through packaging of the necessary enzymes from plants and subsequently by adapting their own genomes to code for these proteins as a safeguard to insure independent reestablishment. We know today that some animal viruses utilize membranes of their hosts as their own protective coat, and it is thought that this aids their reinvasion of other cells. This characteristic is by no means common to all animal viruses. The fact that some fungi can lay down both cellulose and chitin on the same wall indicates that it is not a matter of preference through natural selection that led to the cell wall specificity shown.

The use of the term "cellulose" is misleading because it is an artificial description based on the precipitable property of a glucose polymer. Cellulose is a β-glucan in which the glucose residues are connected through β-1, 4-linkages. The fact that other β-glucans have glucose residues linked as β-1,3;1,6 does not make them radically different entities from an evolutionary standpoint. To hold that viewpoint would be tantamount to saying that nucleotide residues arranged in 2′,5′- rather than the common 3′,5′-linkage do not represent nucleic acid. The evolutionary significance lies only in the enzymes responsible for their respective syntheses. Since both "cellulose" and the noncellulose form of β-glucan are widespread among the fungi, there are no strong reasons to prefer one type of polymer and not the other in phylogenetic considerations. The enzymes responsible for the synthesis of both types of polymers were, undoubtedly, present early in evolution. With respect to the distribution of cellulose and noncellulose β-glucan (hemicellulose) among the fungi, all major groups except the Heterobasidiomycetes have one or the other or both in their walls (see Bartnicki-Garcia, 1968). When the distribution of chitin in fungal walls is surveyed, only the Oomycetes and anascosporogenous yeasts are found to be devoid of this polymer—not considering the exceptions mentioned earlier. Considering the three types of polymers we have mentioned from the standpoint of biosynthesis, chitin is the most elaborate to make, since it requires at least seven enzymes to convert glucose to the precursor, UDP-*N*-acetylglucosamine, that is used in chitin biosynthesis. The β-glucans, on the other hand, are much simpler to make, as glucose is probably directly converted to the nucleotide sugar by a single enzyme (Fig. 2). The exact steps of these biosynthetic paths are not yet known, and what is presented here is an approximation.

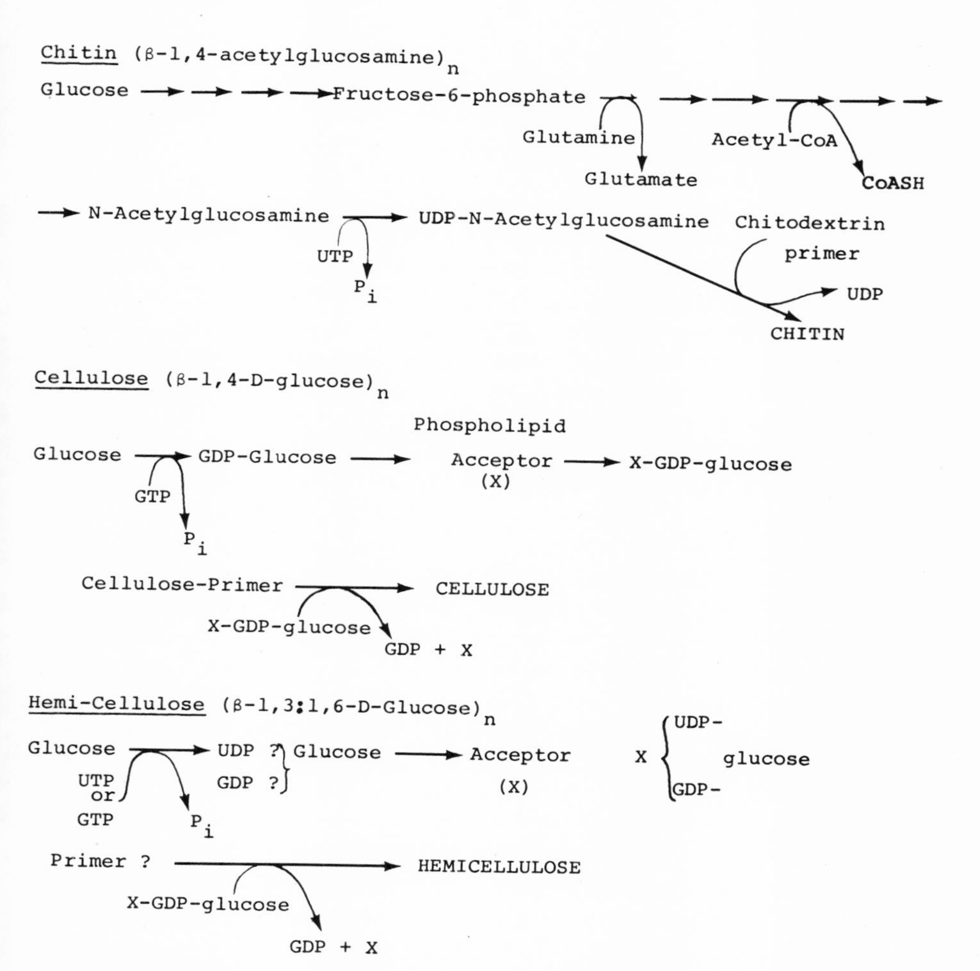

FIG. 2. Probable pathways for the biosynthesis of cell wall polymers chitin, cellulose, and hemicellulose.

One further point worthy of mention is the observation by Bartnicki-Garcia and his associates (Bartnicki-Garcia and Nickerson, 1962; Bartnicki-Garcia and Reyes, 1965; Bartnicki-Garcia, 1968) that although members of the Mucorales do not display the capacity to synthesize β-glucan during vegetative mycelial growth, a β-glucan–polyglucosamine complex is formed in the walls of the sporangiospores. This means that certain genes are permanently turned off during vegetative growth. Laying down of β-glucan in the sporangiospores may have a storage function so that glucose can be acquired easily during germination and early development. The Acrasiales have adapted to the use of glycogen for this storage role.

Returning to the problem of the Oomycetes, one can argue that they constitute a group of phylogenetically related organisms many of which have lost the capacity to synthesize chitin. A host of biochemical traits are found to be common to the Oomycetes and Hyphochytridiomycetes (see later sections) which contain chitin and this would support the concept of similar phylogenetic origins of these two classes. The work of Lin and Aronson (1970) interjects a cautionary note against making generalizations, because *Apodachlya,* a typical Oomycete, contains both chitin and cellulose in its wall. Furthermore, most of the Oomycetes studied in detail have cellulose as a minor component in their walls. The major wall polymer is noncellulose β-glucan (Bartnicki-Garcia, 1968).

Mahadevan and Tatum (1967) have isolated a morphological mutant of *Neurospora crassa* that is incapable of forming chitin and other cell wall structures. A single gene mutation is responsible for this defect, which affects both carbohydrate and cell wall biosynthesis. *Ceratocystis ulmi,* which is related to *Neurospora* on the basis of morphotaxonomy categorization, contains significant amounts of cellulose and chitin. Conceivably, *Ceratocystis* for ecological reasons either developed or retained the capacity to make cellulose, although Bartnicki-Garcia (1968) considers this as a possible case of relatedness between some Ascomycetes and Oomycetes.

The list of fungi in the different taxa that contain cellulose and chitin is growing slowly. In time, a significant number of species will have been examined and it will be no surprise if we find that cellulose is more widespread in fungi than hitherto assumed. It is our feeling that the occurrence of noncellulose β-glucan should be regarded as a more fundamental trait of fungi since it is present in almost all major taxonomic groups.

The present author (LéJohn, 1971*b*) has suggested what could prove to be an easy biochemical screening test to determine whether members of the Oomycetes and Hyphochytridiomycetes contain chitin in their walls. This was based on the data obtained on the stimulatory effects of uridine nucleotides and uridine nucleotide amino sugars on an NAD-linked glutamic dehydrogenase found in these fungi. The few Oomycetes examined and known not to contain chitin in their walls responded to these stimulators, but those containing chitin as well were insensitive. However, several other regulatory facets on the enzyme were retained by both types of Oomycetes including the Hyphochytridiomycetes.

Attempts to phylogenetically relate siphonaceous algae with the Oomycetes on the basis of cell wall composition have met with constant setbacks. A detailed study conducted by Parker *et al.* (1963) comparing the cell wall composition of the siphonaceous alga, *Vaucheria,* with that of several members of the Oomycetes exposed more dissimilarites than similarities. The only useful comparison was that both taxa contained cellulose in their cell walls. But even this similarity was faulted when the properties and the quantity of the celluose were determined. Whereas the fungal walls contained poorly crystalline cellulose polymers that constituted only 25% of the dry weight, *Vaucheria* wall contained highly organized cellulose that made up more than 90% of the dry weight. Furthermore, *Vaucheria* had no hemicelluose, which was the predominant entity in the oomycetous fungi.

Cell wall chemistry, it appears, cannot be considered a reliable indicator of phylogenetic relationships between very different taxa, but it is conceivable that it could serve some useful purpose in delineating species at the generic level. The fungal cell wall is probably too adaptive a structure to depend on for elucidating ancestral lineage, and it may be that it yields relatively easily to selection pressures.

Analysis of Ribosomal RNAs

All living organisms that possess ribosomes have particles that sediment in the ultracentrifuge either at the velocity of 70S, or at 80S, or both; the last are located in cells with cellular organelles such as chloroplasts and mitochondria. It is also clear that the information encoded in the DNA for rRNA production in both eukaryotes and prokaryotes is reiterated, so that there is an extreme redundancy for these genes. Therefore, the mechanism which developed this redundant system must have been selected in primitive cells prior to the evolution of the two major cell types known to us today.

All ribosomes are made up of a pair of unequal subunits referred to as 50S and 30S for the 70S particle and 60S and 40S for the 80S particle. The large particles contain a single RNA polymer which varies in size among the eukaryotes (from 1.25 million to 1.75 million, increasing from algae, higher plants, and protozoa to amphibians, insects, and mammals) (Loening, 1968). The size of this RNA is fairly constant among the prokaryotes. The smaller subunit also contains a single RNA polymer which is smaller than the large-subunit RNA. This polymer, however, is of a constant size in both prokaryotes and eukaryotes, although the eukaryote RNA is about 0.2 million daltons larger (Loening, 1968). Because of the rather specialized function of ribosomes and their large demand for protein synthesis during growth, the rRNAs show a tendency to be conserved throughout evolution. They are, consequently, a very good indicator of phyletic relationships, as are a host of essential enzymes. A third RNA component of ribosomes termed 5S is associated with the large subunit. This RNA may not be present in the ribosomes of mitochondria (Lizardi and Luck, 1971). As very little is known so far about the origin and function of this entity, it will not concern us in this survey.

Loening (1968, 1970) and Wittman (1970) have discussed the evolutionary implications of molecular weight variations of rRNAs in prokaryotes and eukaryotes. Loening's sampling of the fungi was regrettably small, restricted to five members of the Ascomycetes. He provided little information beyond the fact that the fungal rRNAs were similar in size to those of higher plants.

Earlier, Taylor and Storck (1964) had compared the sedimentation behavior of ribosomes from 26 species of fungi and two blue-green algae under identical experimental conditions and had found the 70S and 80S ribosomes true to form, although there was some statistical correlation in the observation that the 70S particles of blue-green algae were slightly smaller than 70S and the 80S particles were fractionally larger then 80S. If we can generalize from this, all major groups of the fungi have similar particles based on size and shape.

Lovett and Haselby (1971) have extended these studies and determined the molecular weights of the ribosomal RNA of 16 fungal species representing all major classes of the fungi except the Deuteromycetes and Discomycetes. As a matter of convenience, we will include the Myxomycetes and Acrasiales when we consider the fungi. They came to the conclusion that the large species of rRNA (which in fungi is about 25S) varied sufficiently between the taxa to serve as a phylogenetic marker. Their data have been modified and summarized in Table I. We estimate that between 100 and 500 nucleotide residues have been removed from the 25S rRNA starting with the taxonomically "primitive" Chytridiomycetes and proceeding to the higher fungal forms such as the Basidiomycetes. This is the opposite of what has been observed in animal cells, where the 28S rRNA tends to a larger size with increase in complexity of the species (Loening, 1968, 1970). Myxomycetes had the largest form of 25S rRNA, and both the Oomycetes and Acrasiales had similar 25S components.

As with several other biochemical properties, the Oomycetes have rRNA that does not show a close affinity to the other fungi on the basis of

TABLE I. Molecular Weights of the Large Ribosomal RNA Species (25S) in Fungi

Organism	Mean average molecular weight ($\times 10^6$)
Acrasiales (1 species)	1.42 ± 0.014
Myxomycetes (1 species)	1.45 ± 0.007
Hyphochytridiomycetes (1 species)	1.36 ± 0.016
Oomycetes (2 species)	1.415 ± 0.007
Chytridiomycetes (4 species)	1.34 ± 0.01
Zygomycetes (2 species)	1.335 ± 0.006
Ascomycetes (2 species)	1.305 ± 0.017
Basidiomycetes (3 species)	1.315 ± 0.01

Modified from Lovett and Haselby (1971).

size. An extra 300–400 nucleotides are estimated to have been added onto their 25S component compared to the Chytridiomycetes and Basidiomycetes. There is, therefore, a breakdown in the pattern of sequential reduction in size on the 25S rRNA as one proceeds up the taxonomic ladder of fungi. Taxonomists who try to preserve order and succession in evolution may consider this an additional criterion for separating the Oomycetes from all other fungi. But the Hyphochytridiomycetes muddle this scheme because, as in many other ways, they display hybrid characteristics typical of both the Chytridiomycetes and Oomycetes, e.g., cell wall composition, thallus structure, enzyme distribution and controls, pathway of lysine biosynthesis, and, here, size of rRNA. Specifically, the size of 25S rRNA of this taxon is closer to that of the Chytridiomycetes, although most of the other properties mentioned above (except thallus structure) are typical of the Oomycetes.

From what is currently known about synthesis and processing of the ribosomal components in eukaryotes, it would be interesting to determine what the nature and size of the rRNA precursors are in the various taxa of fungi. If they are of similar sizes and composition, then one may conclude that size differences result from modifications in the mechanism that controls cleavage of the precursor into the "packaged" form of rRNA.

The %GC content of fungal RNA as determined by Storck (1965) for fungal isolates belonging to the Zygomycetes, Deuteromycetes, Ascomycetes, and Basidiomycetes is relatively constant, varying from 44% to 52%, a difference of 8%. This also matches data obtained by several other investigators, as summarized by Storck. With the observations of

Lovett and Haselby (1971) now available, it would be of interest to determine what the %GC content of the 25S rRNA of these fungi would be. There is a trend of increasing GC content with increase in the molecular weight of the 26–28S rRNA in animal systems (see data provided by Loening, 1968, 1970, and Attardi and Amaldi, 1970).

The available data on rRNA are too sparse to draw reasonable conclusions from in terms of the phylogenetic origins of fungi, but they can be usefully employed with other biochemical parameters, as we show later on.

Base Composition of DNA

As with most biochemical parameters, the base composition of DNA has been used to determine the relationships between different organisms, particularly bacteria (Hill, 1966; Marmur *et al.*, 1963; Sober, 1968) and the blue-green algae (Edelman *et al.*, 1967). Before the impressive study of Storck and Alexopoulos (1970), most of the work on fungal DNA base composition had been carried out by Belozersky and his associates in Russia in 1960 (see Vanyushin *et al.*, 1960). More than 500 fungal isolates were studied by Storck and Alexopoulos (1970) and Storck *et al.* (1971), and their sampling was sufficiently diverse to cover all classes except the Chytridiomycetes, Myxomycetes, and Acrasiales. Their findings should strike a discordant note in the tunes of both morphological and biochemical taxonomists. They came to the conclusion that the phylogenetic "resolving power" of the base composition of fungal DNA is limited even with such an impressive coverage. A summary of their major findings and interpretations is probably useful to reproduce here.

Starting with the Oomycetes, in which 27 isolates selected from two families (Saprolegniaceae and Pythiaceae) were analyzed, the %GC varied from 44.5 to 62.0. A single member of the Leptomitales that had been studied independently and cited by Sober (1968) had a very low %GC content of 27.5. Even within the same genus, *Achlya,* the %GC content fluctuated by 22.5 percentage points. This survey was useful only in delineating the various species of *Achlya* into possible subgenera. The %GC range in the Pythiaceae was somewhat more conserved, varying from 49% to 58%. However, the interesting fact was that this disparity occurred within a single taxonomic species, *Pythium cinnamonii.*

The Mucorales also showed little conservativism in their %GC variation, progressing from a low value of 33.2% to 50.3%. The authors were careful to point out that this chemical parameter of %GC failed to complement the taxonomic relationship suggested by Benjamin (1959) based on morphological criteria. The %GC content of the Mucorales did bring out the fact that the Mycotypha, which have been assimilated with members of the Cunninghamellaceae, should not be categorized as such. The

Mycotypha have a %GC content that varies little between 43% and 77.5%, but the Cunninghamellaceae have a lower %GC value range of 27.5–32.5%.

The Ascomycetes, in general, were more conserved in their %GC variation. A significantly large sample of isolates was studied, and the %GC varied by only 11.5 percentage points among 90 isolates. In fact, most of the %GC values fell between 52% and 54%—a remarkable uniformity! As in most cases in biology, there were exceptions. A single family, the Chaetomiaceae, of which three genera were studied, displayed the complete %GC variation recorded in the Ascomycetes. Furthermore, *C. globosum*, which is a reputable morphological variant, had a constant %GC! This emphasizes the point we made earlier that the bridge between genotype and phenotype, namely proteins in function, should be a useful parameter in the study of microevolution of species.

The Deuteromycetes are considered to be the nonspecialized forms of Ascomycetes, and this conviction was strengthened by the finding that they have %GC content close to that of the Ascomycetes. Some 220 Deuteromycetes were examined, and, with a few exceptions, their %GC content varied little from the range of 48–58%. Fewer than 20 species had %GC content outside this range. Whatever factors cause mutational drifts toward higher or lower %GC content in organisms, they have not affected the Ascomycetes to a significant extent when compared to the Phycomycetes. It is of interest also to point out that the form genus, *Fusarium*, for all its taxonomic problems, has retained its %GC close to 50% (50–53%).

The significant intrageneric variation in %GC observed in some of the Mucorales and the Oomycetes was also seen in the unnaturally grouped "yeasts." The distribution of %GC among 124 yeast isolates ranged from 25% to 70%, although the rather high %GC was witnessed in a single case—*Rhodotorula graminis*. When the %GC content of the anascosporogenous yeast was compared to that of the ascosporogeneous forms in the genus *Candida*, more than 50% of the population sampled showed an overlap. In other words, some *Candida* species would be considered ascosporogenous when they are really not, based on %GC. This bolstered the suspicion long held that some imperfect forms of yeasts may have been erroneously located in the genus *Candida*.

In conclusion, analysis of %GC may be valuable only for phylogenetic comparisons at the intrageneric, not the intergeneric, level.

Metabolic Pathways

Over two decades ago, van Niel (1949) observed that if a biochemical parameter is to be of any value in determining phylogenetic relationships, that property must be distinctively free of either the uniformity witnessed in biochemistry or the diversity displayed in characteristics such as %GC content and cell wall chemistry. Therefore, if a particular end product is reached in metabolism by biochemically distinct routes in different organisms, then the distribution of these pathways in various taxa should reflect to some extent how the organisms did relate to each other in ancestral times. The enzymes, and hence the corresponding genes directing their production, of these biochemically related taxa must have traversed a similar or the same evolutionary causeway at some point in time.

In 1960, Vogel opened the door to a new form of systematics of the fungi by using just this criterion. He showed that the fungi are unique in that they are almost exclusively the only organisms that display a sharp dichotomy in the distribution of two distinct pathways known that lead to the biosynthesis of L-lysine. One path, the α,ϵ-diaminopimelic acid (DAP) route (see Fig. 3), which is the most common among bacteria, blue-green algae, some green algae, the Oomycetes and Hyphochytridiomycetes, and both mono- and dicotyledonous plants, can be regarded as universal. The second path, α-aminoadipic acid (AAA), is confined to the euglenoids and the rest of the fungi. On this basis, most fungi should be regarded as phylogenetically distinct from all other present-day life forms. Not only are the starting materials of the two lysine pathways different, but the intermediates are also unrelated. Seven or eight enzymes are involved in each pathway, which signifies that this difference did not arise through a rather sporadic evolutionary process.

An evolutionary modification of metabolic potential that is quite drastic is the development of these two radically different pathways for lysine biosynthesis. But of interest is the fact that although the AAA lysine *biosynthetic* pathway is restricted to the majority of fungi and is scarcely encountered in other living cells examined so far, yet *degradation* of lysine appears to proceed by a reversal of the AAA route or by a modification of it (see Rodwell, 1969, for a comprehensive review). This is true even for mammals and other organisms that are incapable of synthesizing lysine. However, the exact path of lysine degradation in living cells is in controversy because of the existence of fermentative (anaerobic) and aerobic routes. Differences exist in aerobic paths of lysine degradation among organisms depending on whether nitrogen or carbon is required for growth by the organism. To the writer's knowledge, a degradative route for lysine that involves reversal of the DAP pathway is unknown.

If the idea of Horowitz (1945, 1965) that metabolic pathways evolved through a retrograde evolutionary process is right, then the AAA biosynthetic path in fungi could have developed from metabolic elaborations of the catabolic steps proposed for yeast *Hansenula* (Rothstein and

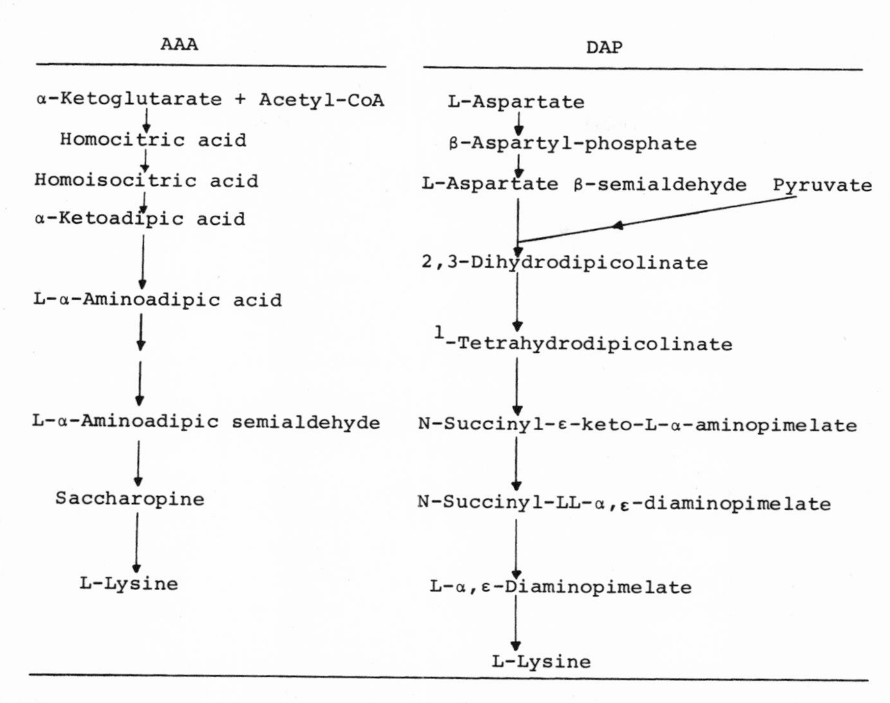

FIG. 3. Pathways of lysine biosynthesis in fungi. DAP pathway in Oomycetes (*Achlya, Sapromyces, Sirolpidium, Pythium,* and *Thraustotheca*) and Hyphochytridiomycetes (*Hyphochytrium* and *Rhizidiomyces*). AAA pathway in Chytridiomycetes (*Rhizophlyctis, Phlyctochytrium, Allomyces,* and *Monoblepharella*), Zygomycetes (*Rhizopus, Cunninghamella,* and *Syncephalastrum*), Hemiascomycetes (*Dipodascus, Tapharina,* and *Saccharomyces*), Euascomycetes (*Penicillium, Neurospora, Gibberella, Venturia, Morchella,* and *Sclerotina*), Heterobasidiomycetes (*Ustilago*), and Homobasidiomycetes (*Polyporus, Coprinus,* and *Calvatia*).

Hart, 1964, 1965) and for mammals such as rats, guinea pigs, and man (Higashino and Lieberman, 1965). The proposed degradation paths are shown in Fig. 4. Why a few fungi such as the Oomycetes and Hyphochytridiomycetes developed the DAP lysine biosynthetic path in place of the AAA path is the real enigma. The DAP lysine biosynthetic pathway appears to be the prevalent route in living cells. It would be interesting to determine whether these fungi with a DAP route degrade lysine at all and, if they do, what the pathway is. As bacteria and other members of prokaryotes conduct lysine synthesis by the DAP pathway, and because DAP is a common constituent of the mucopeptide of many though not all bacterial cell walls, it has been suggested that the conversion of DAP to lysine would be relatively easy to select in nature as it requires only a decarboxylation process.

(Hansenula)

L-lysine ⟶ α-aminoadipic acid ⟶ α-ketoadipic acid

(Mammals)

L-lysine ⟶ saccharopine ⟶ α-aminoadipate-δ-semialdehyde ⟶ α-aminoadipic acid ⟶ α-ketoadipic acid.

FIG. 4. Possible lysine degradation pathways in eukaryotes.

The catabolic and anabolic routes of amino acid metabolism in extant species occur, in most cases, by drastically different paths. This is unlike the central complex reactions of metabolism (glycolysis and the citric acid cycle) that use predominantly the same enzymes reversibly. The weight of evidence supports the notion that catabolic pathways were the initial steps of metabolism during evolution of the primitive cell(s). It is logical, therefore, to assume that lysine catabolism preceded lysine biosynthesis. The AAA pathway or a modification of it should therefore be the ancestral route of the two existing biosynthetic pathways. This is in contradistinction to the proposal made by Vogel *et al.* (1970), an opinion which we have supported (LéJohn, 1971*b*). Vogel based his argument purely on the grounds of lysine biosynthesis and did not consider the important catabolic route as well—probably with good reason since the catabolic pathway is still in doubt. If the AAA lysine biosynthetic path did precede the DAP pathway, then we must ask why some fungi have the AAA pathway and others the DAP pathway. In the prokaryotic bacteria, genetic elements of both pathways presumably exist, although the AAA pathway may be adapted for catabolism. In higher animals, some elements of the AAA pathway exist, although lysine is not made but dismutated.

The vague picture that emerges from all this is that prokaryotes may have a complete DAP lysine biosynthetic pathway and a partial and modified AAA lysine degradative pathway. Eukaryotes (higher animals), on the other hand, may have only a modified AAA pathway of lysine degradation. Eukaryotes (higher plants) possess the DAP pathway for lysine biosynthesis. Eukaryotes (fungi) have one or the other type of pathway, both of which operate in the biosynthetic mode. A possible evolutionary line of descent of the biosynthetic and degradative pathways of lysine is shown in Fig. 5. No matter how these ancestral groups are arranged, the fungi show a definite dichotomy, one stream leading to the Oomycetes, Hyphochytridiomycetes, green algae, and higher plants, the other to all other fungal species.

Evolution of Enzyme Systems

Except for the yeasts, most fungi do not have the tremendously flexible abilities of bacteria to adapt quickly to nutritional fluxes in their environment. Therefore, it seems that fungi have developed the capacity to produce a plethora of resting (dormant) cells such as conidiospores, sporangiospores, and oidiospores that retain the capacity to reinitiate and complete the growth cycle when the environment is favorable. We should therefore not expect to see in fungi enzymatic mechanisms of adaptation commonly found in bacteria. When dealing with aquatic fungi that multiply and disperse via motile cells (zoospores), adaptive controls similar to those found in algae would be expected. Terrestrial fungi should have a different system of adaptation in view of the constantly hostile and changing environment they inhabit. In fact, we see that aquatic molds generally produce motile cells devoid of a cell wall, but terrestrial fungi produce, mostly, nonmotile cells with strong protective envelopes.

Just as the morphology of the fungal cell has evolved in drastically dif-

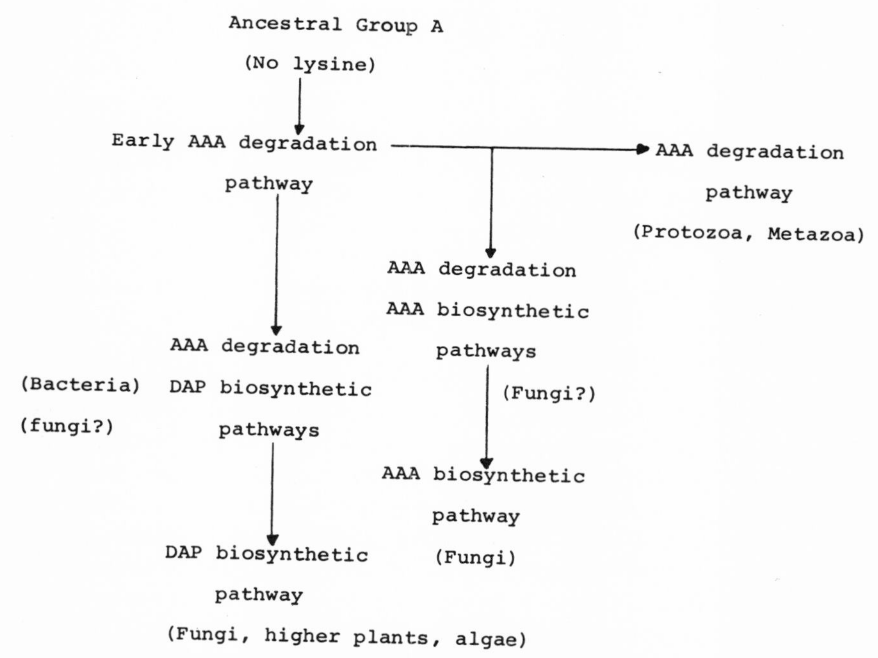

FIG. 5. Integration of catabolic and anabolic pathways of lysine metabolism and phylogenetic implications.

ferent ways in response to the environment, so, it seems, have the metabolic potentialities. In response to its environment, the cell can either (1) lose selectively those metabolic capabilites that become redundant or (2) acquire, elaborate, and improve on existing metabolic abilities until, ultimately, completely new capacities develop. Both mechanisms can operate simultaneously on the same organism so that eventually some enzyme systems can become defunct while novel systems evolve. Enzyme systems can change from inducibility to constitutivity or *vice versa*. Alternatively, the allosteric effects on an enzyme or constellation of enzymes may be modified.

Enzymes of extant fungal species must have undergone a long, arduous evolutionary path. Homologous enzymes from various fungal species must have been subjected to the rigors of selection over and over again so that they are tailored to meet their current-day environmental needs. Therefore, when similarities exist in the sensory abilities of such enzymes with respect to their physiological and biochemical environments, the information obtained tells not only of species character but also of possible parallelism in evolution when different taxa are involved. With the aid of other criteria, the problem of parallelism can be eliminated and a monophyletic tree can be developed for such species.

Dehydrogenases. During our studies on the metabolic controls of enzymes in several fungal isolates, it became clear that the sensory abilities of homologous enzymes from the "lower" fungi formed an excellent taxonomic criterion. Six distinctly different enzymes were selected for careful study. They are NADP- and NAD-linked glutamic dehydrogenases, NAD- and NADP-linked isocitric dehydrogenases, and L(+)– and D(–)-lactic dehydrogenases (NAD-linked). The distribution (Table II) of these enzymes among the various fungal isolates complemented to a great extent the taxa segregation proposed by mycologists on the basis of morphology and other structural criteria.

The first striking feature is that all the fungi belonging to the Acrasiales and Phycomycetes (Chytridiales, Blastocladiales, Hyphochytriales, Saprolegniales, Leptomitales, Lagenidiales, Peronosporales, and Mucorales) possess only an NAD-linked glutamic dehydrogenase. Members of the Ascomycetes, Discomycetes, Basidiomycetes, and Deuteromycetes have the NADP-linked glutamic dehydrogenase as well (Table III). An NAD-linked isocitric dehydrogenase was found only in the Chytridiales, Blastocladiales, and Mucorales of the Phycomycetes (Table II). The limited data available suggest that this enzyme may be widespread among the higher fungi. In this regard, the Oomycetes are an exception. The NADP-linked isocitric dehydrogenase is common to all groups of fungi.

TABLE II. Distribution of Glutamic Dehydrogenases in Fungi

Organism	Order	Class	Coenzyme specificity	
			NAD-linked	NADP-linked
Dictyostelium discoideum	Acrasiales	Myxomycetes	+	–
Polysphondylium violaceum	Acrasiales	Myxomycetes	+	–
Entophlyctis sp.	Chytridiales	Chytridiomycetes	+	–
Rhizophlyctis rosea	Chytridiales	Chytridiomycetes	+	–
Rhizophydium sphaerocarpum	Chytridiales	Chytridiomycetes	+	–
Allomyces cystogenus[a]	Blastocladiales	Chytridiomycetes	+	–
A. arbuscula[a]	Blastocladiales	Chytridiomycetes	+	–
A. javanicus[a]	Blastocladiales	Chytridiomycetes	+	–
Blastocladiella emersonii[a]	Blastocladiales	Chytridiomycetes	+	–
Rhizidiomyces apophysatus[a]	Hyphochytriales	Hyphochytridiomycetes	+	–
Hypochytrium catenoides	Hyphochytriales	Hyphochytridiomycetes	+	–
Aplanopsis terrestris	Saprolegniales	Oomycetes	+	–
Aphanomyces laevis	Saprolegniales	Oomycetes	+	–
A. stellatus	Saprolegniales	Oomycetes	+	–
A. cochlioides	Saprolegniales	Oomycetes	+	–
Thraustotheca clavata	Saprolegniales	Oomycetes	+	–
Achlya sp. (1969)[a]	Saprolegniales	Oomycetes	+	–
A. radiosa	Saprolegniales	Oomycetes	+	–
A. bisexualis (male)	Saprolegniales	Oomycetes	+	–
A. bisexualis (female)	Saprolegniales	Oomycetes	+	–
A. ambisexualis (male)	Saprolegniales	Oomycetes	+	–
A. ambisexualis (female)	Saprolegniales	Oomycetes	+	–
A. americana	Saprolegniales	Oomycetes	+	–
A. flagellata[a]	Saprolegniales	Oomycetes	+	–
Saprolegnia parasitica	Saprolegniales	Oomycetes	+	–

TABLE II. (Continued)

Organism	Order	Class	Coenzyme specificity	
			NAD-linked	NADP-linked
S. megasperma[a]	Saprolegniales	Oomycetes	+	−
S. mixta	Saprolegniales	Oomycetes	+	−
S. terrestris[a]	Saprolegniales	Oomycetes	+	−
S. delica	Saprolegniales	Oomycetes	+	−
Protoachlya paradoxa[a]	Saprolegniales	Oomycetes	+	−
Isoachlya eccentrica[a]	Saprolegniales	Oomycetes	+	−
I. intermedia	Saprolegniales	Oomycetes	+	−
I. unispora	Saprolegniales	Oomycetes	+	−
Aphanomyces euteiches	Saprolegniales	Oomycetes	+	−
Apodachlya brachynema	Leptomitales	Oomycetes	+	−
Sapromyces elongatus	Leptomitales	Oomycetes	+	−
Pythium debaryanum	Peronosporales	Oomycetes	+	−
P. oligandrum	Peronosporales	Oomycetes	+	−
P. undulatum	Peronosporales	Oomycetes	+	−
P. splendens	Peronosporales	Oomycetes	+	−
P. catenulatum	Peronosporales	Oomycetes	+	−
P. butleri	Peronosporales	Oomycetes	+	−
P. splendens	Peronosporales	Oomycetes	+	−
P. ultimum	Peronosporales	Oomycetes	+	−
P. heterothallicum (male)	Peronosporales	Oomycetes	+	−
P. heterothallicum (female)	Peronosporales	Oomycetes	+	−
P. sylvaticum (male)	Peronosporales	Oomycetes	+	−
P. sylvaticum (female)	Peronosporales	Oomycetes	+	−
P. intermedium (+)	Peronosporales	Oomycetes	+	−
P. intermedium (−)	Peronosporales	Oomycetes	+	−
Phytophthora cinnamoni	Peronosporales	Oomycetes	+	−

Phytophthora palmivora	Peronosporales	Oomycetes	+	−
Absidia glauca	Mucorales	Zygomycetes	+	−
Cunninghamella blakesleeana	Mucorales	Zygomycetes	+	−
Mucor hiemalis	Mucorales	Zygomycetes	+	−
Rhizopus stolonifer	Mucorales	Zygomycetes	+	−
Zygorynchus mollieri	Mucorales	Zygomycetes	+	−
Phycomyces blakesleeanus	Mucorales	Zygomycetes	+	−
Piricularia oryzae[b]	Moniliales	Deuteromycetes	+	+
Fusarium oxysporum[b]	Moniliales	Deuteromycetes	+	+
Candida utilis[b]	Moniliales	Deuteromycetes	+	+
Hypomyces rosellus	Hypocreales	Ascomycetes	+	+
Saccharomyces cerevisiae	Endomycetales	Ascomycetes	+	+
Hansenula subpelliculosa[b]	Endomycetales	Ascomycetes	+	+
Sordaria fimicola (+ and −)	Sphaeriales	Ascomycetes	+	+
Neurospora crassa	Sphaeriales	Ascomycetes	+	+
Coprinus lagopus[b]	Agaricales	Basidiomycetes	+	+
Schizophyllum commune[b]	Agaricales	Basidiomycetes	+	+

[a] Cultures obtained from private collection of Drs. J. S. Lovett (Purdue University), R. Emerson (UCLA), R. Beneke (Michigan State), and R. S. Fuller (Georgia). All other cultures were purchased from American Type Culture Collection (USA) and the Centra-albureau voor Schimmelcultures (Baarn, Netherlands).

[b] Data taken from Casselton (1969).

TABLE III. Some Chemical Parameters of Fungal Taxonomy

Taxonomic group	%GC-DNA (range)	"25S" rRNA (× 10^6)	Major cell wall constituent				
			Cellulose	Hemicellulose	Chitin	Chitosan	Mannan
Acrasiales		1.42	+				
Chytridiomycetes		1.32–1.36		+	+		
Hyphochytridio-mycetes		1.36	+		+		
Oomycetes	40.5–62	1.40–1.43	+	+			
Zygomycetes	27.5–59	1.33–1.34					
Mycelia					+	+	
Sporangia				+	+		
Ascomycetes							
Yeasts	25.0–70	1.31		+			+
Others	48.5–60	1.30		+	+		
Deuteromycetes	35.5–64.5			+	+		
Discomycetes							
Heterobasidiomycetes		1.32			+		+
Homobasidiomycetes	50–59.5	1.30–1.32		+	+		
Myxomycetes		1.45					

Interestingly, all the Phycomycetes and Acrasiales harbor only the NAD-linked D(−)-lactate dehydrogenase except for the Zygomycetes that elaborate the NAD-linked L(+) form as well.

The unique distribution of the two coenzyme-specific forms of glutamic dehydrogenase provided us with a valuable criterion to use in phylogenetic comparisons of the lower fungi by studying their catalytic and regulatory properties. Based on the control properties of the enzyme obtained from various sources, three distinct enzyme types and one subtype could be recognized. Type I (unregulated) form of glutamic dehydrogenase is widespread among all Chytridiales and Mucorales with a few exceptions. The Blastocladiales and a genus of the Mucorales, *Absidia,* have a form of the enzyme that shows complex multivalent controls. Bivalent cations such as Ca^{2+} and Mn^{2+} activate the reductive amination reaction leading to the biosynthesis of glutamate but inhibit the reverse reaction, a unidirectional control described in detail earlier (LéJohn, 1968). Adenylates modulate the catalytic activity in such a way that AMP activates and ATP inhibits the enzyme reversibly. This enzyme type is classified as type II. The existence of type II enzyme in *Absidia* was surprising since members of this genus appear to have adapted morphologically to a terrestrial habitat, whereas the Blastocladiales are aquatic forms of fungi. As a start, this raised the interesting possibility that the controls on this enzyme evolved similarly through convergent evolution. But the complexity of the controlling effects observed makes this an unlikely possibility. For one thing, the modulating effects of Ca^{2+} and Mn^{2+} would be expected to evolve among freshwater species where the concentrations of these cations are rather low. In soil, there is an abundance of these cations and there is no apparent advantage to using them as fine modulators of an enzyme such as glutamic dehydrogenase. An alternate explanation is that they may represent the "fossils" of a control system developed in primitive ancestors of *Absidia.* We have yet to encounter these bivalent cation and adenylate effects on other isolates of the Zygomycetes.

The Oomycetes and Hyphochytridiomycetes harbor the type III glutamic dehydrogenase, which is distinguishable by the following features. Although NAD is the coenzyme substrate, NADP (H) interacts with the enzyme only as an allosteric activator (LéJohn *et al.,* 1970; LéJohn and Stevenson, 1970). In contrast, GTP and ATP activate the type III enzyme but inhibit the type II form. AMP, an activator of the type II enzyme, inhibits the type II species. Hyphochytridiomycetes possess a type III enzyme that responds somewhat differently to AMP and ATP, the former activating and the latter and GTP inhibiting; but in many other ways, such as activation by phosphoenolpyruvate and short-chain acyl-CoA derivatives and inhibition by long-chain acyl-CoA derivatives, citrate, Ca^{2+}, and Mn^{2+}, the properties are similar to those found in the enzyme from the Oomycetes. In a way, the type III enzyme of the Hyphochytridiomycetes is intermediate (hybrid?) between the typical type II and type III controls. It is also noteworthy that synthesis of the type III enzyme is sensitive to glucose catabolism, being considerably repressed when glucose is actively metabolized and induced when it is not. This is observed in all of the Oomycetes and Hyphochytridiomycetes studied. On the basis of enzyme controls, lysine biosynthetic pathway, cell wall composition, and molecular size of the 25S rRNA component, the Hyphochytridiomycetes have proved to be a fascinating hybrid of typical Chytridiomycetes and Oomycetes characteristics. A summary of several biochemical parameters on which the Phycomycetes could be distinguished is presented in Table III. It would be of value to determine the %GC and other properties listed that are currently not known for the Hyphochytridiomycetes.

Another aspect of the controls elicited on the glutamic dehydrogenases of the Oomycetes is the activating effects of uridylates and uridine nucleotide amino sugars on enzyme obtained from isolates known not to contain chitin in their cell walls. Whenever chitin is present, as in *Apodachlya* (Leptomitales) and *Rhizidiomyces* (Hyphochytriales), these allosteric ligands were without effect (LéJohn, 1971*b*). One possible physiological implication of this control is clear. It appears that those Oomycetes that can develop chitin in their walls during growth must have constitutive levels of enzymes responsible for chitin biosynthesis. The amino group of the amino sugar precursors is derived from glutamine (glutamate). Therefore, the supply of glutamate is crucial to cells that may have inducible enzyme systems for biosynthesis of amino sugars. Those Oomycetes that do not produce chitin in their cell walls may have some of the terminal enzymes permanently repressed and the only amino sugars synthesized may be those required for membrane and other metabolic functions rather than for cell wall development.

Taken together, the enzymatic controls operative on glutamic dehydrogenases from the Blastocladiales, some Mucorales, Hyphochytriales, Leptomitales, Saprolegniales, and Peronosporales interrelate more at the order level than at the class level. In this regard, this biochemical parameter appears to be more reliable than %GC content of DNA, molecular weights of ribosomal RNA, and even chemical composition of cell walls. It also brings out clearly the possibility of detecting intergeneric relationships. For instance, it is a distinct possibility that some of the Mucorales are related to the Blastocladiales, that the Hyphochytriales are closer to the Oomycetes than hitherto assumed, and that the Blastocladiales and Hyphochytriales may also share some common ancestor or have traversed similar environmental paths at some stage.

TABLE IV. Enzyme Controls Represented as Types[a]

Taxonomic group	NAD-GDH	D(−)-LDH	Enzyme organization	
			Tryptophan	Aromatic cluster
Chytrodiomycetes				
Chytridiales	I	I	I	Common
Blastocladiales	II	I	I	Common
Hyphochytridiomycetes	III (h)	II	?	?
Oomycetes	III	II	IV	Common
Zygomycetes	I & II	I	III	Common
Ascomycetes				
Yeasts	I[b]	III	II, III	Common
Others	I, IV[c]	—	I	Common
Deuteromycetes	I[b]	—	I	Common
Discomycetes	I[b]	—	I	?
Heterobasidiomycetes	I[b]	—	III	Common
Homobasidiomycetes	I[b]	—	I	?
Myxomycetes	?	?	I	?
Acrasiales	III[b]	II[b]	?	?

[a] Definitions: NAD-dependent glutamic dehydrogenases: Type I, No activators or inhibitors yet detected. Type II, Activators (Ca^{2+}, Mn^{2+}, AMP) and inhibitors (citrate, chelons, ATP, fructose-1,6-diphosphate, GTP). Type III, Activators [NADP (H), ATP, acetyl-CoA, and short-chain acyl-CoA derivatives, uridylates and uridine nucleotide amino sugars, GTP, and P-enolpyruvate] and inhibitors (AMP, citrate, long-chain acyl-CoA derivatives, Ca^{2+}, and Mn^{2+}). Type III(h), intermediate, Same as type III except that GTP and ATP are inhibitors while AMP is an activator. Type IV, Requires further resolution in terms of enzyme regulatory patterns. D(−)-NAD-dependent lactate dehydrogenases: Type I, Has GTP as an allosteric inhibitor and ATP as an active site competitive inhibitor. Type II, Has ATP as an active site competitive inhibitor only. Type III, Cytochrome-linked enzyme. Tryptophan pathway enzymes: The definitions of Hutter and DeMoss are (1967) followed. Aromatic cluster: See Ahmed and Giles (1969) for an interpretation.

[b] These systems are not well resolved as yet.

[c] No information available.

Data from the other enzymes studied from a phylogenetic viewpoint also provide interesting though not as compelling information as the glutamic dehydrogenases. The Oomycetes, Hyphochytridiomycetes, and Blastocladiales have an NAD-dependent D(−)-lactic dehydrogenase that is inhibited by ATP (see Table IV). But this is a common feature of most lactic dehydrogenases studied. Only the enzymes from the Oomycetes and Hyphochytridiomycetes are inhibited markedly by GTP in an allosteric manner, indicating a similar evolutionary trend for this locus in the enzyme from the two taxa.

More dramatic is the distribution of NAD- and NADP-linked isocitric dehydrogenases in all major fungal groups (Table II). Only the Hyphochytridiomycetes and Oomycetes are devoid of the NAD-linked variety of these two enzymes. On the surface, this seems like a major departure from the distribution of key enzymes in fungi. Normally, in most living organisms, there is an NAD- and an NADP-linked isocitric dehydrogenase. Together, it is thought (Kaplan, 1963), they interact to maintain an equilibrium between cellular pyridine nucleotides (NAD, NADH, NADP, and NADPH). With the absence of an NAD-linked isocitric dehydrogenase and also of an NADP-linked glutamic dehydrogenase in the Oomycetes and Hyphochytridiomycetes, the problem of pyridine nucleotide balance could only be resolved by the activating effect of NADP elicited on the NAD-linked glutamic dehydrogenase and discussed in great detail elsewhere (LéJohn and Stevenson, 1970).

It is difficult to arrive at any conclusion other than that the evolutionary course has been different for glutamic dehydrogenases of the Oomycetes and Hyphochytridiomycetes from that encountered by the same enzyme from other phycomycetous fungi. The controls on the enzymes are too complex and must have required multiple steps of natural selection. Because of the multivalent nature of the controls (LéJohn *et al.*, 1969, 1970), independent "reactive" sites are probably involved for each type of allosteric ligand. Retention of these complex controls in different genera and other taxa attests to our conviction that a study of enzyme regulation is a more sensitive indicator of phylogenetic relationships than most other biochemical parameters and may be unmatched for its ease and simplicity.

Aldolases. Rutter (1965) was one of the pioneers who used enzymatic characteristics as a criterion of phylogeny. Although his studies were restricted to a few fungi (*Aspergillus niger, Saccharomyces cerevisiae*), they warrant some comments in the hope of precipitating studies of the lower fungi in a similar manner. Rutter recognized two major classes of fructose diphosphate aldolase distributed among metazoa, protozoa, and green algae (class I) and bacteria, fungi, and blue-green algae (class III). *Euglena* and *Chlamydomonas* harbored both classes of aldolase. Some of the physicochemical properties elucidated for the two classes of enzyme are as follows. The molecular weight of class I enzyme is about two times that of class II enzyme. Class II enzyme has a bivalent metal requirement and is activated by K^+, but not class I form. Examination of the mechanism of enzyme action indicates that the class I enzyme utilizes a lysine residue for Schiff base formation (a catalytic intermediate) and has functional carboxy-terminal residues of tyrosine. These properties are not displayed by the class II enzyme.

The important question to answer is whether these enzymes are homologous in the evolutionary sense as are the glutamic dehydrogenases we examined. If they are, it would be worthwhile to study them in greater detail at the taxa level over a wider range of fungal types.

RNA Polymerases. Nonphotosynthetic eukaryotes are endowed with at least three distinct RNA polymerases, two of which are localized within the nucleus and the third present in the mitochondrion. RNA polymerase of the nucleoplasm has been designated as II, that in the nucleolus as I, and the mitochondrial form is III (see *Cold Spring Harbor Symposia,* 1970). The nucleolar RNA polymerase has proved to be worth studying from an evolutionary standpoint. Because this enzyme is probably quite specialized *in vivo*, as are the genes themselves that it acts on, it is less liable to encounter and sustain drastic evolutional changes than the other polymerases. The RNA polymerase I has been found to be inhibited by cycloheximide in two fungi by Griffin and his colleagues (Horgen and Griffin, 1971; Timberlake *et al.,* 1972). RNA polymerase I from *Blastocladiella emersonii* is inhibited by relatively high concentrations (0.2 mM) of the drug, whereas *Achlya ambisexualis* is almost completely inhibited by 2 μM cycloheximide. Cycloheximide has also been implicated as an *in vivo* inhibitor of ribosomal RNA synthesis in *Chlorella* (Wanka and Schrauwen, 1971), *Neurospora* (Fiala and Davis, 1965), other fungi (Haidle and Storck, 1966), yeast (DeKloet, 1965), and mammalian cells (Muramatsu *et al.,* 1970; Willems *et al.,* 1969).

RNA polymerases from prokaryotes are not affected by cycloheximide, and rRNA synthesis proceeds unimpeded *in vivo.* Thus it would appear that the relative sensitivity of the RNA polymerase I of eukaryotes to this agent could be used as a phylogenetic marker.

The available data, which are admittedly sparse and flimsy, show that rRNA synthesis in fungi and algae is more sensitive to cycloheximide than that in mammals and higher plants. Directing our attention to the fungi only, we see that the Oomycetes constitute a group that displays a higher sensitivity to this agent than any other fungi tested to date. All other fungi are about 10–100 times less sensitive to the inhibitory action of this agent on RNA synthesis (Horowitz *et al.,* 1970, and unpublished data). An attempt at determining the differential sensitivities of members of the phycomycetous fungi to cycloheximide in the synthesis of RNA is currently under way in our laboratory.

Enzyme Aggregation and Gene Clusters

In bacteria and fungi, studies have been made by Giles and his associates that indicate possible physical aggregation of enzymes that catalyze the conversion of phosphoenolpyruvate and erythrose-4-phosphate to chorismic acid. Functionally related genes tend to cluster as an operon in some cases in the prokaryotes and are dispersed in eukaryotes (Lewin, 1970; Vogel, 1971). The reverse situation seems to exist for enzymes involved in the prechorismate steps that lead to the biosynthesis of the aromatic amino acids tyrosine, tryptophan, and phenylalanine. As will be shown subsequently, the tryptophan pathway itself has also been subjected to similar scrutiny.

Giles and his associates (Berlyn and Giles, 1969; Ahmed and Giles, 1969) studied the sedimentation behavior of five of the seven enzymes involved in the prechorismate steps (Fig. 6) and found that whereas in all of the bacteria and blue-green algae examined the enzymes do not aggregate as physically related species, without exception the fungal enzymes found in the Blastocladiales, Oomycetes, Zygomycetes, Ascomycetes, and Basidiomycetes aggregate. Data were not presented, but Ahmed and Giles discussed their findings that the enzyme aggregates from the water molds (Phycomycetes) were more susceptible to dissociation into active components than the enzymes from the "higher" fungi. Where data were available, the genes coding for these enzymes in bacteria had been mapped as unlinked cistrons distributed widely over the entire genome. On the other hand, sufficiently detailed genetic maps obtained for *Neurospora crassa* (Rines *et al.*, 1969) and *Saccharomyces cerevisiae* (Ahmed and Giles, 1969) show that the genes for the aromatic amino acids pathway (*arom*) are clustered. A nonsense mutation at one end of the gene cluster region leads to the loss of activity of all five enzymes, indicating that a single polycistronic message may be involved in the synthesis of the enzymes. But as the enzymes can be separated into individually active entities, it is clear that polycistronic transcription simply facilitates formation of a multienzyme complex, not a single giant multifunctional polypeptide.

These studies of Ahmed and Giles cannot resolve taxa problems of the fungi, but they indicate that there may be an underlying interrelationship among all fungi studied in this context. The dissimilarity in this biochemical character between fungi belonging to the Oomycetes and the blue-green algae should be emphasized.

Studies of Hutter and DeMoss (1967), DeMoss and Wegman (1965), DeMoss (1965), and Bonner *et al.* (1965) on the biosynthesis of tryptophan in *Neurospora* and other fungi lend support to the argument that enzyme aggregates do not necessarily reflect linked gene clusters. Four unlinked genes in *Neurospora* specify the enzymes that catalyze the reactions converting chorismic acid to tryptophan (Fig. 6). The resolution achieved in studies of the enzymes of the tryptophan pathway in *Neurospora* has not been matched for any other fungus. Therefore, our discussion on this subject will remain speculative.

In Fig. 7, the gene–enzyme relationships of the tryptophan pathway of *Escherichia coli* are presented. In *Neurospora,* the three enzyme activities

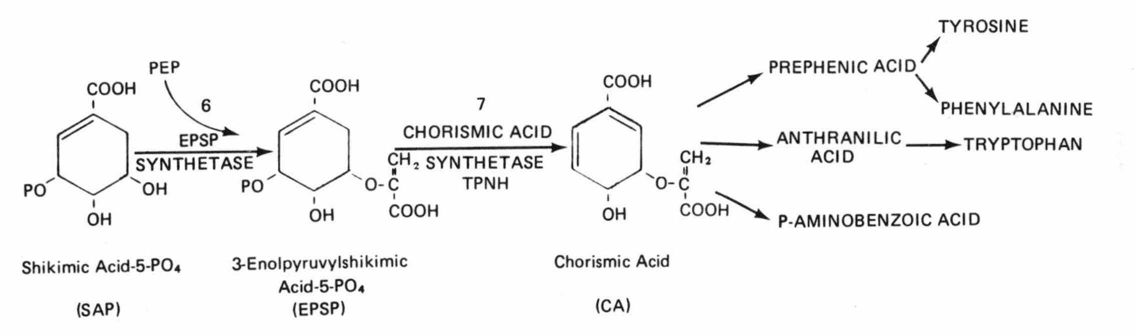

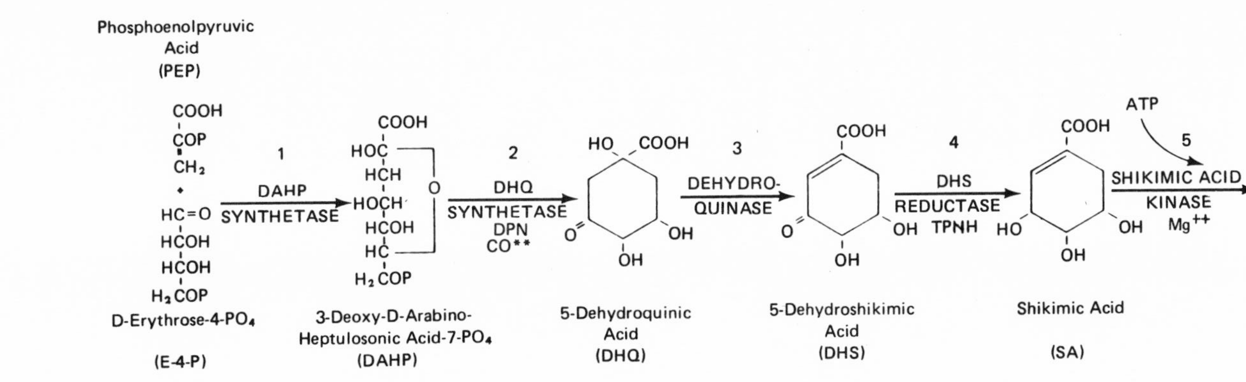

FIG. 6. Reaction steps leading to the biosynthesis of chorismic acid and members of the aromatic amino acids starting from phosphoenolpyruvate and erythrose-4-PO_4. The numbers 1-7 represent the number of key steps involved in the prechorismic acid path.

anthranilate synthetase, PRA isomerase, and InGP synthetase remain associated as a multienzyme complex even with the rigors of purification (Gaertner and DeMoss, 1969). The purified complex sediments at 10.3S and can dissociate to a 7.4S and a 4.4S unit on titration of the sulfhydryl groups with *p*-hydroxymercuribenzoate. The separated 7.4S fragment retains full activity for InGP synthetase and PRA isomerase reactions, but the 4.4S component is inactive unless a sulfhydryl reducing agent is present when some of the anthranilate synthetase activity can be restored. It is not known whether the 7.4S component is specified by a single cistron or by more than one, because the subunits have yet to be separated and characterized. If the subunits are identical, then a single cistron is implied. Of interest, however, is the finding that mutations clustered at one end of the *tryp2* gene of *Neurospora* affect the PRA isomerase activity whereas those located at the opposite end affect the InGP synthetase function (DeMoss *et al.*, 1967).

Hutter and DeMoss (1967) used the technique of sucrose density gradient centrifugation to determine the nature of organization of enzymes of the tryptophan pathway in other fungi. Attempts by these workers to rationalize the sedimentation patterns obtained into the phylogenetic framework of Gaumann (1964) led to uncertainties. Therefore, they proposed a genetic control scheme of evolution of the tyrptophan enzymes starting

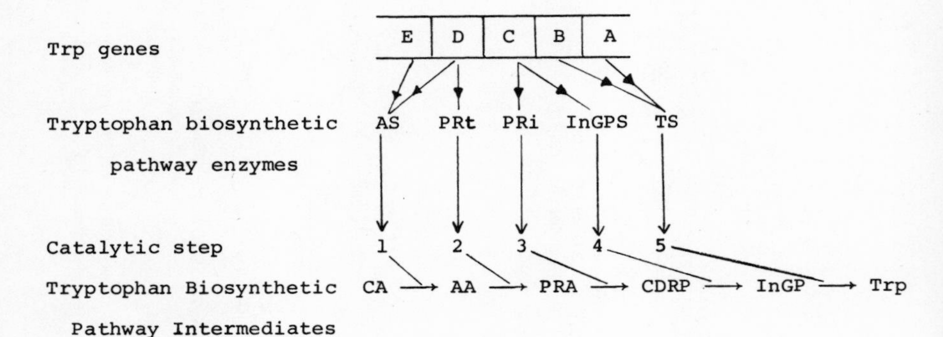

FIG. 7. Gene–enzyme relationships of *E. coli* tryptophan biosynthetic pathway. Abbreviations: As, anthranilate synthetase; PRt, phosphoribosyltransferase; PRi, phosphoribosylisomerase, InGPS, indoleglycerol phosphate synthetase; TS, tryptophan synthetase representing the trivial names of enzymes of the pathway of tryptophan biosynthesis. CA, Chorismic acid; AA, anthranilic acid; PRA, *N*-(5′-phosphoribosyl) anthranilate; CDRP, 1-(*O*-carboxyphenylamino)-1-deoxyribulose-5-phosphate; InGP, indoleglycerol phosphate; Trp, tryptophan. A–E represent the genetic description of the cistrons of the postchorismate tryptophan biosynthetic pathway.

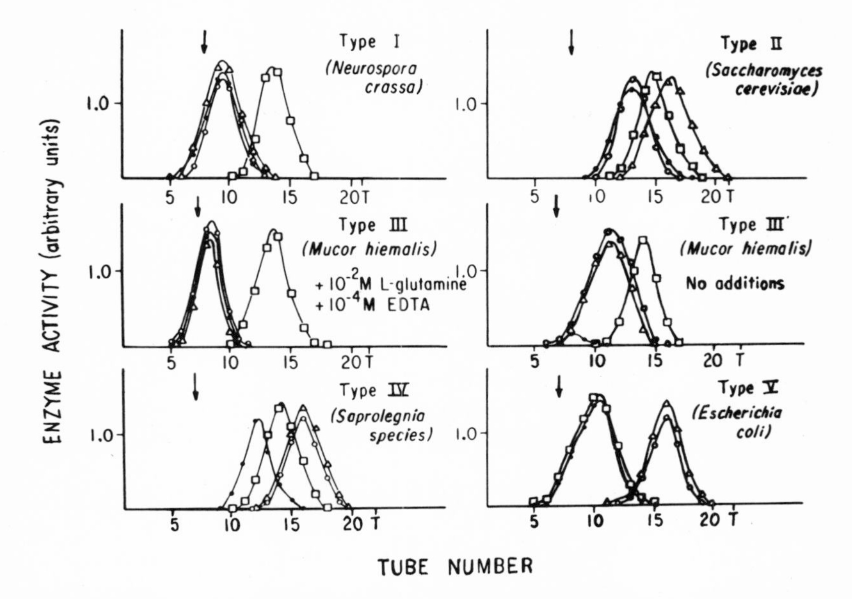

FIG. 8. Distribution of enzyme activities after zone centrifugation in sucrose. The activities of anthranilate synthetase (●), PR transferase (□), PRA isomerase (△), and InGP synthetase (○) are presented in normalized units. Data taken from Hutter and DeMoss (1967) with permission.

with no aggregation and ending with a completely stable state of aggregation.

According to their sedimentation data, Hutter and DeMoss delineated five categories of enzymes of the tryptophan pathway of bacteria and fungi. These categories, shown in Fig. 8, are indicative of the ability of two or more or any of the five enzyme activities to remain associated after sedimentation of the protein mixture in sucrose density gradients. On the evidence presented for *Neurospora* and discussed above, the activities for the PRA isomerase and InGP synthetase reside either in a single protein or else in tightly liganded protein chains. Examination of the data of Hutter and DeMoss reveals that the same may be true for the Mucorales, Saprolegniales, Peronosporales, Hetero- and Homobasidiomycetes, Chytridiales, and Blastocladiales. The only exception are the Endomycetales (yeasts) that have the isomerase and the synthetase activities resident in proteins which no doubt are different since they sediment differently. When the sedimentation coefficients of the various enzyme species are taken into consideration as well as their capacity to remain associated, the scheme displayed in Fig. 9 is obtained. It immediately becomes evident that the PR

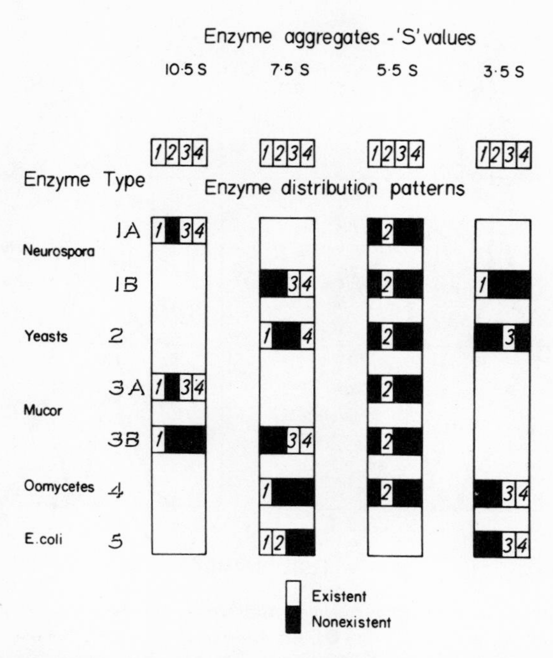

FIG. 9. A reinterpretation of the data of Hutter and DeMoss (1967) shown in Fig. 8. The four rectangles shown enclosing the numbers 1–4 under "Enzyme aggregates" depict the various possible sizes of the entire complex of enzymes or components thereof. Under "Enzyme distribution patterns," these same rectangles display the components that are found and their respective sizes in the various fungi studied. Some of the data on the physical and chemical properties of the *Neurospora* enzyme aggregate reported on by DeMoss's group (1965–1969) have been incorporated into this figure.

transferase protein (enzyme 2) has remained isolated in all species of fungi and has remained the same size (5.5S). Bacteria, on the other hand, are different, for the transferase sediments in association with anthranilate synthetase activity at 7.5S. Without the stabilizing ligands glutamine and EDTA, the isomerase and InGP synthetase of *Mucor* cosediment at the characteristic 7.5S found for all other fungi except the Oomycetes. It should be recalled that in the presence of organomercuric compounds, the 10.5S complex of *Neurospora* dissociates into 7.4S and 4.4S components and the isomerase and synthetase activities remain with the 7.4S product. It could be argued then that the Oomycetes have two separate proteins sedimenting at 3.5S which are responsible for the isomerase and InGP synthetase activities but that these proteins are either covalently linked or else strongly associated in the other fungi. Thus some continuity could be observed in the properties of the tryptophan enzymes of the fungi. When the distribution of the various categories (types I to IV) designated by Hutter and DeMoss among the various taxa of fungi is studied, it becomes clear that a logical phyletic scheme could not be produced. For instance, the type I category, which we call *Neurospora* form, was present in representative members of the Chytridiales, Blastocladiales, Myxomycetes, Ascomycetes, Discomycetes, and Basidiomycetes. The type III or *Mucor* form was located in the Zygomycetes and Basidiomycetes. The Oomycetes were the only fungi that exclusively displayed the type IV enzyme pattern. Similarities exist between the enzyme distribution patterns of the yeasts (type II) and the Oomycetes (type IV). The only difference is in the size of InGP synthetase, which may or may not be linked to anthranilate synthetase in yeasts. The former protein is much smaller in the Oomycetes as it sediments at 3.5S compared to 7.5S for yeasts. In the presence of stabilizing ligands, the enzyme distribution pattern for *Neurospora* (type I) is identical to that for *Mucor* (type III). When the two enzyme systems are exposed to dissociation conditions (see 1B and 3B) in Fig. 9, the two systems show a single difference in the size of enzyme 1, which remains large (10.5S) in the Mucorales but is very small in *Neurospora* (3.5S). Only residual anthranilate synthetase activity can be recovered in *Mucor* without stabilizers, and it is possible that this is due to some loose association between a 3.5S component and the 7.5S isomerase and InGP synthetase complex—the condition under which anthranilate synthetase activity appears in *Neurospora* enzyme system.

Peripheral Biochemical Properties

Cytochrome System. Boulter and Derbyshire (1957), using relatively insensitive spectroscopic methods, studied the visible absorption spectra of the cytochromes of 45 fungal species representing all major classes, and compared their spectra to those obtained for yeasts. They concluded that the cytochromes of all fungi were similar and compared favorably with the cytochromes of mammalian and avian cells. They were quite different from the cytochrome systems of bacteria and green plants.

Gleason and Unestam (1968), using much improved spectrophotometric techniques, detected some spectral differences between the cytochromes of the Oomycetes and Chytridiomycetes. They concluded that the chytrids probably have three *b*-type and two *c*-type chytochromes whereas the Oomycetes have two *b*-type and a single *c*-type as well as cytochrome a-a_3. They also suggested that green algae, the Oomycetes, and higher plants were alike in that they lacked cytochrome c_1 found in

Ascomycetes, Chytridiomycetes, and metazoa. But as the authors stated, no great significance can be attributed to these findings until the various cytochromes are isolated, purified, and characterized.

Two aspects of cytochrome studies that in our opinion should give us a better resolution of the phylogenetic relationships of fungal cytochromes are (1) regulation of biosynthesis of the cytochrome complex and (2) primary amino acid sequences of cytochromes from a wide variety of fungi. Unfortunately, the information available on either of these two aspects is rather limited. As the cytochrome system represents the final common path of oxidation and is coupled to oxidative phosphorylation in the eukaryotic cell, the components of this system can interact with each other reversibly and constitute a constellation of apoproteins each with its own typical heme prosthetic group (King, 1971). This complex, by virtue of its specialized function, has not undergone radical evolutional changes in structure. Support for this notion comes from the studies of Margoliash and associates (Nolan and Margoliash, 1968; Margoliash and Schejter, 1966; Smith, 1970) on cytochrome *c*, whose primary structure has been determined for a wide range of taxonomic forms including yeasts and *Neurospora*. The conclusion reached from their studies is that all extant eukaryotic species have a common ancestor and that the gene for this protein has survived drastic mutational alterations for some 1 billion years. Cytochrome *c* is the only hemoprotein of all the cytochromes that displays a constancy of physical and chemical properties. It seems, therefore, that it is the variations (both qualitative and quantitative) of the other components of the cytochrome complex that may yield invaluable information on the nature of inter- and intrageneric relationships. Thus as the complexities in the induction and repression mechanisms operative on the cytochrome complex become unraveled for different taxa, and their different constitutions are determined, we will be in a better position to incorporate this parameter into phylogenetic studies of a single phylum.

Polyunsaturated Fatty Acids. A wide variety of fatty acids with more than two double bonds exist in nature, and they are distinguished by differences in the length of their carbon chain, degree of unsaturation, and position of the double bonds (Bloch, 1965). Two distinct pathways of their biosynthesis are recognized (von Klenk, 1961; Mead, 1960), one leading to the production of α-linolenic acid (9,12,15-octadecatrienoic acid) and the other to γ-linolenic acid (6,9,12-octadecatrienoic acid) and arachidonic acid (5,8,11,14-eicosatetraenoic acid) (Fig. 10). Interestingly, the α-linolenic acid appears to be synthesized predominantly by higher plants and algae, not by protozoa and metazoa. The γ-linolenic acid can be synthesized by all algae (except the blue-greens and the greens) and also by animals (Bloch, 1965). It is now known to be erroneous to describe these as "plant"

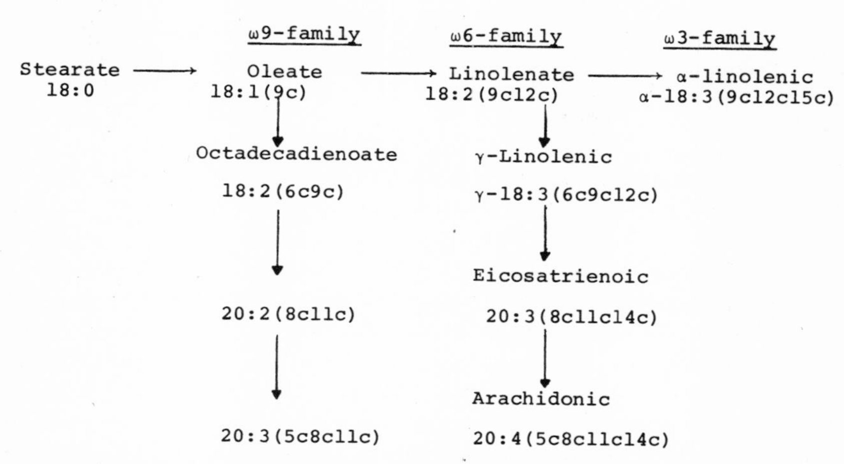

FIG. 10. Biosynthesis of polyenoic acids. ω6-Family found in Oomycetes (Pythiales and Saprolegniales) and Zygomycetes (Entomophthorales and Mucorales). ω3-Family found in Ascomycetes (Taphrinales, Myringiales, Helotiaceae, Pezizales, Hypocreales, Sphaeriales, and Euascomycetes or Deuteromycetes), Heterobasidiomycetes (Auriculariales, Phelogenaceae, and Ustilaginales), and Homobasidiomycetes (Exobasidiaceae and Thelophoraceae). Data from Shaw (1965). (ω3+ω6)-Families found in four marine fungi and one freshwater form of the "lower" fungi (Phycomycetes). Data from Ellenbogen and Aaronson (1969).

and "animal" pathways since the metabolic sequence, including aerobic desaturation, that is followed is common to both phyla. The "plant" pathway is just a manifestation of a specific metabolic block which denies access of substrate to the desaturation site (Hitchcock and Nichols, 1971). What is of interest to us is the fact that Shaw (1965) showed in his studies of 31 fungal species, covering four orders of the "Phycomycetes," seven orders of the Ascomycetes, and five orders of the Basidiomycetes, that all of the "Phycomycetes" (Oomycetes and Zygomycetes) produce γ-linolenic acid, not the α-linolenic acid. The Ascomycetes and Basidiomycetes display the reverse ability. The capacity for synthesis of both forms of polyunsaturated fatty acid can coexist and be expressed depending on the growth condition in the same organism as observed in red algae, chrysomonads, *Euglena,* and *Chlamydomonas* (Hitchcock and Nichols, 1971). It is therefore difficult to rationalize the phylogenetic significance of the dichotomy of the pathways shown by fungi as did Vogel (1965) for the DAP and AAA lysine biosynthetic pathways. Although Bartnicki-Garcia (1970) dismisses this pathway homology of the "Phycomycetes" lightly by laying greater weight on cell wall characteristics, tryptophan pathway, and lysine biosynthesis pathway, we feel it requires more serious consideration. Certainly, the suggestion of convergent evolution among "Phycomycetes" for biosynthesis of γ-linolenic acid, while convenient, is unconvincing.

Sterols

The occurrence of various sterols and their patterns of biosynthesis can be examined for evidences of their value as phylogenetic markers in fungi, but the primary conclusion that can be drawn is with respect to the major sterol form found in these organisms. Blue-green algae contain no sterols; cholesterol is found in the red algae and animals; phytosterol is typical of green plants and ergosterol of fungi (Goad, 1967).

Examination of the mechanism(s) of sterol biosynthesis indicates that all isoprenoid derivatives arise from the mevalonate to squalene sequence. This probably occurred very early in evolution since blue-green algae appear to produce nonsteroid derivatives of mevalonate, the quinones (Danielsson and Tchen, 1968). Beyond squalene, the key intermediates in steroid biosynthesis diverge, with lanosterol being a key intermediate in fungi and vertebrates, while cycloartenol is the precursor of the phytosterols of green plants (Fig. 11).

Some studies have been made of the biosynthesis of the specific forms of ergosterol and phytosterol. Despite their similarities in structure, it appears that they are formed from different biosynthetic paths, as suggested by the opposite stereochemical steps in Δ^{22} double bond formation, which would then divide the algae and fungi into two distinct groups (Goad, 1970).

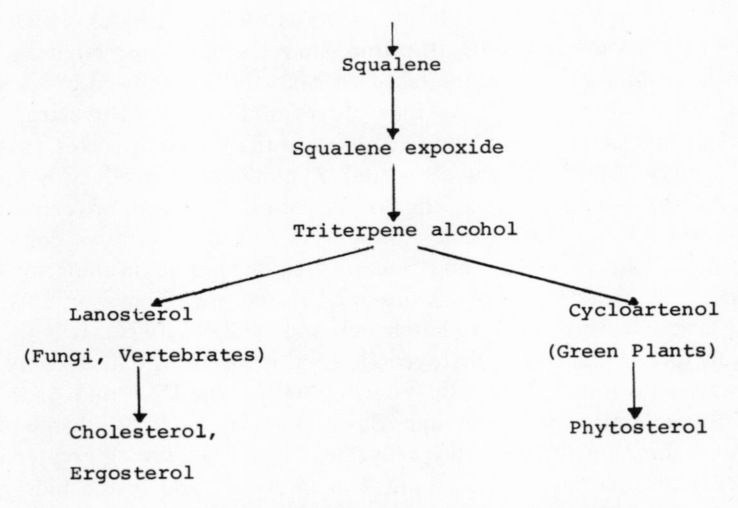

FIG. 11. An abbreviated pathway for sterol biosynthesis in eukaryotes.

The fungi also appear to vary among themselves in patterns of biosynthesis. It has been suggested, for example, that the alkylation step at C-24 occurs at different stages in the biosynthetic process of different fungi. In some Ascomycetes it is in a later step than in the Phycomycetes and Basidiomycetes. The alkylation step occurs late in Ascomycetes and early in the other fungi (Goad, 1970). However, information about the precise steps and sequences of sterol biosynthesis is not yet adequate to provide a useful grouping of the fungi.

Apart from the biosynthetic patterns of sterols, it has also been suggested that the occurrence of certain types of sterols in fungi may have phylogenetic implications. Goad (1967) reported that a variety of chitin-walled fungi contained ergosterol or closely related sterols. McCorkindale *et al.* (1969) extended this work to the cellulose-walled fungi (Saprolegniales, Leptomitales, and Peronosporales) and showed that while the chitin-walled forms (Mucorales) contained ergosterol and related sterols, the Oomycetes did not but rather contained demosterol, fucosterol, and cholesterol derivatives found in algae and pollen. Of interest is the fact that Peronosporales do not produce sterols, although these compounds influence the growth and reproductive events of these organisms. It is possible that some crucial early step in sterol biosynthesis is missing through mutation in the Peronosporales. But until the proper biosynthetic steps are elucidated, the occurrence of similar or different sterols in various fungi is not adequate to indicate phylogenetic relationships. Besides, a greater number and variety of species would have to be examined for reliable comparisons to be made.

BIOCHEMICAL EVOLUTION OF THE FUNGI—COMMENTARY

Marcel Florkin (1944) was the first to broadcast the idea that evolutionary descent of living things proceeds with a corresponding modification at the molecular level. This view has since been supported by observations that although there is a fundamental chemical unity of life, yet still there are subtle biochemical variations along each branch of the phylogenetic tree. Generally, each branch contains a biochemical homology that is unmatched at the phenotypic level. Such biochemical homology can reflect itself in enzyme distribution patterns and in an adaptive feature as metabolic control of enzymes. But the question of biochemical homology through enzymes should be approached with caution as the mere presence or absence of enzyme(s) could produce spurious evolutionary correlations that are meaningless for intertaxa relationships. Take, for example, the distribution of eight types of dehydrogenases in the fungi that superficially could be considered as four erstwhile twins (Table V). The enzymes show either distinct coenzyme (NAD or NADP) specificity or substrate isomer [D(−)- or L(+)-lactate] specificity. Whereas no difference is observed in the enzyme distribution patterns for the Chytridiales, Blastocladiales, and Mucorales, when adaptive and physicochemical properties of these enzymes are considered, homology breaks down and the Blastocladiales can be seen as more closely allied to some of the Mucorales. The controls elicited on the NAD-specific glutamic dehydrogenases of the Phycomycetes indicate that Hyphochytridiomycetes and the Oomycetes are close relatives, and this supports the data on the distribution of dehydrogenases among these two groups. Furthermore, the enzyme control systems of lactic and glutamic dehydrogenases clearly outline the vast gulf that exists between the other members of the Phycomycetes and the so-called cellulose fungi.

The precision and resolving power of enzyme regulation as a phylogenetic parameter are realized when enzyme organization is employed instead (Table IV) to delineate groups. Hutter and DeMoss's classification of the enzymes of the tryptophan pathway relegates the type I tryptophan system to the Chytridiales, Blastocladiales, Aspergillales, Pezizales,

TABLE V. NAD- and NADP-Dependent Dehydrogenases[a]

	GDH		IDH		NAD-LDH		MDH	
Taxonomic group	NAD	NADP	NAD	NADP	D(−)	L(+)	NAD	NADP
Chytridiomycetes								
Chytridiales	+		+	+	+		+	
Blastocladiales	+		+	+	+		+	
Hyphochytridiomycetes	+			+	+		+	+
Oomycetes	+			+	+		+	+
Zygomycetes	+		+	+	+		+	
Ascomycetes								
Yeasts	+	+	+	+	+[b]	+	+	?[c]
Others	+	+	+	+		+	+	?
Deuteromycetes	+	+	+	+		+	+	?
Discomycetes	+	+	+	+		+	+	?
Heterobasidiomycetes	+	+	+	+		+	+	?
Homobasidiomycetes	+	+	+	+		+	+	?

[a] GDH, Glutamic dehydrogenase; IDH, isocitric dehydrogenase; LDH, lactic dehydrogenase; MDH, malic dehydrogenase.

[b] Cytochrome-linked enzyme.

[c] No information available.

Sphaeriales, Myxomycetes, and Agaricales. The type III tryptophan system is also widely dispersed among the Mucorales, Tremellales, and Ustilaginales. Only the types II and IV tryptophan systems are confined to individual fungal classes. Organization of enzymes of the aromatic cluster studied by Ahmed and Giles (1969) is even less definitive, because the *arom* system only delineates the lower fungi from the higher forms. There is a hint that the water molds may have a slightly different *arom* enzyme aggregate than the other Phycomycetes because their enzyme aggregates are more susceptible to dissociation.

It seems as though a good approach to understanding fungal phylogeny would be to trace the probable evolutionary course of fungal dehydrogenases by sequence studies and through study of the distribution of these enzymes in algae, higher plants, protozoa, and metazoa. Unfortunately, the available information on algal and higher plant glutamic dehydrogenases is rather sparse and indecisive. All reliable reports indicate that algae have a single enzyme that can use either NAD or NADP as coenzyme. Furthermore, the glutamic dehydrogenases found in the cytoplasm and chloroplast of *Chlorella* are both nonspecific for the two coenzymes (Kivic *et al.*, 1969).

Russian workers have carried out most of the fine analyses of the kinetic properties of the glutamic dehydrogenases of *Chlorella* (Shatilov *et al.*, 1969, 1970; Tomava *et al.*, 1969). They report that the enzyme in *Chlorella* is similar to the bovine liver enzyme with regard to its response to adenylates and guanylates and its nonspecificity.

Bacteria have either an NAD- or an NADP-specific glutamic dehydrogenase—never both. Animals have a nonspecific form of the enzyme just as algae do. A phylogenetic lineage for glutamic dehydrogenases starting from prokaryotes to eukaryotes is consequently not evident from the distribution of the enzyme forms in the various phyla. However, within the fungal species, an organized pattern exists in the distribution of the enzyme and it is this point to which we will now address ourselves.

EVOLUTION OF FUNGAL GLUTAMIC DEHYDROGENASES

Glutamate dehydrogenase catalyzes the reversible reaction involving the oxidative deamination of glutamate and reductive amination of α-ketoglutarate. The enzyme is strategically placed in cellular metabolism connecting the central complex of aerobic oxidative metabolism (citric acid cycle) with the biosynthesis of the glutamate family of amino acids. The reaction it catalyzes is crucial from another point of view since the coenzyme reactants are vital in energy metabolism and the substrates glutamate and

α-ketoglutarate participate in many transaminating reactions that lead to the biosynthesis of several other amino acids. The enzyme must therefore be one of the very early catalysts to accede development of aerobiosis in early cells. To perform all of these functions, the enzyme must be well regulated so that the activity will respond rapidly to the prevailing concentrations of its substrates. For these reasons, the diversity in enzyme controls elicited by glutamic dehydrogenases of fungi and described here and elsewhere is regarded as a compelling phylogenetic parameter. We find that the lower fungi have only an NAD-specific glutamic dehydrogenase and the higher forms two enzymes, one NAD-specific and the other NADP-specific. What about the transition between these two extremes? This transitory enzyme may be either nonspecific for the two pyridine nucleotide coenzymes or else specific for one coenzyme form that is used as substrate while the second coenzyme acts as an allosteric regulator. To date, no fungus has been found to contain a glutamic dehydrogenase that is nonspecific for the coenzyme used. But the Hyphochytridiomycetes and Oomycetes harbor an enzyme that uses NAD as coenzyme and interacts with NADP (NADPH), which acts as "pacemaker." It therefore appears that a monophyletic scheme that would account satisfactorily for the phylogenetic origins of glutamic dehydrogenases in fungi cannot really be deduced without strong bias. I have proposed a complex scheme in which I assumed that the enzyme found in the Hyphochytridiomycetes may be a forerunner of those found in the Chytridiomycetes and "higher" fungal forms (LéJohn, 1971*b*). This was based not just on the properties of the glutamic dehydrogenases found in fungi, but also on the consideration of other parameters such as cell wall chemistry, lysine biosynthetic pathways, and distribution of other enzymes among the various taxa of the fungi. But if the Ascomycetes and Basidiomycetes are phylogenetically related to some of the Phycomycetes, as is currently believed by several mycologists, then it becomes extremely difficult to come to any other conclusion than that some ancestral phycomycete developed the capacity to elaborate both an NAD- and an NADP-dependent glutamic dehydrogenase since all "higher" fungi so far examined possess both forms. But despite our large survey of the Phycomycetes, as shown in Table II, a single organism is yet to be detected that has these twin forms of the enzyme. The closest to this possibility is found among the Oomycetes and Hyphochytridiomycetes that harbor an NAD-dependent glutamic dehydrogenase allosterically regulated by NADP. This particular enzyme displays a kinetic mechanism that is typical of only NADP-dependent glutamic dehydrogenase from other life forms (Stevenson and LéJohn, 1971). All of the other NAD-dependent glutamic dehydrogenases of the Phycomycetes tested to date have a typical "NAD-type" kinetic mechanism. One is justified in suspecting that this unique enzyme type present in the Oomycetes and Hyphochytridiomycetes may represent an evolutionarily important protein. It could bridge the vast chasm that exists in our knowledge of the evolutionary progress of Phycomycetes to the Ascomycetes and Basidiomycetes.

Present-day Phycomycetes are supposed to derive from ancestral algae. But extant species of algae tested so far have only a single nonspecific glutamic dehydrogenase that can use both NAD and NADP as coenzyme. It is hardly likely that all Phycomycetes lost this ability to utilize NADP as substrate for their glutamic dehydrogenase and, on evolving into the "higher" fungal forms, reacquired the capacity, this time in the form of a separate protein. The Oomycetes are therefore of signal importance because their unique glutamic dehydrogenase can be interpreted to be related to the other fungal glutamic dehydrogenases in two ways. First, the enzyme could represent a protein that is evolutionarily more sophisticated than that of the Chytridiomycetes and is evolving toward the development of a new catalytic function in which NADP would serve as substrate. The "higher" fungi achieved this through the Oomycetes. Second, the enzyme may represent a diversionary form with its ancestral root from algae. The regulatory property of NADP simply developed out of the catalytic locus. This is easy to comprehend from the hypothetical scheme of Fig. 1. In this case, the Oomycetes and Hyphochytridiomycetes can be considered to be totally unrelated to all other fungi. The fact that they happen to contain only an NAD-dependent glutamic dehydrogenase, like all Phycomycetes, would therefore be coincidental.

Can we use some of the other biochemical parameters to resolve this paradox of the Phycomycetes? I think not, for we see that, irrespective of what parameter is selected, major and minor difficulties are encountered. Considering cell wall chemistry, superficially, a correlation is difficult to see. On closer examination, however, it is established that chitin and cellulose, which are regarded as the fabrics of fungal walls, are present simultaneously in several taxa. Traditionally, cellulose is considered to be the wall polymer of Oomycetes and chitin the polymer of most other fungi. But in actual fact cellulose is only a minor component of the walls of Oomycetes, and it is a poorly crystalline structure at that. Chitin and cellulose are occasionally detected in Oomycetes in significant quantities (Lin and Aronson, 1970); in Hyphochytridiomycetes (Fuller and Bashad, 1960); in the ascomycetes and deuteromycetes *Ceratocystic ulmi* (Rosinski and Campana, 1964), *Aspergillus* (Farr, 1954), *Fusarium* (Marchant, 1966), *Venturia* (Jaworski and Wang, 1965); and in the basidiomycete *Schizophyllum* (Wang and Miles, 1966). Because X-ray analytical methods were not used in most of these studies, except those of Rosinski and Fuller and Farr, one must draw cautious conclusions. It is possible, though, that these cellulose–chitin coexistences in certain taxa are evolutionary signals which should not be ignored. They may represent properties that have been dismissed through selection pressures in most other extant species but, for special environmental reasons, did not disappear in the existing species in which they are found. Some would argue against this and propose in place parallel evolution of distinct ancestral forms of fungi. I am of the opinion that not enough species have been examined and, in due course, more fungal isolates of different taxa will be found to harbor cellulose and chitin in their walls. I reiterate the belief that cellulose is only a special form of those diverse forms of polymers generically termed β-glucans that is ubiquitous in all fungi except some of the "yeasts." Consequently, the Oomycetes may not be particularly different in cell wall construction from most other fungi.

What about the lysine biosynthetic pathway? Here, the disparity is more dramatic and difficult to reconcile. As nothing is known about the lysine degradation pathway in the Oomycetes and most fungi, the problem remains unresolvable at this time.

Other parameters considered in this review, such as separability of constituent enzymes from an enzyme complex, cytochrome systems, fatty acid biosynthetic pathways, sterol biosynthesis, %GC content, and molecular weights of ribosomal RNAs, display even less resolving power in fungal phylogenetics than enzyme distribution, enzyme regulation, and cell wall chemistry.

It is hoped that more and detailed studies on these latter parameters will be initiated and extended so that the biochemical reasoning of these enigmatic organisms may ultimately provide us with the "footprints" needed to trace the ancestral origins and relationships of the fungi.

Acknowledgments

I thank Miss R. Stevenson most warmly for her unflinching help with the literature survey. I also extend my appreciation to several colleagues and students who assisted directly or indirectly with some of the experiments reported here. This work was supported by a grant from the National Research Council of Canada.

REFERENCES

Ahmed, S. I., and Giles, N. H., 1969, Organization of enzymes in the common aromatic synthetic pathway: Evidence for aggregation in fungi, *J. Bacteriol.* **99**:231–237.

Alexopoulos, C. J., 1962, *Introductory Mycology,* Wiley, New York.

Attardi, G., and Amaldi, F., 1970, Structure and synthesis of ribosomal RNA, *Ann. Rev. Biochem.* **39:**183–226.

Bartnicki-Garcia, S., 1968, Cell wall chemistry, morphogenesis and taxonomy of fungi, *Ann. Rev. Microbiol.* **22:**87–108.

Bartnicki-Garcia, S., 1970, Cell wall composition and other biochemical markers in fungal phylogeny, in: *Phytochemical Phylogeny* (J. B. Harborne, ed.), pp. 81–103, Academic Press, New York.

Bartnicki-Garcia, S., and Nickerson, W. J., 1962, Isolation, composition, and structure of cell walls of filamentous and yeast-like forms of *Mucor rouxii, Biochim. Biophys. Acta* **58:**102–119.

Bartnicki-Garcia, S., and Reyes, E., 1965, *Bacteriol. Proc.* **26:**(abstr.).

Benjamin, R. K., 1959, The megasporangiferous Mucorales, *Aliso* **4:**321–433.

Berlyn, M. B., and Giles, N. H., 1969, Organization of enzymes in the polyaromatic synthetic pathway: Separability in bacteria, *J. Bacteriol.* **99:**222–230.

Bloch, K., 1965, Lipid patterns in the evolution of organisms, in: *Evolving Genes and Proteins* (V. Bryson, and H. J. Vogel, eds.), pp. 53–65, Academic Press, New York.

Bonner, D. M., DeMoss, J. A., and Mills, S. E., 1965, The evolution of an enzyme, in: *Evolving Genes and Proteins* (V. Bryson, and H. J. Vogel, eds.), pp. 305–319, Academic Press, New York.

Boulter, D., and Derbyshire, E., 1957, Cytochromes of fungi, *J. Exp. Bot.* **8:**313–318.

Cantino, E. C., 1955, Physiology and phylogeny in the water molds—A reevaluation, *Quart. Rev. Biol.* **30:**138–149.

Casselton, P. J., 1969, Concurrent regulation of two enzymes in fungi. *Sci. Progr.* (Oxf.) **57:**207–227.

Cook, P. W., 1966, *Entophylctis reticulospora* sp. nov.: A parasite of *Closterium, Trans. Brit. Mycol. Soc.* **49:**545–550.

Danielsson, H., and Tchen, T. T., 1968, Steroid metabolism, in: *Metabolic Pathways,* Vol. 2, 3rd ed. (D. M. Greenberg, ed.), Academic Press, New York.

DeKloet, S. E., 1965, Ribonucleic acid synthesis in yeast: The effect of cycloheximide on the synthesis of ribonucleic acid in *Saccharomyces carlsbergensis, Biochem. J.* **9:**566–581.

DeMoss, J. A., 1965, The conversion of shikimic acid to anthranilic acid by extracts of *Neurospora crassa, J. Biol. Chem.* **240:**1231–1235.

DeMoss, J. A., and Wegeman, J., 1965, An enzyme aggregate in the tryptophan pathway of *Neurospora crassa, Proc. Natl. Acad. Sci.* **54:**241–247.

DeMoss, J. A., Jackson, R. W., and Chalmers, J. H., Jr., 1967, Genetic control of the structure and activity of an enzyme aggregate in the tryptophan pathway of *Neurospora crassa, Genetics* **56:**413–424.

Edelman, M., Swinton, D., Schiff, J. A., Epstein, H. T., and Zeldin, B., 1967, Deoxyribonucleic acid of the blue-green algae (Cyanophyta), *Bacteriol. Rev.* **31:**315–331.

Ellenbogen, B. B., and Aaronson, S., 1969, Polyunsaturated fatty acids of aquatic fungi: Possible phylogenetic significance, *Comp. Biochem. Physiol.* **29:**805–811.

Farr, W. K., 1954, Structure and composition of the walls of conidiophores of *Aspergillus niger* and *A. Carbonarius, Trans. N.Y. Acad. Sci.* **16:**209–214.

Fiala, E. S., and Davis, F. F., 1965, Preferential inhibition of synthesis and methylation of ribosomal RNA in *Neurospora crassa* by actidione, *Biochem. Biophys. Res. Commun.* **18:**115–118.

Fitch, W. M., and Margoliash, E., 1970, The usefulness of amino acid and nucleotide sequences in evolutionary studies, in: *Evolutionary Biology,* Vol. 4 (T. Dobzhansky, M. K. Hecht, and W. C. Steere, eds.), pp. 67–109, Appleton-Century-Crofts, New York.

Florkin, M., 1944, *L'Evolution Biochimique,* Masson, Paris.

Fuller, M. S., and Bashad, I., 1960, Chitin and cellulose in the cell walls of *Rhizidiomyces* sp., *Am. J. Bot.* **47**:105–109.

Gaertner, F. H., and DeMoss, J. A., 1969, Purification and characterization of a multienzyme complex in the tryptophan pathway of *Neurospora crassa, J. Biol. Chem.* **244**:2716–2725.

Gaumann, E., 1964, *Die Pilze, Grundzuge ihrer Entwicklungsgeschichte und Morphologie,* Birhauser, Basel.

Gleason, F. H., Unestam, T., 1968, Cytochromes of aquatic fungi, *J. Bacteriol.* **95**:1599–1603.

Goad, L. J., 1967, Aspects of phytosterol biosynthesis, in: *Terpenoids in Plants* (J. B. Pridham ed.), pp. 159–190, Academic Press, New York.

Goad, L. J., 1970, Sterol biosynthesis, in:*Natural Substances Formed Biologically from Mevalonic Acid* (Biochemical Society Symposium 29) (T. W. Goodwin, ed.), pp. 45–77, Academic Press, New York.

Haidle, C. W., and Storck, R., 1966, Inhibition by cycloheximide of protein and RNA synthesis in *Mucor rouxii, Biochem. Biophys. Res. Commun.* **22**:175–180.

Higashino, H., and Lieberman I., 1965, Lysine catabolism by liver after partial hepatectomy, *Biochim. Biophys. Acta* **111**:346–348.

Hill, L. R., 1966, An index to deoxyribonucleic acid base compositions of bacterial species, *J. Gen. Microbiol.* **44**:419–437.

Hitchcock, C., and Nichols, B. W., 1971, *Plant Lipid Biochemistry,* Academic Press, New York. pp. 387.

Horgen, P. A., and Griffin, D. H., 1971, Specific inhibitors of the three RNA polymerases from the aquatic fungus *Blastocladiella emersonii, Proc. Natl. Acad. Sci.* **68**:338–341.

Horowitz, N. H., 1945, On the evolution of biochemical syntheses, *Proc. Natl. Acad. Sci.* **31**:153–157.

Horowitz, N. H., 1965, The evolution of biochemical syntheses—Retrospect and prospect, in: *Evolving Genes and Proteins* (V. Bryson and H. J. Vogel, eds.), pp. 15–23, Academic Press, New York.

Horowitz, N. H., Feldman, H. M., and Pall, M. L., 1970, Derepression of tyrosinase synthesis in *Neurospora* by cycloheximide, actinomycin D and puromycin, *J. Biol. Chem.* **245**:2784–2788.

Hutter, R., and DeMoss, J. A., 1967, Organization of the tryptophan pathway: A phylogenetic study of the fungi, *J. Bacteriol.* **94**:1896–1907.

Jaworski, E. G., and Wang, L. C., 1965, Gross cell wall composition of *Venturia inaequalis, Phytopathology* **55**:401–403.

Joshi, J. G., Hashimoto, T., Hanabusa, K., Dougherty, H. W., and Handler, P., 1965, Comparative aspects of the structure and function of phosphoglucomutase, in: *Evolving Genes and Proteins* (V. Bryson and H. J. Vogel, eds.), pp. 207–219, Academic Press, New York.

Kaplan, N. O., 1963, Multiple forms of enzymes, *Bacteriol. Rev.* **27**:155–169.

Kaplan, N. O., 1965, Evolution of dehydrogenases, in: *Evolving Genes and Proteins* (V. Bryson and H. J. Vogel, (eds.), pp. 243–277, Academic Press, New York.

King, M. E., 1971, Regulation of cytochrome biosynthesis in some eukaryotes, in: *Metabolic Pathways,* Vol. 5, 3rd ed. (H. J. Vogel, eds.), pp. 55–76, Academic Press, New York.

Kivic, P. A., Bart, C., and Hudock, G. A., 1969, Isozymes of glutamate dehydrogenase in *Chlamydomonas reinhardtii, J. Protozool.* **16**:743–744.

Klein, R. M., and Cronquist, A., 1967, A consideration of the evolutionary and taxonomic significance of some biochemical, micromorphological, and physiological characteristics in the thallophytes, *Quart. Rev. Biol.* **42**:105–296.

LéJohn, H. B., 1968, Unidirectional inhibition of glutamic dehydrogenase by metabolites: A possible control mechanism, *J. Biol. Chem.* **243**:5126–5131.

LéJohn, H. B., 1971*a*, Enzyme regulation, lysine pathways and cell wall structures as indicators of major lines of evolution in fungi, *Nature* **231**:164–168.

LéJohn, H. B., 1971*b*, Relationship between uridine nucleotide sugar activation of glutamic dehydrogenases in fungi and existence of chitin and cellulose in their walls, *Biochem. Biophys. Res. Commun.* **42**:538–544.

LéJohn, H. B., and Stevenson, R. M., 1970, Multiple regulatory processes in nicotinamide adenine dinucleotide-specific glutamic dehydrogenases. Catabolite repression; NADP(H) and P-enolpyruvate as activators; allosteric inhibition by substrates, *J. Biol. Chem.* **245**:3890–3900.

LéJohn, H. B., Jackson, S. G., Klassen, G. R., and Sawula, R. V., 1969, Regulation of mitochondrial glutamic dehydrogenase by divalent metals, nucleotides and α-ketoglutarate, *J. Biol. Chem.* **244**:5346–5356.

LéJohn, H. B., Stevenson, R. M., and Meuser, R., 1970, Multivalent regulation of glutamic dehydrogenases from fungi: Effects of adenylates, guanylates and acyl coenzyme A derivatives, *J. Biol. Chem.* **245**:5569–5576.

Lewin, B. M., 1970, *Gene Expression,* Interscience, New York.

Lin, C. C., and Aronson, J. M., 1970, Chitin and cellulose in the cell walls of the oomycete, *Apodachlya* sp., *Arch. Mikrobiol.* **72**:111–114.

Lizardi, P. M., and Luck, D. J. L., 1971, Absence of a 5S RNA component in the mitochondrial ribosomes of *Neurospora crassa, Nature New Biol.* **229**:140–142.

Loening, U. E., 1968, Molecular weights of ribosomal RNA in relation to evolution, *J. Mol. Biol.* **38**:355–365.

Loening, U. E., 1970, The mechanism of synthesis of ribosomal RNA, in: *Organization and Control in Prokaryotic and Eurkaryotic Cells* (H. P. Charles and B. C. J. G. Knight, eds.), pp. 77–106, Cambridge University Press, London.

Lovett, J. S., and Haselby, J. A., 1971, Molecular weights of the ribosomal ribonucleic acid of fungi, *Arch. Mikrobiol.* **80**:191–204.

Mahadevan, P. R., and Tatum, E. L., 1967, Localization of structural polymers in the cell walls of *Neurospora crassa, J. Cell Biol.* **35**:295–302.

Marchant, R., 1966, Wall structure and spore germination in *Fusarium culmorum, Ann. Bot.* **30**:821–830.

Margoliash, E., and Schejter, A., 1966, Cytochrome *c, Advan. Protein Chem.* **21**:113–286.

Margoliash, E., and Smith, E. L., 1965, Structural and functional aspects of cytochrome *c* in relation to evolution, in: *Evolving Genes and Proteins* (V. Bryson and H. J. Vogel, eds.), pp. 221–242, Academic Press, New York.

Marmur, J., Falkow, S., and Mandel, M., 1963, New approaches to bacterial taxonomy, *Ann. Rev. Microbiol.* **17**:329–372.

McCorkindale, N. J., Hutchinson, S. A., Pursey, B. A., Scott, W. T., and Wheeler, R., 1969, A comparison of the types of sterols found in species of the Saprolegniales and Leptomitales with those found in some other Phycomycetes, *Phytochemistry* **8**:861–867.

Mead, J., 1960, in: *Lipide Metabolism* (K. Bloch, ed.), Wiley, New York.

Mez, C., 1929, Versuch einer Stammesgeschichte des Pilzreichs, *Schr. Königsb. Gelehrt. Ges. Naturwiss. Klasse* **6**:1–58.

Muramatsu, M., Shimada, N., and Higashinakagawa, T., 1970, Effect of cycloheximide on the nucleolar RNA synthesis in rat liver, *J. Mol. Biol.* **53**:91–100.

Nolan, C., and Margoliash, E., 1968, Comparative aspects of primary structures of proteins, *Ann. Rev. Biochem.* **37**:727–790.

Parker, B. C., Preston, R. D., and Fogg, G. E., 1963, Studies of the structure and chemical composition of the cell walls of Vancheriaceae and Saprolegniaceae, *Proc. Roy. Soc. Lon. Ser. B* **158**:435–445.

Raper, J. R., and Flexer, A. S., 1970, The road to diploidy with emphasis on a detour, in: *Organization and Control in Prokaryotic and Eukaryotic Cells* (H. P. Charles and B. C. J. G. Knight, eds.), pp. 401–432, Cambridge University Press, London.

Rines, H. W., Case, M. E., and Giles, H. H., 1969, Mutants in the *arom* gene cluster of *Neurospora crassa* specific for biosynthetic dehydroquinase, *Genetics* **61**:789–800.

Rodwell, V. W., 1969, (1) Carbon catabolism of amino acids, pp. 191–235, (2) Biosynthesis of amino acids and related compounds, pp. 317–373, in: *Metabolic Pathways*, 3rd ed. (D. M. Greenberg, ed.), Academic Press, New York.

Rosinski, M. A., and Campana, R. J., 1964, Chemical analysis of the cell wall of *Ceratocystis ulmi*, *Mycologia* **56**:738–744.

Rothstein, M., 1965, Intermediates of lysine dissimilation in the yeast, *Hansenula saturnus*, *Arch. Biochem. Biophys.* **111**:467–476.

Rothstein, M., and Hart, J. L., 1964, Products of lysine metabolism in yeast, *Biochim. Biophys. Acta* **93**:439–441.

Rutter, W. J., 1965, Enzymatic homology and analogy in phylogeny, in: *Evolving Genes and Proteins* (V. Bryson and H. J. Vogel, eds.), pp. 279–291, Academic Press, New York.

Shatilov, V. R., Evstigneeva, Z. G., and Kretovich, V. L., 1969, Glutamate dehydrogenase of *Chlorella*, *Biochimiya* **34**:409.

Shatilov, V. R., Kolashina, G. S., and Kretovich, V. L., 1970, Induction of NADP-specific glutamate dehydrogenase in *Chlorella*, *Dokl. Akad. Nauk SSSR Ser. Biol.* **194**:969.

Shaw, R., 1965, The occurrence of γ-linolenic acid in fungi, *Biochim. Biophys. Acta* **98**:230–237.

Smith, E. L., 1970, Evolution of enzymes, in: *The Enzymes: Structure and Control*, Vol. 1, 3rd ed. (P. D. Boyer, ed.), pp.267–339, Academic Press, New York.

Sober, H. A. (ed.), 1968, *Handbook of Biochemistry: Selected data for Molecular Biology*, Chemical Rubber Co., Cleveland, Ohio.

Stevenson, R. M., and LéJohn, H. B., 1971, Glutamic dehydrogenases of Oomycetes: Kinetic mechanism and possible evolutionary history, *J. Biol. Chem.* **246**:2127–2135.

Storck, R., 1965, Nucleotide composition of nucleic acids of fungi. I. Ribonucleic acids, *J. Bacteriol.* **90**:1260–1264.

Storck, R., and Alexopoulos, C. J., 1970, Deoxyribonucleic acid of fungi, *Bacteriol. Rev.* **34**:126–154.

Storck, R., Nobles, M. K., and Alexopoulos, C. J., 1971, The nucleotide composition of ribonucleic and deoxyribonucleic acids in some species of the Hymenochaetaceae and Polyporaceae, *Mycologia* (*N.Y.*) **63**:38–49.

Taylor, M. M., and Storck, R., 1964, Uniqueness of bacterial ribosomes. *Proc. Nat. Acad. Sci*, **52**:958–965.

Taylor, M. M., Glascow, J. E., and Storck, R., 1967, Sedimentation coefficients of RNA from 70S and 80S ribosomes, *Proc. Natl. Acad. Sci.* **57**:164–169.

Timberlake, W. E., McDowell, L., and Griffin, D. H., 1972, Cycloheximide inhibition of the DNA-dependent RNA polymerase I of *Achlya bisexualis*, *Biochem. Biophys. Res. Commun.* **46**:942–947.

Tomava, N. G., Evstigneeva, Z. G., and Kretovich, V. L., 1969, Effect of nitrate and ammonium nitrogen on nitrate reductase and some dehydrogenases of *Chlorella*, *Biochimiya* **34**:249.

van Niel, C. B., 1949, The "Delft school" and the rise of general microbiology, *Bacteriol. Rev.* **13**:164–174.

Vanyushin, B. F., Belozersky, A. N., and Bogdanova, S. L., 1960, A comparative study of the nucleotide composition of ribonucleic acid and deoxyribonucleic acids in some fungi and myxomycetes, *Dokl. Akad. Nauk SSSR* **134**:1222–1225.

Vogel, H. J., 1960, Two modes of lysine synthesis among the lower fungi: Evolutionary significance, *Biochim. Biophys. Acta* **41**:172–173.

Vogel, H. J., 1964, Distribution of lysine pathways among fungi: Evolutionary implications, *Am. Naturalist* **98**:435–446.

Vogel, H. J., 1965, Lysine biosynthesis and evolution, in: *Evolving Genes and Proteins* (V. Bryson and H. J. Vogel, eds.), pp. 25–40,Academic Press, New York.

Vogel, H. J. (ed.), 1971, Metabolic regulation, in: ***Metabolic Pathways,*** Vol. 5, Academic Press, New York. [A cluster of articles on the structure, organization, and function of prokaryotic and eukaryotic operons.]

Vogel, H. J., Thompson, J. S., and Shockman, G. D., 1970, Characteristic metabolic patterns of prokaryotes and eukaryotes, in: *Organization and Control in Prokaryotic and Eukaryotic Cells* (H. P. Charles and B. C. J. G. Knight, eds.), pp. 107–119, Cambridge University Press, London.

von Klenk, E., 1961, Chemie und Stoffwechsel der Polyenfettsauren, *Experentia* **17**:199–204.

Wang, C. S., and Miles, P. G., 1966, Studies of the cell walls of *Schizophyllum commune, Am. J. Bot.* **53**:792–800.

Wanka, F., and Schrauwen, P. J. A., 1971, Selective inhibition by cycloheximide of ribosomal RNA synthesis in *Chlorella, Biochim. Biophys. Acta* **254**:237–240.

Watts, D. C., 1971, Evolution of phosphagen kinases, in: *Molecular Evolution: Biochemical Evolution and the Origin of Life,* Vol. 2 (E. Shoffeniels, ed.), pp. 150–173, North-Holland, Amsterdam.

Watts, R. L., 1970, Proteins and plant phylogeny, in: *Phytochemical Phylogeny* (H. J. Harborne, ed.), pp. 145–178, Academic Press, New York.

Willems, M., Penman, M., and Penman, S., 1969, The regulation of RNA synthesis and processing in the nucleolus during inhibition of protein synthesis, *J. Cell Biol.* **41**:177–187.

Wittman, H. G., 1970, A comparison of ribosomes from prokaryotes and eukaryotes, in: *Organization and Control in Prokaryotic and Eukaryotic cells* (H. P. Charles and B. C. J. G. Knight, eds.), pp. 55–76, Cambridge University Press, London.

4

The Phylogeny of Angiosperms—A Palynological Analysis

P. K. K. NAIR
National Botanic Gardens
Lucknow, India

INTRODUCTION

One of the exciting features of the living system is its potential to evolve and progress, or else succumb to death and extinction. The facts of organic evolution are vividly exemplified in the multifarious manifestations of the living units, e.g., the species, each of which constitutes a compromise of diverse morphosystems at various levels of evolutionary advancement. In the fabric of morphosystems of any one species, the key role is played by the reproductive system, the biological efficiency of which is the chief factor that has governed the very existence of life through the endless stream of time and space.

If fossil records are an indication of the primitive plant form, the sporelike single-celled plant body should be considered to indicate the beginning of morphological evolution in plants (Engel *et al.*, 1968; Kremp, 1972). It is therefore imperative that all the morphological diversifications of the plant body are in effect built around the fundamental spore body and that the morphological evolution of spores (in cryptogams) and pollen (in phanaerogams) is a true reflection of the tendencies of plant evolution (Nair, 1972). As Banks (1970, p. 38) stated, "Pollen grains seem to provide more continuity and a greater stratigraphic control than do leaves and wood." This is the fundamental principle on which palynophylogeny is based (see also Nair, 1970*a*) and is applied here with regard to the problem of angiosperm phylogeny.

PHYLOGENY AND EVOLUTION IN "PREANGIOSPERMS"

The angiosperms are the latest evolved group in the evolutionary hierarchy of the plant kingdom; all other plant groups that preceded it constitute the "preangiosperms," at some level of the evolution of which the origin and organization of the unique angiosperm morphostructure occurred (Puri, 1967; Axelrod, 1970; Chaloner, 1970; Muller, 1970; Nair, 1970*a*; Schopf, 1970; Schuster, 1972; Meeuse, 1972). Morphologists and phylogenetic taxonomists reckon the preangiosperms as being composed of the Thallophyta, Bryophyta, Pteridophyta, and Gymnospermae in the order of phylogenetic succession, but the origin and early evolution of each group are shrouded in mystery. The gaps in phyletic history represent the abyss of evolutionary activity, death, and extinctions which inevitably preceded the successful emergence of the fittest in the struggle for survival.

Thallophyta

The Thallophyta consists of delicate, aquatic gametophytic plant forms, each with a life cycle that includes more than one asexual spore type in the gametophytic phase and the spore (zygospore) sporophyte alone constituting the whole sporophytic phase. The occurrence of more than one spore type in the life cycle of the same plant is evidently a measure designed by nature to insure the security of the life span of the plant form, which being gametophytic is highly susceptible to inclement environments that could lead to its death and decay. Evidently, it is this potential for successive sporulation at any occasion in any one life cycle and the capacity of the spore walls to resist decay that have accounted for the survival of thallophytic plant forms through the long span of geological time to the present.

The spore morphology of the Thallophyta in general is characterized by the absence of any definite and clearly organized occurrence of the germinal aperture or mark, which is of primary importance in considerations of the morphological evolution of spores and pollen grains (Nair, 1965). However, the "Riss lineae" observed in some spores, and described as an inconspicuously organized triletous mark, might be considered to signal the beginning of organized apertural evolution. Because of the lack of a clear apertural organization in the spores of the Thallophyta, the author (Nair, 1970*b*) has characterized such spores as "amorphous" and has designated the group as "Amorphosporophyta."

Bryophyta

The dominance of the aquatic Thallophytes slowly diminished with the emergence of land masses out of the past seas, which contributed to the origin and organization of the amphibian plant forms, the Bryophyta, with a new pattern and trend of structural evolution. In general, the plant body in bryophytes is thalloid, constituted of parenchymatous cells, and the thallus produces the sexual units, the antheridia and archegonia. The zygote, resulting from sexual union, develops into a vegetative sporophyte, the capsule, different from the spore sporophyte of the Thallophyta. Evidently, the bryophytes represent a new phase in plant evolution, during which the fundamental organization of the plant body for leading a terrestrial existence was accomplished.

The spore morphology also reflects the new trend of structural organization and evolution, with three well-defined spore types—the trilete, monolete, and alete. That the trilete form is primitive and that the monolete and alete are independent derivatives from the trilete have been clearly explained by Mehra (1968) in his treatise on the phylogeny of the Archegoniatae. In contrast to the thallophytes, there are no asexual spores in the bryophytes and the only spore type that occurs in the life cycle is the result of sexual activity involving the formation of the zygote and the consequent process of reduction division. The attainment of such a new bioform with a unique structural organization (including the trimorphous condition of the spores) and functional mechanism could not have been reached all of a sudden, but must have been the result of a long span of evolutionary activity beginning at the level of the chlorophycean Thallophyta. Perhaps it means that in every evolutionary phase there are two stocks, one representing *instant evolution*, accounting for the establishment of the base phylum, and another representing *slow evolution*, accounting for the origin of the succeeding phylum (Fig. 1).

Pteridophyta

The Pteridophyta, with a well-organized sporophytic vegetative structure differentiated into the root and shoot, mark the next succeeding phase in structural evolution. In Pteridophyta, compared to Bryophyta, the spores are trimorphous, being trilete, monolete, and alete as well as heterosporous, owing to dimorphism in size (see also Pettit, 1970). In the Psilophyta, the extinct primitive group of Pteridophyta, the spores were triletous, which again is evidence of the primitiveness of the trilete spore form (Eggart, 1974). The occurrence of the trimorphous tendency of morphological evolution in the group may be considered to indicate the close phylogenetic link of the Pteridophyta with the Bryophyta, although there is no clear evidence of fossil or living link forms (see Proskauer, 1960) to serve in the reconstruction of phyletic history. At this point, it may be of interest to mention that Schopf (1970) advanced evidence from Precambrian algal fossils to support the view of the chlorophycean origin of vascular plants.

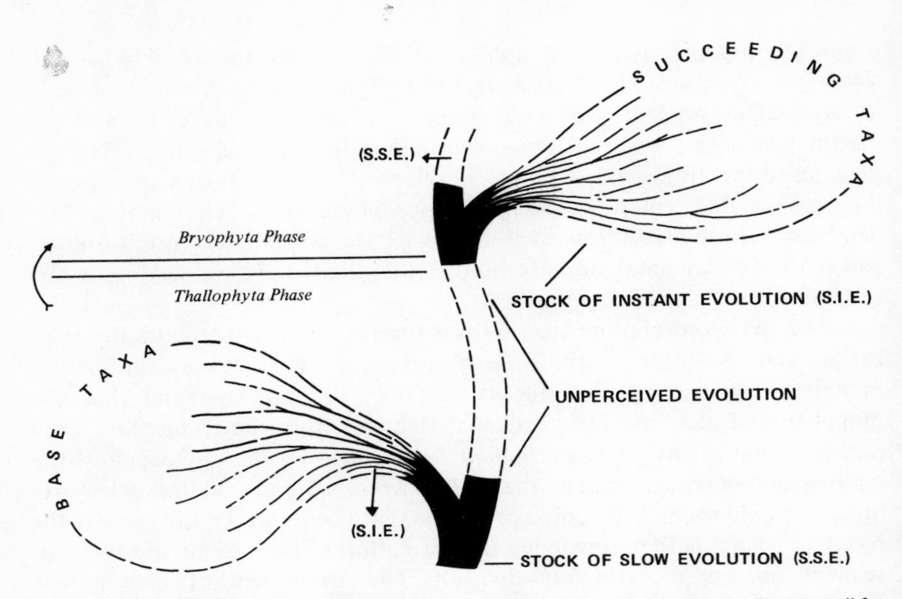

FIG. 1. Possible mechanism of plant evolution. The "succeeding taxa" become the "base taxa" for the group that emerges in succession; at the level of species and lower microtaxa, the evolution may be reticulate and not parallel as shown (see Grant, 1972).

Gymnospermae

The Gymnospermae mark the beginning of the seed habit and are also characterized by the tree habit, with a plant body that is well suited to lead a terrestrial existence. With the accomplishment of a certain degree of perfection in the plant body construction, the gametophyte assumed a new role in the life cycle of the gymnosperms. The female gametophyte remains within the sporophytic plant body, whereas the male gametophyte moves or is moved to the female to effect fertilization. The mobility of the male gametophyte, the pollen grain, is facilitated by a variety of factors: (1) reduction of the cellular organization to a single-cell state, equivalent to the spore stage (microspore) in the Pteridophyta; (2) the tree habit, raising the plant high in the air, so that the microspore movement is uninterrupted; and (3) special structural mechanisms (e.g., wings) and physical properties of pollen (e.g., dry, falls loose, simpler external morphology, apertural structure suited to harmomegathy) (Wodehouse, 1935).

With respect to pollen morphology, the trimorphous condition occurs in gymnosperms as well as in bryophytes and pteridophytes, the gymnosperm apertural types being the trichotomocolpate, monocolpate, and inaperturate (equivalent to the trilete, monolete, and alete, respectively). The aperture, whenever present, is situated on the distal pole, as opposed to the proximal position in cryptogamic spores. Further, the development of the wings in the pollen of the pines and their allies was a new and unique event in the morphological evolution of gymnosperm pollen. Taking note of the fact that the trimorphous trend of morphological evolution in the Bryophyta continues to occur in the Pteridophyta and Gymnospermae, the author (Nair, 1970*b*) has proposed the placement of the above groups in the phylum "Trimorphosporophyta."

Evolutionary Trends in Preangiosperms

The facts of structural organization and trends of morphological evolution of the preangiosperms should serve to provide a clue to the possible nature and direction of the evolutionary process in the angiosperms. The individual identity of the Thallophyta, Bryophyta, Pteridophyta, and Gymnospermae within the preangiosperms is clear, each group representing a new phase in plant evolution. The organization of every new form (e.g., a new species) with a new combination of characters and functional synchrony must have inevitably been preceded by a long chain of fast evolutionary processes, involving variation, recombination, and natural selection. The nature of the evolutionary activity in the organization and establishment of a new plant body system as exemplified, for example, by a whole group such as the Bryophyta, is hidden by the phylogenetic gap that precedes the appearance of every plant group. For each perceivable group, the picture is one of sudden appearance of a new plant structure with a new biological process, and this holds true for the angiosperms, too.

The point of interest is that every plant group apparently has a long phyletic history, perhaps reaching back to the base level of its preceding group (see Fig. 1), at which point the process of more rapid diversification was set in motion, soon after the primary organization of the *ideal* primitive structure suited to a new situation. It may therefore be assumed that the ancestry of the Bryophyta reaches back to the Chlorophyta, that of the Pteridophyta to the Marchantiales-Anthocerotales complex, and that of the Gymnospermae to the Psilophyta-Lycopodiales complex. The origin and evolution of the vascular sporophytes, comprised of the pteridophytes and gymnosperms (among preangiosperms), are of special significance in tracing the phylogeny of angiosperms, as presently attempted.

The early history of the pteridophyte plant body may be reflected in extinct plant forms such as *Rhynia* and *Hornia* on the one hand, representing herbaceous plants, and the lepidodendrons on the other, representing arboreal plants. The herbaceous as well as the arboreal forms (e.g., *Pleuromoia*) possessed tuberous stem structures for the conservation of food materials and for supplying them to the growing shoot or for use in times of need. The fact that the arboreal habit was not immediately successful in the early stages of the land flora is evident from the occurrence of a large number of arboreal genera as fossil forms only. Further, the arboreal habit is confined to the Cyatheaceae alone among the present-day pteridophytes. All the same, it was essential that the reproductive structures be held high and away from the ill effects of the new lands that continued to appear through geological time, as well exemplified in the gymnosperms. The complete dominance of the arboreal habit in the gymnosperms suggests their filial link with the extinct arboreal pteridophytes (i.e., stachyosporous stock).

PHYLOGENY AND EVOLUTION OF ANGIOSPERMS

The origin of the angiosperms has often been described in the words of Darwin as an "abominable mystery," because there are no or perhaps only a few authentic pre-Cretaceous fossil records (see Muller, 1970; Tidwell *et al.*, 1970) and there is no conclusive morphological evidence among present-day plants, with the result that phylogenetic affinity to the various groups of preangiosperms, even reaching the level of the Thallophyta (Chlorophyceae; Schopf, 1970), has been proposed. However, Meeuse (1972) believes that there is "no inexplicable 'gap' in the fossil record, no baffling or 'abominable' mystery as far as the origin of flowering plants is concerned if *living* taxa can be shown to link all higher cycadopsids from the gymnospermous to the angiospermous level of evolution of the reproductive region."

The perceivable new characteristics that provide the angiosperms with a distinctive identity, as different from any one group of the preangiosperms, involve both vegetative and reproductive structures, and their functional mechanisms, of which may be mentioned the reproductive system (the flower with its floral leaves, the closed ovule, the phenomenon of fertilization) and the vascular system (the occurrence of vessels and libriform fibers; see Esau, 1960). The unique structural organization of the angiospermous plant body system is reflected vividly in the morphology of pollen grains.

The taxonomic system of Cronquist (1968) is considered here as being the nearest to the principles of morphological evolution of pollen grains (Nair, 1972). According to this system of classification, Angiospermae (i.e., Magnoliophyta) belong to the class Magnoliatae (Dicotyledones) and the class Liliatae (Monocotyledones). The author (Nair, 1972) has proposed the division of the primitive Magnoliatae (belonging to the subclass Magnoliidae of Cronquist, 1968) into the subclasses Magnoliidae (orders Magnoliales, Piperales, Aristolochiales, and Nymphaeales, barring Nelumbonaceae of Cronquist, 1968) and Ranunculiidae (orders Ranunculales and Papaverales and perhaps a few more of Cronquist, 1968), the two groups being characterized by monocolpate and tricolpate pollen, respectively, and this amended proposal is followed in the following discussion.

The pollen grains of the Angiospermae consist of an array of apertural types which may be classified broadly into the two fundamental groups monocolpate and tricolpate. The monocolpate group contains the fundamental trichotomocolpate (trilete) and the specialized inaperturate, apart from the monocolpate type itself, which together suggest the phylogenetic alliance of the monocolpate group with the preangiosperms, in which the trimorphous condition is the rule. The tricolpate group contains various apertural types, namely, colpate, colporate, porate, and pororate, in varying numbers (two, three, or more) and disposition (equatorial or global), and, further, this group is devoid of the monocolpate type, except perhaps as abormalities in such marginal taxa as the Nelumbonaceae (Chopra, 1972).

Comparing the trends of spore–pollen evolution in the preangiosperms and the angiosperms, it is evident that the trend of morphological evolution, as reflected in the occurrence of the trimorphous condition of pollen grains in the preangiosperms suddenly changed in the angiosperms (see also Chaloner, 1970; Muller, 1970), in which new apertural forms represented by the tricolpate group appeared. The distribution of the two pollen apertural groups, the monocolpate and the tricolpate, in the angiosperms is suggestive of their phylogeny and evolutionary trends. The monocolpate group dominates the Liliophyta (Monocotyledones) and is the only form occurring in the Magnoliidae of the Dicotyledones, whereas the tricolpate group is characteristic of the Ranunculiidae.

On the basis of palynological evidence, it is logical to assume that groups having monocolpate pollen, comprised of the Liliatae and Magnoliidae (dicots), being possessed of the trichotomocolpate, monocolpate, and acolpate forms, are evidently related to the preangiosperms, which are characteristically trimorphous. It was explained earlier that every succeeding group might have originated from the basal taxa of the immediately preceding group, and applying such a principle to the angiosperms, it becomes apparent that the origin of angiosperms perhaps occurred at the level of the pteridosperms, as proposed by Eames (1961).

The earliest, possibly extinct, taxa of angiosperms may have been those showing the trimorphous trend of pollen evolution, as that in the preangiosperms, which among the present-day taxa occurs in the Liliatae (i.e., monocots) and the Magnoliidae (dicots). The phylogenetic link between the pteridospermous preangiosperms and the monocolpate group of present-day angiosperms is further evident from the fact that the aperture is distal in position in both (although rarely not so in some pteridosperms; Potonie, 1967).

The tree habit being dominant and successful in the gymnosperms, it is possible that the same habit prevailed in the primitive angiosperms, too, and therefore the magnolian group among the present-day angiosperms, by virtue of its tree habit and its possession of the monocolpate type of pollen, merits a primitive status. Consequently, the hypothetical protangiosperm was possibly magnolian in its plant-body structure and biology; from the magnoliaceous protangiosperms diverged the monocot stock on the one hand and the magnolian stock on the other as represented by the taxa of present-day angiosperms comprised in the Magnoliidae. The pollen grains of *Schisandra* (Schisandraceae), having a trichotomocolpate aperture on the distal face, the arms of which extend to the proximal face, and also three independent zonal colpi (Erdtman, 1952), suggest the possible origin of the fundamental tricolpate pollen characteristic of the ranunculean stock. Based on these facts, it is supposed that the Magnoliidae (including perhaps the protangiosperms) constitute the most primitive group that emerged in the wake of the origin of the angiosperms.

The families of plants belonging to the present-day Magnoliidae (as modified by Nair, 1972, on the proposal of Cronquist, 1968) are as follows: Austrobaileyaceae, Lactoridaceae, Magnoliaceae, Winteraceae, Degeneriaceae, Himantandraceae, Annonaceae, Myristicaceae, Canellaceae, Illiciaceae, Schisandraceae, Eupomatiaceae, Amborellaceae, Trimeniaceae, Monimiaceae, Gomortegaceae, Calycanthaceae, Lauraceae, Hernandiaceae (order Magnoliales), Chloranthaceae, Saururaceae, Piperaceae (order Piperales), Aristolochiaceae (order Aristolochiales), Nymphaeaceae, and Ceratophyllaceae (order Nymphaeales, excluding Nelumbonaceae). The flowers of the plants constituting the above families are either magnolioid or strobiiloid, providing thereby a united picture of both the classical concept and the Englerian concept of the origin of the flower. It is imperative that the taxa of Magnoliidae consist of two evolutionary stocks, represented by the magnolioid flower (named here as phyllomosporous stock) with many floral parts, and by the piperacean monochlamydous flower (stachyosporal stock) with reduced floral members, that are independent entities which diverged early in the evolution of the magnolian monocolpate group. With the onset of new evolution, as represented by new pollen morphotypes and floral types of the Ranunculiidae (as defined by Nair, 1972), the evolution of magnolian taxa might have been slowed down, and the fewer number of families belonging

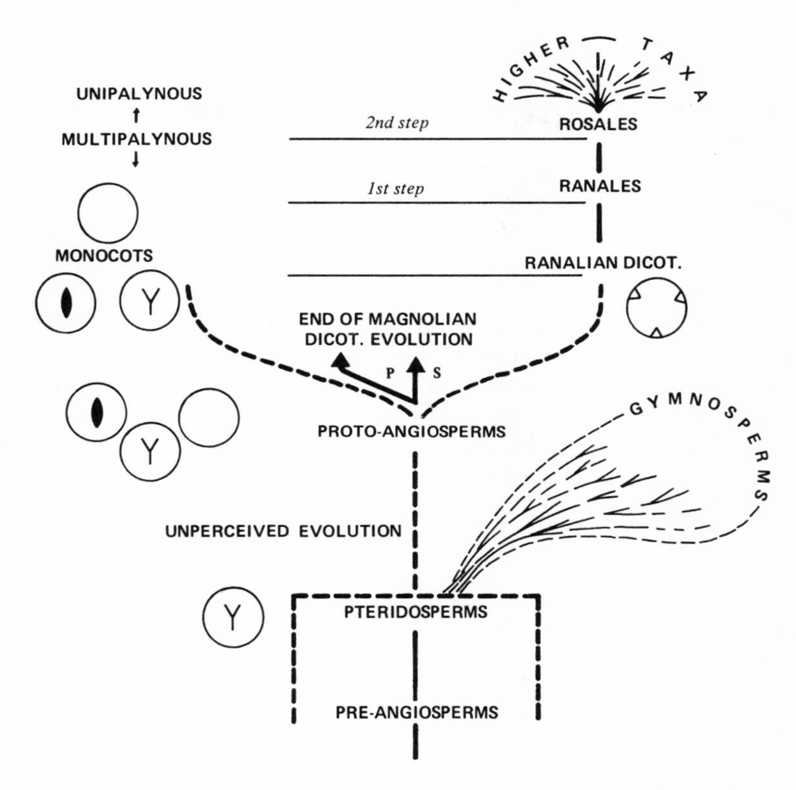

FIG. 2. The phylogeny of angiosperms. P, Phyllomosporal stock; S, stachyosporal stock.

to the group should be taken as evidence that magnolian evolution ended at the primitive level of angiosperm evolution (Fig. 2).

The Liliatae, by virtue of their dominance of monocolpate pollen forms, are clearly related to the magnolian stock of dicots, but the ancestry of Liliatae has been variously interpreted. The view of Hutchinson (1959) represents one line of opinion, that the monocots originated from dicots, as did the ranunculian dicots. The dominance of monocolpate pollen in Liliatae and the lack of such pollen in the Ranunculiidae suggest that the possibility of a phylogenetic relationship between the two groups is remote, indeed. On the other hand, the possibility of a phylogenetic connection between the Magnoliidae and Liliatae is apparently greater, palynologically, but cannot be fully supported by floral and vegetative characters. The totality of plant morphology suggests an early divergence of the monocots and magnolian dicots from the protangiosperms, which might have been magnolian in general morphology with arboreal and polyfloral habits and monocolpate pollen.

The ranunculian stock of dicots dominates the angiospermous vegetation of the present day. The Ranunculaceae, accepted as the most primitive family by a majority of morphologists and taxonomists, alone contains almost all the pollen types known among the ranunculian stock of dicots. The occurrence of multipalyny at the primitive level is evidence of relatively rapid evolutionary activity, prior to the resolution of taxa characterized by unipalyny. The Rosales also are considered to be primitive and ancestral with regard to the evolution of the Sympetalae (Wernham, 1911), and this order also is characterized by multipalyny. It may therefore be presumed that the early evolution of the ranunculean dicots could have been completed in two steps, at the levels of the Ranales (Ranunculaceae and their allies) and the Rosales (Nair, 1972), and that further evolution occurred along unipalynous lines polyphyletically (Fig. 2).

REFERENCES

Axelrod, D. I., 1970, Mesozoic palaeogeography and early angiosperm history, *Bot. Rev.* **36**:277–320.

Banks, H. P., 1970, Major evolutionary events and the geological record of plants, *Biol. Rev.* **45**:317–318.

Chaloner, W. G., 1970, The rise of the first land plants, *Biol. Rev.* **45**:353–378.

Chopra, S., 1972, Palynological investigations of some vegetable crops, Ph.D. thesis, Agra University, India.

Cronquist, A., 1968, *The Evolution and Classification of Flowering Plants,* Nelson, London.

Eames, A. J., 1961, The morphological basis for a Palaeozoic origin of the angiosperms, *Rec. Advan. Bot. Res.,* pp. 722–726.

Eggart, D. A. 1974. The sporangium of Horneophyton lignieri (Rhyniophytina) *Amer. J. Bot.* **61(4)**:405–413.

Engel, A. E. L., Nagy, B., and Nagy, L. A., 1968, Alga-like forms in the Onverwacht series, South Africa: Oldest recognized lifelike forms on earth, *Science* **161**:1005–1008.

Erdtman, G., 1952, *Pollen Morphology and Plant Taxonomy/Angiosperms,* Almqvist and Wiksell, Stockholm.

Esau, C., 1960, *Anatomy of Seed Plants,* Wiley, New York and London.

Grant, V., 1972, *Plant Speciation,* Columbia University Press, New York.

Hutchinson, J., 1959, *The Families of Flowering Plants,* 2nd ed., 2 vols., Clarendon, Oxford.

Kremp, G. O. W., 1972, Advancing organisation, time, and the orderly progressions of life, *J. Palynol.* **8**:1–27.

Meeuse, A. D. J., 1972, Palm and pandan pollination: Primary anemophily or primary entomophily? *Botanique* (*Nagpur*) **3**(**1**):1–6.

Mehra, P. N., 1968, Conquest of land and evolutionary patterns in early land plants (15th Sir Albert Charles Seward Memorial Lecture), Birbal Sahni Institute of Paleobotany, Lucknow.

Muller, J., 1970, Palynological evidence on early differentiation of angiosperms, *Biol. Rev.* **45**:417–450.

Nair, P. K. K., 1965, *Pollen Grains of Western Himalayan Plants*, Asia Publishing House, New York.

Nair, P. K. K., 1970*a*, *Pollen Morphology of Angiosperms: A Historical and Phylogenetic Study*, Scholar Publishing House, Lucknow/Vikas Publishing House, Delhi.

Nair, P. K. K., 1970*b*, A palynological basis for plant classification, *J. Palynol.* **6**:111–112.

Nair, P. K. K., 1972, Pollen morphology and phylogenetic classification of primitive angiosperms, *Advan. Plant Morphol.* (*Meerut*) (Prof. V. Puri Commemorative Volume), pp. 255–263.

Pettit, J., 1970, Heterospory and the origin of seed habit, *Biol. Rev.* **45**:401–416.

Potonié, R., 1967, New phylogenetic facts on fossil spores. *Rev. Paleobot. Palynol.* **1**:71–82.

Proskauer, J., 1960, Studies in Anthocerotales, VI. *Phytomorphology* **10**:1–19.

Puri, V., 1967, The origin and evolution of angiosperms, *J. Indian Bot. Soc.* **46**:1–14.

Schopf, J. W., 1970, Precambrian micro-organisms and evolutionary events prior to the origin of vascular plants, *Biol. Rev.* **45**:319–352.

Schuster, R. M., 1972, Continental movements, "Wallace's line" and Indo-Malaya Australasian dispersal of land plants: Some eclectic concepts, *Bot. Rev.* **38**(1):3–86.

Tidwell, W. D., Rushforth, S. R., Reveal, J. L., and Behunin, H., 1970, *Palmoxylon sumperi* and *P. pristina*: Two pre-Cretaceous angiosperms from Utah, *Science* **168**:835–840.

Wernham, H. F., 1911, Floral evolution with particular reference to the sympetalous dicotyledons, *New Phytol.* **10**:73–83, 109–120, 145–159, 217–226, 293–305.

Wodehouse, R. P., 1935, *Pollen Grains*, Macmillan, New York.

5

Gene Flow in Seed Plants

DONALD A. LEVIN
Department of Botany
University of Texas
Austin, Texas

and

HAROLD W. KERSTER
Environmental Studies Center
California State University
Sacramento, California

INTRODUCTION

Gene dispersal (flow, or migration) within and between plant populations has been of continuous interest to plant breeders and seed producers for many decades. Economic considerations have stimulated studies of gene flow as a function of distance, breeding system, pollinating agent, and planting design in numerous domestic plants. Only during the past two decades have a large body of plant evolutionists become interested in information accruing from these studies, and in the rates of gene flow in wild populations. Their efforts have concentrated primarily on related problems such as adaptations for and mechanics of pollen and seed (or fruit) dispersal, plant–pollinator coevolution, adaptive radiation in pollination and seed dispersal mechanisms, and colonization and the alteration of species boundaries. Early in this century, anecdoctal evidence on the movement of pollen and seed vectors, dispersal of pollen by wind, and the range extensions of weed species led to the casual assumption that gene flow must be extensive, and that it must play a major role in the cohesion of populations and population systems. This view eroded as more information became available and was more critically interpreted (e.g., Grant, 1958, 1971; Ehrlich and Raven, 1969; Stebbins, 1970*a*; Bradshaw, 1972).

There is considerable evidence that levels of gene exchange within and between populations may be orders of magnitude below those previously accepted. Information on potential and actual gene flow in plants is scattered throughout the crop, forestry, genetics, and evolution literature. Wolfenbarger (1946, 1959) compiled extensive and thorough reviews of the literature prior to 1960. This chapter in a sense is an extension of that survey and a development of the Ehrlich and Raven (1969) thesis on gene flow restriction. It is intended to organize a body of information on the nature and regulation of pollinator foraging and attendant pollen flow in animal-pollinated plants, pollen flow in anemophilous plants, seed dispersal patterns, and gene flow within and between populations. In addition, it is intended to show what the observed gene flow levels mean in terms of the breeding structure of plant populations and population systems.

POTENTIAL GENE FLOW

"Potential gene flow" refers to the deposition of pollen and seeds from a source as a function of distance. In contrast, "actual gene flow" refers to the incidence of fertilization (in the case of pollen) and establishment of reproductive individuals (in the case of seeds) as a function of the distance from a source. As will be developed below, there may be a great disparity between actual and potential gene flow. Nevertheless, the movement of pollen and seeds affords valuable insight into the nature of actual gene dispersal in the absence of appropriate genetic markers and knowledge of gene frequencies.

POLLEN FLOW

The pollen of most species is dispersed by animals or by the wind. Pollination by animals is more efficient than wind pollination, in that it increases the probability of seed set and reduces the number of pollen grains which need be produced per ovule (Baker, 1963; Faegri and van der Pijl, 1966). Pollination by animals is most prevalent in relatively closed communities, especially tropical rainforests (Baker, 1959, 1963, 1970). On the other hand, wind pollination is relatively common in open, often arid regions of diverse climates, i.e., in prairies, temperate forests, and savannas, which are characterized by relatively low rainfall, deciduous habit, low species diversity, and unambiguous stimuli for synchronized flowering (Baker, 1959, 1963; Whitehead, 1969; Stebbins, 1970*b*).

Animal-Mediated Pollen Flow

The pollen dispersal curves of entomophilous plants are dependent on the foraging habits of pollinators, which in turn are responsive to properties of the plant population and local flora in which it is embedded. Unfortunately, the pollen dispersal schedule typically cannot be measured directly, because distinguishable pollen phenotypes are lacking, or if they are present, as in species with heteromorphic incompatibility, the phenotypes are not segregated in space. As an alternative, we may estimate the dispersal pattern from knowledge of foraging behavior and pollen replacement schedules, or the dispersion of dye particles applied to anthers.

The foraging areas of bees are best understood and will be considered first. H. Müller (1882) was the first to report that honeybees have narrow foraging areas and that many bees return on successive trips to the same plants on which they have fed earlier. Other early observations of this behavior were reported by Giltay (1904) and Bonnier (1906). Minderhoud (1931) marked honeybees foraging on plots of *Brassica juncea, Trifolium pratense, Taraxacum officinale,* and *Reseda lutea,* and observed that bees return time after time, even day after day, to an area less than 10 m². Buzzard (1936) noted that each plant of *Cotoneaster horizontalis* appeared to have its own population of bees which strayed from one bush to another only when their branches were interlaced. Butler *et al.* (1943) studied foraging in nature and in artificial populations. They found that honeybees foraging in small plots of *Echinops sphaerocephalus* isolated by 17 m seldom switched plots. They also noted that 90% of the bees marked near the center of a patch of *Epilobium angustifolium* were foraging there 24 hr later and that 10% could be found within the adjacent 5 m. The fixation of bees to localized feeding sources was observed in an artificial population composed of 112 petri dishes containing sugar syrup. An individual bee tended to confine its foraging to a single dish, and when it strayed it was to a neighboring dish (Fig. 1). Foraging areas of less than 18 m² were observed by Singh (1950), who plotted the course of individual bees on monocultures of legumes and composites. The restricted movement of honeybees in dense arrays is also apparent in the distribution of flight distance. In turnips (Bateman, 1947*a*) and vetch (Weaver, 1957), the frequency distributions of flight lengths are highly leptokurtic; i.e., they depart from a normal distribution in having an excess of short- and long-distance flights.

The foraging behavior of honeybees on trees is similar to that described for herbs. Bees in fruit orchards restrict their activity to single or adjacent trees (MacDaniels, 1931; Roberts, 1956). Singh (1950) followed 64 honeybees and found that 45 of them continued to forage on one tree, 16 on

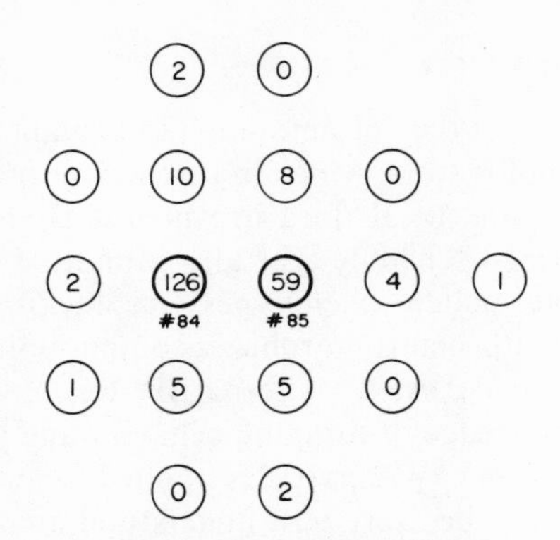

FIG. 1. The number of times bees marked on dishes No. 84 and 85 were seen foraging on these and neighboring dishes. After Butler *et al.* (1943).

two trees, two on four trees, and one on five trees. Free (1960) notes that on the average bees visited two fruit trees per trip and that flights were between neighboring trees. The foraging areas of individual vectors remained small over consecutive trips.

Relatively little information is available on the foraging areas of bumblebees. It appears that their foraging areas on herbs are similar to those of honeybees, although bumblebees may visit twice as many flowers per trip (Ribbands, 1955; Free, 1964, 1968). In fields of cotton, bumblebees work methodically from plant to plant in the immediate vicinity of the first plant visited (Thies, 1953; Simpson and Duncan, 1956). On fruit trees, bumblebees tend to work less methodically than honeybees, and fly more frequently from tree to tree (Brittain and Newton, 1933).

Attempts have been made to estimate the distribution of pollen within a crop by dusting the anthers of selected flowers with methylene blue and examining other flowers for the stain 24 hr later. In cotton, Thies (1953) treated 21 flowers in the edge row of a ten-row block and found that after 2 hr every flower had been dusted. In other studies on cotton (Stephens and Finkner, 1953; Simpson, 1954), dye was found on nearly all flowers in the vicinity of the donors and up to 27 m away. In one experiment with methylene blue, Sindu and Singh (1961) determined that the percentage of recipient flowers decreased from 48% at 6 m to 18% at 12.2 m, 6% at 15.2 m, and 2% at 16.7 m. Although the technique is simple and has appeal, a compelling relationship between dye dispersal and pollen dispersal has not been established. Moreover, such studies suffer from the fact that the flowers, but not stigmas *per se*, were scored for the presence of dye.

The foraging behavior of bees in natural populations is compatible with that observed in synthetic populations. Actively feeding bees move from a plant to one of its nearest neighbors, and the frequency distribution of flight distances is markedly leptokurtic. The distribution of honeybee and butterfly flight distance (collectively) in a colony of *Lithospermum caroliniense* shows a preponderance of very short flights; fewer than 5% of all flights are beyond 10 ft (Fig. 2). The distribution is remarkable for its leptokurtosis (kurtosis = 50.48). Honeybees foraging in a dense array of *Liatris aspera* have a mean flight distance of 0.58 m and a kurtosis of 26.11 (Levin and Kerster, 1969*b*). The foraging behavior of bees working dense colonies of *Echinops sphaerocephalus, Vernonia fasiculata, Eupatorium maculatum, Lythrum salicaria, Lythrum alatum, Monarda fistulosa, Pychnanthemum virginianum,* and *Veronicastrum virginicum* were similar *inter se* and to behavior for *Liatris aspera* (Levin and Kerster, 1969*b*).

Mark and recapture studies on colonial butterflies indicate that often they are relatively sedentary and forage within a small area. Dowdeswell *et al*. (1949) reported that 98% of the *Maniola jurtina* individuals marked at a site were recaptured there. Restricted foraging areas have been reported for several other temperate zone species including *Polyommatus incarus* (Dowdeswell *et al*., 1940), *Papilio glaucus* (Fales, 1959), *Philotes sonorensis* (Mattoni and Seiger, 1963; Keller *et al*., 1966), and *Cercyonis oetus* (Emmel, 1964). Tropical species with ranges of only hundreds of square meters include *Euptychia hermes* (Emmel, 1968), *Victorina epaphus* (Young, 1972*a*), *Heliconius erato* (Turner, 1971), *Anartia amalthea* (Fosdick, 1973),

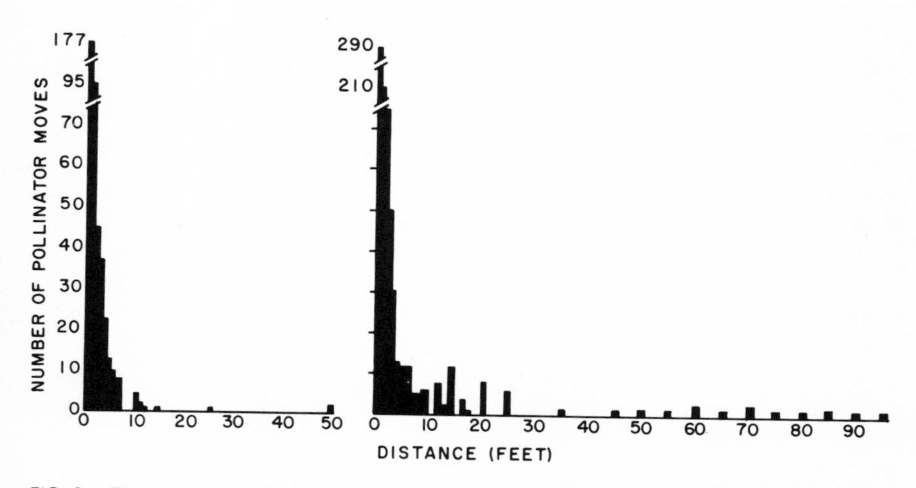

FIG. 2. The distribution of pollinator flight distances in populations of *Lithospermum caroliniense* (left) and *Phlox pilosa* (right). After Kerster and Levin (1968) and Levin and Kerster (1968).

and *H. ethilla* (Ehrlich and Gilbert, 1973). Ehrlich (1965) reported that 97% of *Euphydryas editha* specimens released at different localities were recaptured within less than 1 ha from the point of origin. However, individuals in some populations of this species may travel 500 m or more from the site of release (Gilbert and Singer, 1973). Dispersal distances exceeding 500 m also have been reported in the temperate zone *Pieris raphae* (Emmel, 1972), *Danaus plexippus* and *D. gilippus* (Brower, 1961), and *Erebia epipsodea* (Brussard and Ehrlich, 1970), and in the tropical *Anartia fatima* (Young, 1972*b*). Unfortunately, we do not know whether or to what extent pollen is transferred between populations serving as food resources by these and other highly vagile species.

Little is known about the foraging flight distances of butterflies within populations. Two studies are most notable in this regard. Levin and Kerster (1968) reported the flight distances of the species of *Colias, Pieris,* and *Polites* servicing *Phlox pilosa.* The distribution of distances for three populations (collectively) is depicted in Fig. 2. Most flights were among plants that were near neighbors. The mean flight distance was approximately 1.5 m, the maximum observed being about 35 m. Plant densities for the populations were 5, 8, and 13 plants/m^2. A pattern of near-neighbor visitation also was documented for the butterfly *Heliconius ethillia* (Ehrlich and Gilbert, 1973). In this case, food plants comprised several genera and often were located 10 m or more apart. Moreover, this pollinator established "traplines" which were patrolled with some regularity. The outcome for the food plants of *Heliconius* and for *Phlox pilosa* is assortative pollination by virtue of spatial affinity.

The restricted movements of butterflies should be accompanied by restricted pollen dispersal within and between the plants on which they are feeding. The only information on pollen dispersal by lepidopterans comes from two studies on interspecific pollen exchange in *Phlox*. The first, heretofore unpublished, involves the dispersal of *P. glaberrima* pollen in a population of *P. pilosa* located near Lansing, Cook County, Illinois. The pollen grains of *P. glaberrima* are much larger than those of *P. pilosa* and are readily scored on stigmas of the latter. Ten plants of *P. glaberrima* were introduced as a single cluster within a *P. pilosa* population containing over 1000 plants and having approximately 10 plants/m^2. Ten 1-m-wide concentric rings were established around the introduced species, and stigmas were collected from each ring in order to determine the rate at which stigma contamination and alien (= interspecific) pollen loads declined as a function of distance. Our observations are presented in Table I. The ring adjacent to the pollen source had the largest percentage of contaminated stigmas and alien pollen loads. In the second ring (1.1–2.0 m) there were only 14% as many contaminated stigmas as in the first, and

TABLE I. Intrapopulation Pollen Flow as a Function of Distance in *Phlox*

Distance (m)	Number of stigmas examined	Contaminated stigmas (%)	Mean alien load
0–1.0	284	35.71	1.60
1.1–2.0	340	5.71	0.16
2.1–3.0	215	6.05	0.09
3.1–4.0	615	4.23	0.05
4.1–5.0	601	3.66	0.05
5.1–6.0	515	1.17	0.01
6.1–7.0	606	1.82	0.02
7.1–8.0	251	1.20	0.01
8.1–9.0	221	2.31	0.02
9.1–10.0	411	1.22	0.01

pollen receipt per stigma was only 10% that of the first. From 1 m to 10 m there was a steady but gradual decline in pollen receipt. Although we were observing the consequences of interspecific pollinator flights, the pattern which emerged is similar to that which would accrue from intraspecific pollinator flights, because the food sources are so similar. The data are consistent with the short flight distances of lepidopterans in *P. pilosa* populations (Levin and Kerster, 1968).

The second investigation deals with pollen exchange between *Phlox divaricata* and *P. bifida* as a function of distance (Levin, 1972*a*). The spatial association of the two species varied from confluent (mixed) to a separation of 35 m. A summary of the results in Table II reveals that pollen flow is most extensive between confluent populations. This is consistent with the sedentary behavior of the pollinators, which include species of *Papilio*, *Polites*, and *Colias*. Pollen flow is much reduced between contiguous populations and continues to decline as distance increases.

Foraging behavior in hummingbirds is best known in territorial species. Most or all feeding takes place within the territory so that pollen dispersal is largely confined to the area defended by an individual. The size of the area is probably a function of the supply of nectar per unit area, the number of other hummingbirds, and the distribution and species composition of the major food plants. The territories may be short-term because of asynchronous flowering of members of the same species, or because a migrant leaves the area en route to its breeding or wintering grounds. Territory size may be as large as 1000 m^2 (*Panterpe insignis*,

TABLE II. Interpopulation Pollen Flow in *Phlox*

Proximity	Number of alien grains (average)	Contaminated stigmas (%)	Domestic/alien grain ratio
	P. divaricata		
Confluent (3)[a]	0.51	22.8	144:1
Contiguous (2)	0.18	11.7	280:1
5–15 m (4)	0.14	9.3	439:1
20 m (2)	0.13	10.8	540:1
30–35 m (3)	0.06	6.7	823:1
	P. bifida		
Confluent	0.74	20.7	90:1
Contiguous	0.28	13.0	247:1
5–15 m	0.24	10.9	234:1
20 m	0.10	6.8	611:1
30–35 m	0.06	4.6	920:1

Summarized from Levin (1972*b*).
[a] Number of populations studied is given in parentheses.

Wolf, 1969), or only a few square meters (*Calypte anna*, Stiles, 1972). Grant and Grant (1968) report that 25 *Delphinium cardinale* plants were included in the territory of male Costa hummingbirds (*Calypte costae*).

The dispersal of *Delphinium cardinale* pollen by hummingbirds was analyzed by Schlising and Turpin (1971). They applied radioactive iodine (I^{131}) in liquid buffer solution to stamens of plants in the field, and then traced and recorded radioactivity on flowers elsewhere in the population. In one population, I^{131} was applied over 8 days to 76 flowers on six plants in the center of a circular zone with a radius of 11 m. After 30 hr, one sector (quarter) of the zone had radioactive pollen on 9% of its flowers. After 8 days, 58% of the flowers in another sector had radioactive pollen on them The receipt of radioactive pollen was greatest by those plants near the pollen source, and diminished abruptly beyond 4 m (Fig. 3). Nevertheless, some pollen was carried to the periphery of the population. No male territoriality was detected during this study. Linhart (1973) has shown that *Heliconia* species which form dense populations are fed on primarily by territorial hummingbirds. Consequently, pollen dispersal is over a few meters, with an upper limit of about 25–120 m depending on population density. In species where plants tend to be scattered, pollination is accomplished by nonterritorial hummingbirds, and pollen movement may approach 200 m, the number of flowers receiving marked pollen decreasing in a linear fashion from the source.

The pattern of foraging in natural populations and crops may be described as "opportunistic." A vector makes frequent trips each day to one or a few substantial pollen and nectar sources, a form of behavior that is not very efficient in transferring pollen between plants with a multitude of flowers (e.g., trees) or between populations. Although this pattern is the most prevalent one, it is not universal. Female euglossine bees of the neotropical lowlands visit the same large number of small and widely scattered pollen sources on a daily basis for many months (Dressler, 1968; Williams and Dodson, 1972; Janzen, 1971*a*). They may be described as "traplining" bees, the trapline consisting of as many as ten divergent species including those within the Apocynaceae, Leguminosae, Convolvulaceae, and Rubiaceae. The males also pollinate widely scattered plants; however, those observed do not run a specified trapline (Dodson, personal communication). These vectors provide effective outcrossing for plants at low densities in spite of the fact that a significant portion of the visits are interspecific. Long-distance pollination may also be experienced regularly by species serviced by lepidopterans and hummingbirds (Janzen, 1971*a*; Ehrlich and Gilbert, 1973, Dodson, personal communication).

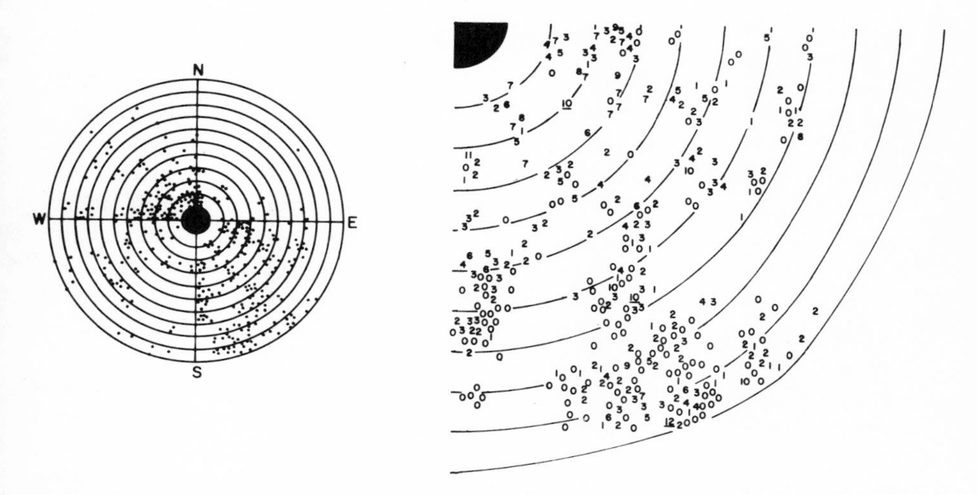

FIG. 3. The approximate location of *Delphinium cardinale* flowering stems relative to the source of radioactive flowers depicted as a black core. Stems bearing radioactivity on at least one flower 8 days after the application of radioactive iodine are shown on the left. The number of flowers per stem in the southeast sector that contained radioactive pollen is shown on the right. After Schlising and Turpin (1971).

Thus far we have emphasized general properties of groups of pollinators. This is not to imply that pollinators do not alter their foraging behavior in response to environmental variables. The foraging behavior of pollinators is plastic, and in part dependent on plant density. The relationship between plant density within continuous arrays and honeybee flight parameters was determined in 27 populations, three for each of nine forb species (Levin and Kerster, 1969*a*). The species, their densities, and pollinator flight parameters are summarized in Table III. Within each species, an increase in density was accompanied by a decrease in flight distance mean (Fig. 4) and variance. When all species were considered collectively, the correlation coefficient for plant spacing mean vs. flight distance mean was 0.83, and that for flight vs. spacing variation was 0.78. Thus we may conclude that bees are very sensitive to plant density and respond in a similar fashion regardless of the plant species involved. Density-dependent foraging distances and pollen dispersal may be a common feature for bees and bee-pollinated plants. This mode of foraging is the most efficient in terms of nutrition gathered vs. energy expended per unit time. Density-dependent foraging also is evident in hummingbirds (e.g., Pitelka, 1942; Grant and Grant, 1968; Wolf, 1969) and butterflies (Levin and Kerster, 1968; Ehrlich and Gilbert, 1973).

Plant density per unit area and the size of population should also influence the incidence of interpopulation pollinator flights. The more plants present, the greater the probability that a pollinator will forage exclusively in a single population. What is important ultimately is not the number of plants per population, but the number of flowers and the nectar production per flower (Heinrich and Raven, 1972). The more flowers there are per unit area or the more nectar per flower, the more worthwhile the food source. The numbers of vectors feeding in a population also may affect the propensity for interpopulation flights. Populations receiving much pollinator attention will be depleted of their "resources" more rapidly than those receiving little attention, so that we may expect vectors from the former to move to other populations more frequently.

At some point, reduction in plant density will not only affect the foraging behavior of the pollinators, but it will also affect the desirability of the species as a food source. When this point is reached, there will be a reduction in pollinator constancy and in the number of pollinator visits per flower (*cf.* Free, 1970). A self-incompatible species will experience a decline in seed set in consonance with a decline in pollinator reliability, but pollen dispersal distance will remain relatively large. A self-compatible, self-pollinating species with a penchant for cross-fertilization would not suffer a decline in seed set, but the incidence of self-fertilization would rise as the propensity to feed on the species diminished (Bateman, 1956). In terms of pollen dispersal, self-pollination is equivalent to zero dispersal. Thus as density declines the effect of selfing may counterbalance the wider dispersal of pollen by animals.

The foraging behavior of a pollinator species is conditioned not only by the contemporary stage of plant species in an area but also by the long-term reliability of particular food resources (Southwood, 1962), pollinator density (Dethier and MacArthur, 1964; Johnson, 1969; Shapiro, 1970), and efficiency of adults in locating larval and adult resources or the ability of dispersing adults to assess the quality of larval resources (Gilbert and Singer, 1973). In general, the vagility of butterflies is related to the predictability and availability of their biotic environments. Butterflies which are K-strategists (MacArthur and Wilson, 1967) or occupy mature communities with limited food resources tend to maintain small sedentary populations, whereas species which are r-strategists or exploit transient habitats rich in foodstuffs tend to be more mobile (Young, 1972*b*; Ehrlich and Gilbert, 1973).

The specific relationship between pollinator flight distances and pollen dispersal distances remains to be determined. A disparity between these statistics may stem from self-compatibility, pollen carryover beyond the first recipient plant, and the removal or loss of pollen from those parts of the vector's body which would touch the stigma.

Pollen carryover in *Phlox* and *Lithospermum* probably is small. *Phlox* is pollinated primarily by butterflies, which transport the pollen on their proboscis. *Lithospermum* is pollinated by bees and lepidopterans, which transport the pollen on the proboscis or face. Both vectors usually work

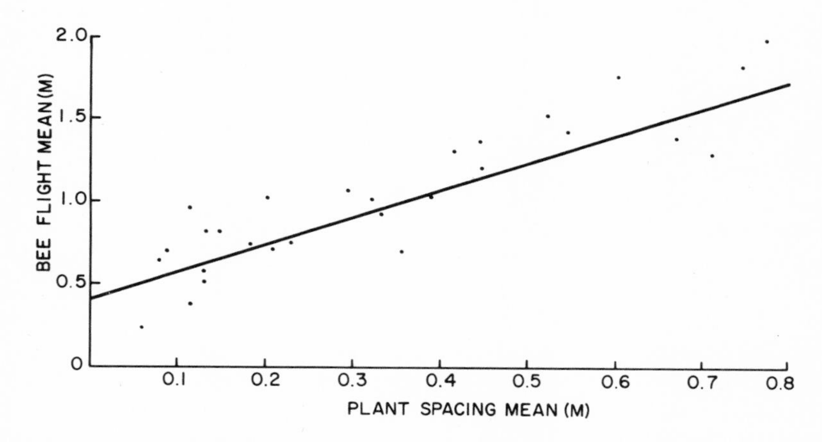

FIG. 4. Relationship between bee flight mean and plant spacing mean per population. Each dot represents a population. After Levin and Kerster (1969*a*).

TABLE III. Summary of Spacing and Flight Parameters in Study Colonies

Species	Population structure	Spacing mean	Spacing variance	Number of bee flights[a]	Flight mean	Flight variance
Echinops sphaerocephalus	Sparse[b]	0.602 ± 0.218	0.038 ± 0.006	215 (10)	1.77 ± 0.122	3.208 ± 0.554
	Intermediate	0.211 ± 0.013	0.034 ± 0.007	370 (11)	0.717 ± 0.048	0.849 ± 0.152
	Dense	0.089 ± 0.009	0.014 ± 0.004	243 (6)	0.686 ± 0.054	0.708 ± 0.156
Vernonia fasciculata	Sparse	0.523 ± 0.019	0.048 ± 0.006	201 (13)	1.534 ± 0.081	1.330 ± 0.033
	Intermediate	0.299 ± 0.011	0.023 ± 0.003	219	1.109 ± 0.074	1.185 ± 0.210
	Dense	0.139 ± 0.007	0.011 ± 0.002	393 (7)	0.815 ± 0.040	0.617 ± 0.095
Eupatorium maculatum	Sparse	0.446 ± 0.016	0.024 ± 0.005	201 (17)	1.375 ± 0.096	1.851 ± 0.382
	Intermediate	0.205 ± 0.010	0.013 ± 0.003	199 (12)	1.026 ± 0.088	1.542 ± 0.390
	Dense	0.108 ± 0.007	0.007 ± 0.001	185 (1)	0.389 ± 0.033	0.197 ± 0.101
Liatris aspera	Sparse	0.67 ± 0.05	0.16 ± 0.03	393	1.41 ± 0.074	2.61 ± 0.52
	Intermediate 1	0.39 ± 0.04	0.06 ± 0.02	337	1.04 ± 0.065	1.41 ± 0.43
	Intermediate 2	0.23 ± 0.02	0.01 ± 0.002	346	0.75 ± 0.048	0.78 ± 0.26
	Dense	0.13 ± 0.01	0.003 ± 0.000	538	0.58 ± 0.027	0.38 ± 0.08
Lythrum alatum	Sparse	0.741 ± 0.041	0.183 ± 0.026	127 (10)	1.84 ± 0.132	2.224 ± 0.622
	Intermediate	0.714 ± 0.004	0.116 ± 0.031	170 (8)	1.31 ± 0.109	2.000 ± 0.358
	Dense	0.357 ± 0.017	0.016 ± 0.006	230 (5)	0.720 ± 0.051	0.592 ± 0.169

Lythrum salicaria	Sparse[b]	0.260 ± 0.185	0.021 ± 0.004	179 (9)	1.84 ± 0.157	4.390 ± 0.702
	Intermediate	0.117 ± 0.005	0.004 ± 0.009	292 (8)	0.959 ± 0.057	0.950 ± 0.177
	Dense	0.082 ± 0.005	0.003 ± 0.0003	297	0.622 ± 0.028	0.230 ± 0.045
Monarda fistulosa	Sparse	0.544 ± 0.031	0.095 ± 0.030	209 (4)	1.493 ± 0.068	0.965 ± 0.137
	Intermediate	0.325 ± 0.020	0.059 ± 0.013	233 (11)	1.024 ± 0.073	1.23 ± 0.215
	Dense	0.131 ± 0.564	0.003 ± 0.001	287 (3)	0.535 ± 0.047	0.602 ± 0.196
Pycnanthemum virginianum	Sparse	0.450 ± 0.024	0.063 ± 0.013	192 (8)	1.202 ± 0.078	1.159 ± 0.197
	Intermediate	0.138 ± 0.008	0.010 ± 0.001	181 (6)	0.736 ± 0.072	0.942 ± 0.229
	Dense	0.062 ± 0.003	0.001 ± 0.0001	207 (3)	0.287 ± 0.041	0.344 ± 0.153
Veronicastrum virginicum	Sparse	0.774 ± 0.049	0.238 ± 0.038	186 (11)	2.00 ± 0.106	2.103 ± 0.358
	Intermediate	0.417 ± 0.018	0.032 ± 0.005	184 (4)	1.388 ± 0.092	1.552 ± 0.362
	Dense	0.134 ± 0.007	0.004 ± 0.001	284 (7)	0.814 ± 0.056	0.892 ± 0.196

From Levin and Kerster (1969*a*).

[a] In parentheses are the number of "lost" flights assigned a distance of 15 m, which have been included in the preceding number.

[b] These colonies were omitted from multispecies calculations because of their manifestly heterogeneous spacing patterns.

several flowers per plant, and then move to one of the plant's near neighbors. The between-plant variances of alien pollen loads in a population of *P. pilosa* and *P. glaberrima* (Levin and Kester, 1967) and of legitimate loads in a population of the heterostylic *L. caroliniense* (Levin, 1968) are rather high, suggesting the absence of substantial between-plant carryover. Not only do neighboring plants differ markedly in alien or legitimate pollen loads, but so do adjacent flowers on the same plant. Since pollinators tend to work adjacent flowers on a plant, we may surmise that most of the pollen is deposited on the first flower visited. This view is supported by experiments involving *P. pilosa* and *P. glaberrima* and the pollinator *Colias eurytheme* (Levin and Berube, 1972). In laboratory trials, heterospecific pollen loads were deposited primarily on the first stigma, and little pollen was deposited beyond the third stigma encountered. Specifically, *P. pilosa* pollen on *P. glaberrima* stigmas declined from a mean of 45 grains on the first flower pollinated to 25 grains on the second, 17 on the third, 5 on the fourth, and 3 on the fifth. In the reciprocal experiment, *P. glaberrima* pollen on *P. pilosa* stigmas declined from a mean of 1.4 grains on the first flower pollinated to 1.2 on the second and 0.02 on the third. No pollen was carried to the fourth and fifth flowers.

Wind-Mediated Pollen Flow

It will be helpful to describe the mechanism of pollen flow prior to discussing flow in relation to distance. As summarized by Tauber (1965), pollen grains are lifted into the air by turbulence and follow the wind currents, which most often move in a horizontal direction. Since pollen grains have relatively slow terminal velocities, they are unlikely to settle under the influence of gravity, and their chief motion will be the result of turbulent diffusion processes in the atmosphere. The eddying of the atmosphere carries pollen in vertical and lateral directions in addition to the prevailing downwind direction. In this way, pollen is mixed into the air current above the vegetation or in the vegetation and tends to follow these currents until it is incorporated by the boundary layer around the surfaces of vegetation. The boundary layer is a more or less stagnant layer of air extending from a fraction of millimeter to a few millimeters above smooth surfaces. Turbulence is suppressed in this layer, and nearly laminar movements parallel to the surfaces predominate. The final departure from the air to surfaces such as a stigma is chiefly by turbulent impaction. The collection efficiency of the stigma, i.e., the ratio of the number of grains captured by a stigma that would have passed through the space occupied by that structure had it not been there, is directly proportional to the density and diameter of pollen grains and to the wind velocity, and inversely proportional to the diameter of the stigma or its projections.

The deposition schedule of particulate matter at points from the source depends on the vagaries of atmospheric circulation and the properties of the particles. Several workers have generated equations to describe the expected behavior of particles in terms of areal and deposition density as a function of distance. These equations may be applied with varying degrees of success in predicting the dispersal of pollen, because once airborne it is subject to the same principles that govern the dispersion of other small particles. A brief review of the prime theoretical considerations will shed light on the factors which influence this mode of dispersal and serve as a prelude to the data on wind-mediated pollen dispersal.

Schmidt (1918) made one of the first attempts to predict the concentration of pollen as a function of distance from the source. He found that pollen grains and other particles of similar size, if subjected to horizontal air currents, fall to the ground at distances computable by Stokes' law, which takes into account wind velocity and the velocity of fall of the pollen grain. Empirical analysis showed that the formulas of Sutton (1932, 1947) for the dispersal of a smoke cloud by eddy diffusion were better predictors than Stokes' law. According to the eddy diffusion theory, the dilution of the suspended pollen cloud by eddies is regarded as the major factor controlling the deposition gradient, while the terminal velocity of pollen and wind velocity are relegated to second-order interactions.

Sutton's formulas for point sources are presented below because they give some quantitative ideas of the rate of pollen dispersion. In the case of a continuously emitting point source located at height h above the ground, the approximate concentration immediately above the ground on an axis downwind is given by the expression

$$\chi(x) = \frac{Q}{\pi C_y C_z u x^m} e^{-h^2/C_z^2 x^m} \tag{1}$$

where χ is the pollen concentration, x is the distance downwind from the point of the source, Q is the source strength in number of grains released per second, C_y and C_z are diffusion coefficients for lateral and vertical directions, u is the mean wind velocity, and m is the degree of turbulence of the air. Values for C_y and C_z are judged to be about 0.5–1.0(meter)$^{1/8}$ and 0.1–0.2(meter)$^{1/8}$, respectively (Gregory, 1973). The turbulence parameter is m = 1.75–2.00, but it cannot exceed 2.0 (Sutton, 1947). From this formula, it follows that there would be zero pollen at ground level below a plant, but that there is a slow and later rapid rise with distance, until a maximum is reached at a distance $d_{\max}$ from the plant equal to

$$d_{\max} = \left(\frac{h^2}{C_z^2}\right)^{1/m} \tag{2}$$

The theoretical treatments of pollen dispersal have dealt with the concentration of pollen reaching the ground at some distance from the source. This value, however, is less important than the amount of pollen present in the air at the level of the stigmas. Sutton's formula for a continuously emitting point source situated at ground level is useful in this regard, if we assume that anthers and stigmas in a population occur in the same plane, some distance above the ground. Pollen concentration at stigma height as a function of distance downwind from the source is

$$\chi(x) = \frac{Q}{\pi C_y C_z u x^m} \tag{3}$$

The formulas do not apply to dispersion within the trunk space of a forest, a level through which high concentrations of pollen are dispersed. The trunk space is characterized by a temperature inversion and suppressed turbulence, because the warmest layer in the forest is usually at the crown. Dispersion distances may be approximated by calculating the hypothetical trajectories which would be followed if the movement of the pollen grains were determined by wind velocity and terminal velocity alone (Tauber, 1965).

The following inferences may be drawn from the theoretical considerations of Sutton (1947). Among the biotic factors affecting the deposition pattern, the nature of the source is of prime importance. The gradient of decreased deposition with distance will be steepest for a point source because of lateral diffusion. This factor is much diminished if pollen is emanating from a continuous crosswind line source, and accordingly the slope of the gradient is less steep. For an areal source, i.e., a rectangular source with the long axis in the direction of the wind, the slope of the gradient will be even less steep than for a linear source. This is due to the fact that pollen traveling beyond the boundary of an areal source represents in a large measure the extended tails of the summed leptokurtic distributions from many plants. Thus the dispersal parameters of single plants and of the population which they comprise may be quite disparate. The height of the source also affects the character of the deposition gradient. The maximum rate of deposition occurs downwind from the source, roughly proportional to the square of the elevation of the source. Pollen grains may be expected to be deposited at distances greater than 10 times the source height (Green and Lane, 1957). Beyond this point, the rate of deposition will decline at a pace almost inversely proportional to the square of the distance (Tauber, 1965). Thus the greater the elevation, the greater will be the mean and variance of the dispersal distance and the longer will be the tail of the distribution.

A second biotic factor is the terminal velocity of the pollen grain. The terminal velocity is proportional to the difference between the density of the pollen grain and air, and is correlated with grain diameter. Empirically determined terminal velocities for several angiosperm and gymnosperm pollens are presented in Table IV. McCubbin (1944) provisionally suggested a method of calculating this value on the assumption that surface drag accounts for most of the retardation. He showed that observed terminal velocities of most spherical and ovoid spores fitted the approximate formula V_s = (length × width)/40, where velocity is in mm/sec and pollen dimensions are in microns. By definition, the loss due to deposition (and the deposition rate) of large, heavy grains is greater than for small, light ones. Accordingly, heavy grains will be dispersed over much shorter distances regardless of the height of the source. Another reason to expect this pattern is that the collection efficiency of the vegetation increases rapidly with the size and mass of the pollen grain. Thus at a given wind velocity, small, light grains tend to follow the air flow around potential collecting objects whereas large, dense grains are diverted appreciably from the airstream and tend to impact. Big pollen grains like those of *Fagus*,

TABLE IV. Observed Terminal Velocities of Pollens

Species	V_s (cm/sec)
Abies pectinata	38.7
Alnus viridis	1.7
Ambrosia trifida	1.56
Ambrosia artemisii folia	1.56
Betula alba	2.4
Carpinus betulus	2.2–6.8
Corylus avellana	2.5
Dactylis glomerata	3.1
Fagus sylvatica	5.5
Larix decidua	9.9–22.0
Larix polonica	12.3
Picea excelsa	8.7
Pinus cembra	4.5
Pinus sylvestris	2.5
Quercus robur	2.9
Salix caprea	2.16
Secale cereale	6.0–8.8
Tilia cordata	3.24
Tilia platyphylla	3.2
Ulmus glabra	3.24

After Gregory (1973).

Ulmus, and *Tilia* will be caught 3–4 times more efficiently than small pollen grains like those of *Alnus*, *Corylus*, and *Betula* (Tauber, 1967).

Several measurements of pollen dispersion from crop fields, single trees, and forest edges are reported in the literature. The methods and equipment employed are quite diverse, reflecting the varied purposes for which experiments were conducted and their degree of sophistication. The data taken from several investigations have been plotted in Fig. 5 by using the first sampling distance outside the source or the position of the maximum count as the initial point.

In spite of the variation introduced by different experimental procedures, and pollen traits, there is good agreement in the shape and slope of the relative pollen concentration curves with distance (Fig. 5). The distributions for the most part are linear on a log/log plot and have similar slopes. On linear/linear scales, the curves are highly leptokurtic with similar levels of kurtosis. Most of the pollen is deposited near the source, but a small fraction of the grains are transported over 250 m. The experimental values for the pollen concentration gradients depicted in Fig. 5 are

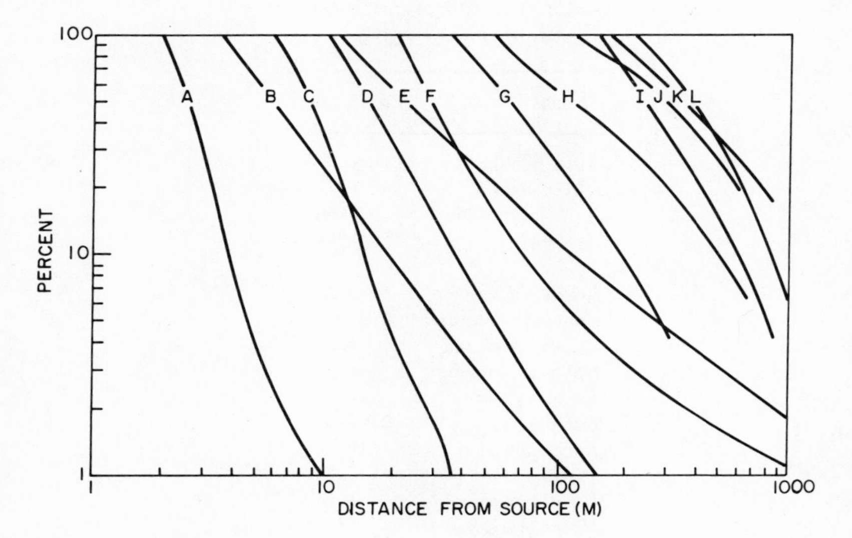

FIG. 5. Change in relative pollen concentration with distance using the first sampling distance outside the source as the initial point. A, *Thalictrum* (Kaplan, 1970); B, pinyon pine (Wright, 1952); C, ragweed (Dingle *et al.*, 1959); D, ash (Wright, 1952); E, slash pine (Wang *et al.*, 1960); F, atlas cedar (Wright, 1952); G, bromegrass (Jones and Newell, 1946); H, sugarbeet (Jensen and Bogh, 1941); I, timothy (Jensen and Bogh, 1941); J, rye (Jensen and Bogh, 1941); K, guayule (Gardner, 1946); L, cocksfoot (Jensen and Bogh, 1941).

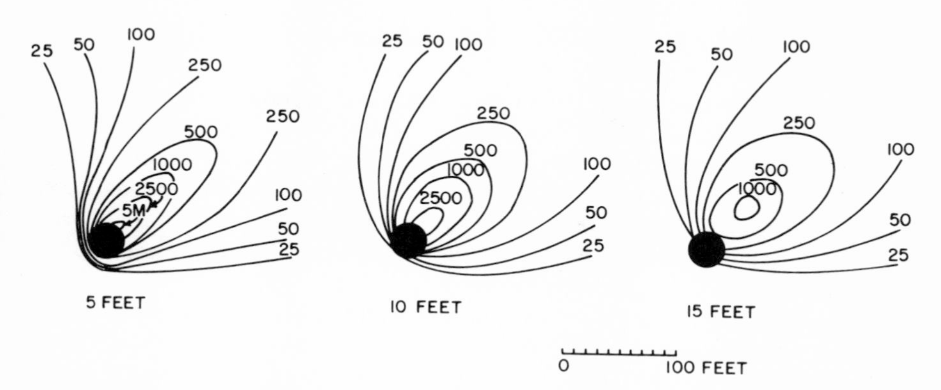

FIG. 6. Ragweed pollen concentration patterns at heights of 5, 10, and 15 ft in grains/m^3. After Raynor and Ogden (1965).

in fair agreement with those predicted by the formulas of Sutton and Gregory. The data do not fit theoretical curves based on the assumption that the trajectory of each pollen grain can be calculated from information on wind velocity, rate of fall, and distance of fall.

Pollen concentration decreases rapidly with distance, pollen falling largely within 50 m of the source. For herbs the point of maximum deposition is adjacent to the source, whereas for trees it is a short distance downwind. The narrow distributions of pollen noted in Fig. 5 are based on measurements made downwind from the source at a time when the wind direction was primarily unidirectional. Pollen flow is restricted if the wind blows equally in all directions. Jensen and Bogh (1941) calculated that such variation in direction would mean a reduction in pollen exchange between two given fields by 15–20% of what occurs in the downwind direction. The reduction is greatest for long distances.

Most studies on pollen dispersion deal only with the concentration gradient along a line radiating from the source in a downwind direction. This treatment is valuable in determining an upper limit for pollen flow, but it fails to provide information on the pollen cloud or plume width and height at various distances from the source. Isoleths of pollen concentration at three heights above a ragweed source are depicted in Fig. 6. It is apparent that there is significant lateral dispersion, and that the point of maximum concentration increases with distance from the source as the sampling points rise in elevation. The width of the plume increases as the source size increases, but more rapidly close to the source than at greater distances. In timothy, the width of the plume has been shown to be an inverse function of wind speed (Raynor *et al.*, 1970*b*). Plume height in ragweed and timothy increases slowly with distance. Area sources have slightly larger heights than point sources.

Since pollen is carried primarily in the direction of the wind with some lateral diffusion, it follows that the pollen concentration above the windward side of a population in which the pollen was produced would be much less than that over the leeward side. This relationship has been demonstrated by Pedersen *et al.* (1961*a*) with pollen distribution in rye fields. The pollen concentrations in the interior and at the leeward edges are 4–5 times greater than those on the windward side. What effect does this pattern of indigenous pollen concentration have on the danger of pollination with extraneous pollen? In order to answer this question, Pedersen *et al.* (1961*b*) established a series of population pairs, each containing a diploid and a tetraploid strain of rye at a different distance from one another. The atmosphere immediately above the tetraploid populations was analyzed for indigenous and extraneous pollen concentrations. The results of this study are summarized in Table V. Where the fields were 50 m apart, 52% of the pollen 0.5 m from the edge of the tetraploid field was diploid, but only about 30% was diploid in the interior of the field. A similar pattern was repeated when the isolation distance was increased, although the overall contribution from the diploid field decreased with distance. We may thus conclude that the windward edge of a population is more subject to alien gene receipt than the remainder of the population.

TABLE V. Percentage of Pollen from a Neighboring Field with 2*n* Rye in the Pollen Mixture over a Field with 4*n* Rye

	2*n* rye pollen (%)			
Distance from neighboring field (m)	Distance from edge of field			
	0.5 m	5 m	15 m	35 m
50	51.7	32.8	27.2	27.6
150	33.0	18.8	15.7	17.2
250	24.3	13.1	11.1	12.5
350	19.2	10.1	8.6	9.9
450	15.8	8.2	7.0	8.0
550	13.5	6.9	5.9	6.9
650	11.8	6.0	5.1	6.0
750	10.4	5.3	4.5	5.3

From Pedersen *et al.* (1961*b*).

The effect of atmospheric and biotic variables on pollen dispersion was considered in a few investigations. Jensen and Bogh (1941) found that for several crop plants the dispersion distance was nearly as great on days of low average wind velocity (7 mph) as on days of higher wind velocity (12 mph). A similar impression emerged from J. W. Wright's (1952) study of forest trees. Pollen traveled as far or slightly farther during relatively calm periods as during windy periods. These data lend support to Sutton's contention that turbulence and not wind velocity is the prime atmospheric parameter.

If pollen dispersal is proportional to the turbulence in the air, the species in sites experiencing greater turbulence is less, all other factors being equal. The amount of turbulence at a given air speed is a function of the roughness of natural surfaces. A roughness parameter has been determined for several vegetation surfaces (Geiger, 1950). In order of decreasing roughness, they are forest, grain field, beet field, tall. grass, prairie grass, and short grass. The development of turbulent elements is facilitated by the less pliable or more irregular vegetation surfaces. The minimum air speed for turbulent flow varies from 0.0001 m/sec for forests to 0.0008 m/sec for tall grass, 0.09 m/sec for prairie grass, and 0.14 m/sec for short grass.

Studies on the dispersal of particulate matter indicate that particles with higher terminal velocities are deposited sooner than those with lower terminal velocities. There is evidence that this relationship may apply to pollen. A correlation between dispersion distance and rate of fall has been observed in trees. Wright (1952) reported that spruce, ash, and pine have relatively high terminal velocities (> 5 cm/sec) and relatively small dispersion distances, whereas poplars, elms, and hazel have large dispersion distances and low terminal velocities (< 5 cm/sec). Raynor *et al.* (1970*a,b*) compared the dispersion of corn, timothy, and ragweed pollen from sources of similar dimensions. The terminal velocity of these species is 30, 5.5, and 2.0 cm/sec, respectively. The decline in airborne pollen with distance was the greatest in corn and least in ragweed. At 10 m from the source edge, corn pollen deposition per unit area was only 8% that at 1 m, in contrast to 20% in timothy and 30% in ragweed. Although pollen size was an important factor, the data were collected at different times so that part of the relationship between terminal velocity and the dispersion schedule may have been obscured by other variables. This problem was essentially eliminated in the simultaneous investigations of pollen dispersion in diploid and tetraploid rye and beets (Pedersen *et al.*, 1961*a*). Pollen diameters in diploid beets varied from 16 to 21 μm, as opposed to 21–28 μm in the tetraploids. Pollen diameters in diploid rye varied from 48 to 64 μm and in tetraploid rye from 57 to 77 μm. The curves for pollen spreading from both diploid and tetraploid rye have quite the same shape, as do the curves for the two beet strains. In rye, the curves for pollen spreading upward from the fields also were little affected by pollen size. Apparently, the size differential was insufficient to influence dispersion.

As noted above, we may expect pollen deposition to decrease more rapidly from an isolated plant (point source) than from a population (areal source). This relationship was found in ragweed and timothy (Raynor *et al.*, 1970*a,b*). The results of the ragweed study are shown in Fig. 7, where the curves separate in order of source diameter, with concentrations from the largest source decreasing most slowly with distance and those from the point source most rapidly.

There have been no definitive experiments dealing with the effect of single biological variables on pollen dispersion. Comparison of data from different studies may afford certain general impressions, but much noise is introduced by the large number of meteorological influences which vary in time and space. In order to study the dispersion and deposition of different pollen grains which are released simultaneously from a single or multiple source, Raynor *et al.* (1966) devised a method whereby pollen of different species may be stained different colors and released into the atmosphere by compressed air–operated atomizing nozzles. The application of this tech-

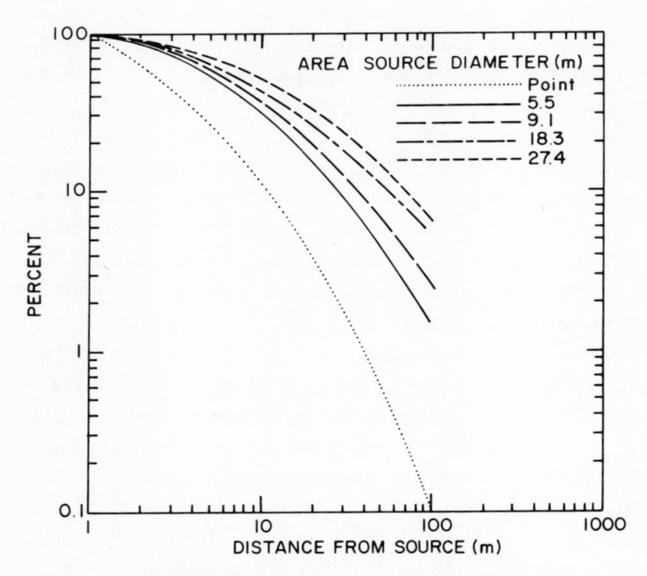

FIG. 7. Change of mean relative source-height concentration of ragweed pollen with distance from an experimental point and area sources. After Raynor *et al.* (1970*a*).

nique offers the opportunity to determine the dispersion and deposition characteristics of several species under one meteorological regime while at the same time eliminating the hazards of background pollen over the study site.

The equations for calculating pollen dispersion assume a free atmosphere over a smooth field, and investigations of pollen dispersal have been conducted in open terrain. For many species, however, this condition is artificial and results in the pollen deposition gradient being less steep than would be experienced in a natural setting. In many communities, conspecific and heterospecific plants intercept significant quantities of pollen on their leaves and branches. The denser the vegetation, the more restricted is the dispersion of that fraction of the pollen which would pass through it. An increase in density also is accompanied by a decline in wind velocity and turbulence, which compounds the effect of the structural barriers. Accordingly, the dispersal of pollen may be dependent on the density of the vegetation—the greater the density, the more retarded the dissemination of pollen.

The effect of a specific type of vegetation (a forest with open trunk space) on the dispersal of ragweed pollen has been investigated by Raynor (1967). A comparison of dispersal in open terrain and into a forest revealed that the forest significantly retarded pollen flow. Less than 30% of the pollen entering the plane of the forest edge was airborne at 60 m, and extrapolation suggested that only about 10% would remain at 100 m. In open terrain, about 50% would be airborne at that distance.

The data on pollen dispersion considered in this section have been obtained from real sources in a natural environment under conditions normally prevailing at the time of anthesis. Although populations in nature are seldom as uniform and compact as many cultivated plots are, pollen flow from them can be more readily judged with the experimental data in mind. Since pollen from anemophilous plants clearly disperses in a fashion similar to that of small particles, calculation of pollen flow by use of existing diffusion models with realistic parameters and with due allowance for source and pollen properties should be practical and may afford the only possible solutions where experimentation is not possible.

Water-Mediated Pollen Flow

Many species of angiosperms grow in water, but of these only a small fraction are hydrophilous, i.e., rely on water as the agent of pollen transfer (Faegri and van der Pijl, 1966). Water pollination may occur either on the surface of the water or below it. In the case of the former, pollen grains float on the surface membrane and diffuse rapidly, especially if they are hydrophobic. This type of pollen dispersal is unique in that it takes place in a two-dimensional medium. Where pollination is below the surface, pollen is not apt to be hydrophobic and accordingly will not diffuse rapidly from the anthers of its own accord. In some plants (e.g., *Vallisneria spiralis,* Kerner, 1898; *Lemna trisulca,* den Hartog, 1964), the entire male flower is released instead of individual pollen grains, and the pollen itself does not touch the surface of the water. Hydrophily does not preclude autogamy. There are no data on pollen dispersion in water, and no discussions of the role of current velocity and chop and dispersion. Equations for the diffusion and sedimentation of pollen-size particles in liquid have been formulated (*cf.* Calde, 1965), but none of these seems appropriate for predicting pollen flow in water. It is apparent, however, that dispersion in water is less random than in air, and that it occurs within a smaller volume of medium. Moreover, the pollen will flow in the direction of the current.

SEED FLOW

The seed is an organ which performs a wide variety of not obviously related functions: multiplication, perennation, dormancy, and dispersal. Some of the features of seeds which are specifically or incidentally related to dispersal and which afford alternative adaptive strategies have been reviewed by Ridley (1930), Salisbury (1942), Ehrendorfer (1965), van der Pijl (1969), Harper *et al.* (1970), and Stebbins (1971). These publications contain a fund of information on seed dispersal mechanics and dispersability. They also reveal the vast anecdoctal evidence about the movement of seeds. Unfortunately, much of this information deals with unusual, long-distance dispersal events and offers little insight into the proportion of seeds deposited various distances from the source. The effect of seed transport on the breeding structure of species is decided by the relation between local seed production and that at greater distances. This is in contrast to the effect of seed transport on the rate of species extension, which is decided by the long-distance events.

A theoretical treatment of optimal dispersal strategies under different levels of temporal heterogeneity led Gadgil (1970) to predict that environmental regimes which are characterized by high variability in time and low variability in space would be best exploited by organisms with high dispersability. In plants, small seed size and high dispersability often go hand in hand (Harper *et al.,* 1970). Thus seed size may be used as a gauge, albeit crude, of the dispersability of species; and we would expect seed size to be smallest in weedy plants or those of open habitats and to increase either as the habit became more reliable or the plant became woodier. Salisbury

(1942) has shown that the weight of windborne fruits and seeds is correlated with features of the habitat in Britain. He found that the mean weight of propagules of species growing in open habitats was 0.0013 g, as compared to 0.0019 g for species of semiclosed or closed nonshaded habitats, 0.0040 g for species of woodland margins, 0.0113 g for shaded herbs, 0.0854 g for shrubs, and 0.6534 g for trees. It would appear that species occupying disturbed or transient habitats have greater dispersability than those in more advanced or stable habitats. A progression in seed weight accompanying increasing woodiness was reported for a sample of the California flora by Baker (1972), and for a sample of the worldwide flora by Levin (1974). Mean seed weights for herbs, shrubs, and trees from these studies are compared with those provided by Salisbury in Table VI. The shift in seed weight is most striking in Britain and least so in California. This difference may reflect differences in the duration of the growing season or the time allotted to seed development (Baker, 1972). Neither the study of Baker nor that of Levin reinforced Salisbury's impression that, among herbs, species of open habitats have the smallest seeds. Baker (1972) found that seed weight varies in relation to moisture availability as well as habit, with seeds in xeric habitats being heavier than those in mesic habitats. Plants of open habitats may actually have less available moisture than those in more mature habitats. Seed size in open habitats would then be a compromise to meet conflicting demands placed on the propagule.

A relationship between the habit of species and adaptations for dispersability by species has been described for single genera and assemblages of genera in the Compositae, Dipsacaceae, and Rubiaceae (Stebbins, 1950; Ehrendorfer, 1965). As one moves from a woody to a herbaceous perennial to an annual habit, specialization for the degree of dispersability is greatly

TABLE VI. Seed Weight in Relation to Plant Habit

	Mean seed weight (g)		
Habit	Britain[a]	California[b]	Worldwide[c]
Herbs	0.0020	0.0057	0.0070
Shrubs	0.0854	0.0075	0.0691
Trees	0.6534	0.0096	0.3279

[a] From Salisbury (1942).
[b] From Baker (1972).
[c] From Levin (1974).

enhanced, although the means by which such is achieved may vary from phylad to phylad. In a similar vein, Stebbins (1950) has shown that in the Compositae and Gramineae there is an inverse correlation between the persistence and ability for vegetative reproduction and the degree of development of a species dispersal mechanism.

Dansereau and Lems (1957) and Harper *et al.* (1970) enunciated ecological correlations between seed dispersal mechanisms and habitat. Weedy herbs tend to have wind-dispersed seeds or explosive mechanisms for seed ejaculation, whereas herbs of later successional communities may employ similar, but less well-developed, mechanisms. Within woodlands, different synusia often have different seed dispersal mechanisms. For example, in the rainforests of south Nigeria, 46% of the canopy species have wind-dispersed seeds and another 46% have animal-dispersed seeds, whereas in the lower stories the values are 7 and 71%, respectively (Jones, 1956). A decline in wind dispersal from the upper to lower strata in a Nigerian forest also was reported by Keay (1957); a similar relationship is evident in Costa Rican forests (Baker, 1973). Herbs commonly display hairs or barbs which facilitate animal dispersal; shrubs often display conspicuous fleshy fruits which attract small animal vectors; trees commonly produce large seeds whose dispersal is accomplished deliberately by animals which collect and store seed crops. Other relationships include larger seed size and air cavities in water-dispersed seeds compared to wind- or animal-dispersed seeds (Harper *et al.,* 1970) and more elaborate chemical defenses in tropical seeds compared to temperate seeds (Janzen, 1971*b*).

Wind-Mediated Seed Flow

Most of the definitive evidence on seed and fruit dispersal relates to wind-dispersed species. This may be attributable to the fact that these species are most amenable to propagule counts at various distances from the source. Dispersal by wind is numerically important in seed plants, and is predominant in regions which are biotically depauperate and subject to high winds and turbulence. This is also true for oceanic islands, even though the devices which promote wind dispersal may be reduced (Carlquist, 1966). Wind is the most significant agent for the dispersal of lighter seeds and fruits.

The distances over which windborne seeds and fruits travel are influenced by their settling rate (= terminal velocity), the height and area of the source, turbulence, and wind velocity. The settling rate is proportional to the weight of the propagule, but is modified by plumes, wings, and seed shape. The settling rate seems to be the limiting factor in dispersal. Most seeds either are not airborne long enough to experience the turbulence that will carry them to high altitudes or are simply too heavy to respond dramatically to the turbulence they experience. The rates of fall of a variety of species are presented in Table VII. The values vary widely within and among the dispersal types. The dust and plumed species are the slowest to settle, followed by the winged species. Ordinarily, small seeds fall at a rate of 75–500 cm/sec.

Most considerations of wind-dispersed diaspores have dealt with plumed fruits and seeds, especially those of the Compositae. In an aerodynamic sense, the principal function of the pappus and other plumelike structures is to increase the resistance between the air and falling propagule, thereby prolonging the time of fall and increasing the distance the propagule may be transported by wind currents (Haberlandt, 1914). After release from the plant, the motion of a plumed propagules has two phases, which may overlap (Burrows, 1973). First, the initial motion is one of rotation and translation with changing speed relative to the air. Shortly thereafter, the propagule reaches an equilibrium flight configuration in speed and orientation relative to the air. Two kinds of plumed propagules can be recognized. Some have a relatively large pappus (or similar structure) of small porosity which allows the propagules to achieve a low terminal velocity, i.e., high buoyancy. Such a pappus acts as a drag parachute. On the other hand, some propagules have small and very porous pappus which functions like a guide parachute. This type of pappus serves to orient the seed in the air and on the ground without adding much in the way of dispersal distance. Once on the ground, propagule orientation plays an important role in determining the nature of the contact with the substrate and the probability of germination (Harper *et al.*, 1970).

The terminal velocities of achenes of the Compositae have been determined for several species (Sheldon and Burrows, 1973), and are

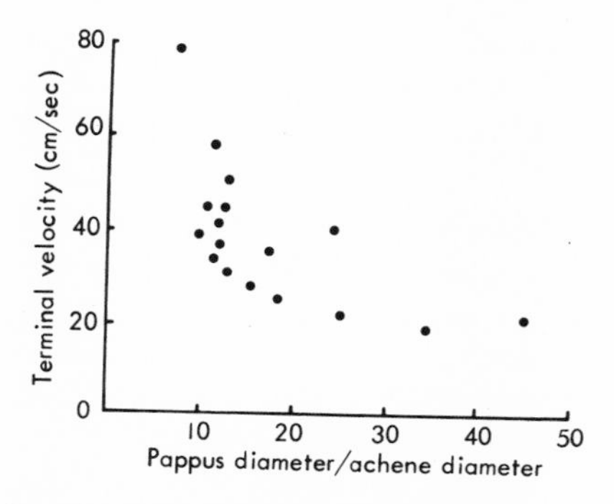

FIG. 8. Relationship between the pappus diameter/achene diameter ratio and the terminal velocity of fruits of selected Compositae. After Sheldon and Burrows (1973).

TABLE VII. Rate of Fall of Seeds or Fruits Through Still Air

Species	Terminal velocity (cm/sec)
Plumed propagules	
Tussilago farfara	19
Cirsium arvense	22
Erigeron acer	23
Sonchus arvensis	24
Senecio vulgaris	28
S. viscosus	32
Cirsium palustre	34
Taraxacum officinale	36
Sonchus oleraceus	36
Eupatorium cannabinum	39
Hypochoeris radicata	40
Senecio jacobaea	42
S. squalidus	46
Tragopogon porrifolius	46
Leontodon autumnalis	51
Carlina vulgaris	58
Carduus tenuiflorus	79
Centaurea scabiosa	220
Epilobium montanum	18
Salix repens	16
Populus tremula	11
Dust propagules	
Orchis latifolia	28
Cypripedium calceolus	25
Corallorhiza innata	21
Epipactus palustris	20
Coryanthes macrantha	14
Winged propagules	
Fraxinus excelsior	200
Carpinus betulus	120
Acer platanoides	107
Abies alba	106
Ailanthus altissima	91
Pinus silvestris	83

After P. Müller (1955) and Sheldon and Burrows (1973).

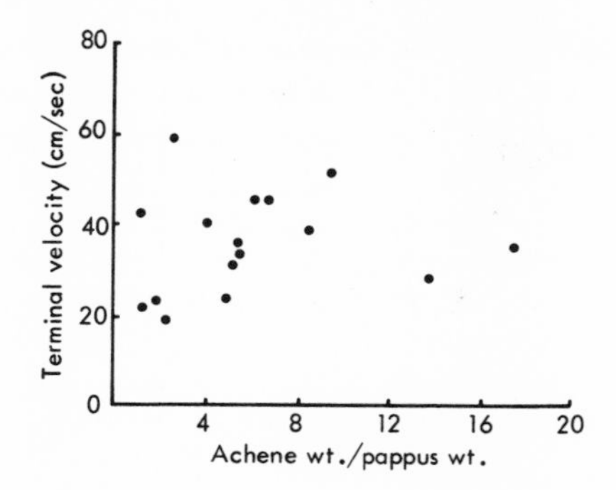

FIG. 9. Relationship between the achene weight/pappus weight ratio and the terminal velocity of fruits of selected Compositae. After Sheldon and Burrows (1973).

enumerated in Table VII. The terminal velocity of composite achenes generally decreases with an increase in the ratio of the pappus diameter to achene diameter (Fig. 8). It is also generally true that an increase in the weight of the pappus in relation to the achene weight decreases the terminal velocity (Fig. 9). Taking into account terminal velocity and the resistance coefficient (which was also empirically determined) and assuming that each achene–pappus unit moves with the local horizontal wind speed at all times after release, Sheldon and Burrows (1973) calculated the maximum dispersal distance for several composites under three wind velocities. These distances are presented in Table VIII, which also includes the resistance coefficient for each species. It is evident that maximum dispersal distances are highly correlated with wind speed, a threefold increase in speed being accompanied by about a threefold increase in distance for all species. However, note that the distances are not great.

The height of presentation also plays a major role in the dispersal profile of plumed fruits and seeds. For *Senecio jacobaea,* Sheldon and Burrows (1973) have shown analytically that the maximum distance at a wind speed of 5.47 km/hr varies as a function of height in the following manner: 30.5 cm tall yields 1.15 m distance; 61.0 cm tall yields 2.26 m distance; 91.4 cm tall yields 3.37 m distance; 121.9 m tall yields 4.48 m distance. The degree of distance differential remains constant at higher wind speeds, when all achenes are moving farther. The role of height difference counterbalancing differences in pappus structure is illustrated in *Liatris*. The float distance of *L. aspera* and *L. cylindracea* achenes was measured from a linear array of plants set crosswind in an experimental garden (Levin and Kerster, 1969*b,* and unpublished). With a wind speed of 10–20 mph, the mean dispersal distance of both species was about 2.5 m, with a maximum distance of 9 m. The interesting point is that our *L. aspera* was 9–10 dm

TABLE VIII. Maximum Dispersal Distance at a Convection Speed of 3.05 cm/sec and Taking into Account the Boundary Layer of the Ground

Species	Resistance coefficient (k/cm^{-1})	Mean plant height to nearest 30 cm	Wind speed (km/hr)		
			5.47	10.94	16.41
Centaurea scabiosa L.	0.02	60	0.52 m	1.04 m	1.56 m
Cardus tenuiflorus Curt.	0.16	60	0.70	1.14	2.12
Carlina vulgaris L.	0.29	30	0.49	0.98	1.47
Leontodon autumnalis L.	0.37	30	0.55	1.09	1.64
Tragopogon porrifolius L.	0.47	30	0.60	1.20	1.80
Senecio squalidus L.	0.47	30	0.60	1.20	1.80
S. jacobaea L.	0.55	90	1.84	3.69	5.53
Hypochoeris radicata L.	0.60	30	0.67	1.33	2.00
Eupatorium cannabinum L.	0.64	90	1.96	3.93	5.89
Sonchus oleraceus L.	0.77	90	2.19	4.37	6.56
Taraxacum officinale Weber	0.77	30	0.76	1.52	2.27
Cirsium palustre (L.) Scop.	0.83	90	2.27	4.54	6.81
Senecio viscosus L.	0.98	30	0.85	1.72	2.57
S. vulgaris L.	1.25	30	0.97	1.94	2.90
Sonchus arvensis L.	1.69	90	3.34	6.67	10.00
Erigeron acer L.	1.87	30	1.20	2.40	3.61
Cirsium arvense (L.) Scop.	2.10	90	3.79	7.57	11.35
Tussilago farfara L.	2.66	30	1.48	2.95	4.43

After Sheldon and Burrows (1973).

tall, whereas *L. cylindracea* was only 4–5 dm tall. The achenes of the latter are dispersed as far as the former because *L. cylindracea* has a large plumose pappus in contrast to the small barbellate pappus of *L. aspera*.

The *Liatris* study is one of the few where the distribution of plumose propagules were studied. As seen in Fig. 10, most achenes of *L. aspera* fall near the seed source (within 2.5 m). A comparable dispersal pattern was described in *Senecio jacobaea*. Poole and Cairn (1940) (cited in Salisbury, 1961) found that about 60% of the disseminules were deposited around the base of the source plants, 39% 3 m away, 08% 6 m away, 02% m away, and 0.005% 40 m away. Brownlee (1911) reported that *Leontodon hispidus* deposition declined markedly with distance, and that few seeds were carried more than 8 m. Returning to *Senecio jacobaea,* Poole and Cairns also reported that only about 0.5% of the total achene production was actually windborne. Similarly, Plummer and Keever (1963) estimated that only one camphor-weed seed in 93,000 actually ascended vertically with wind current. Thus in some species only a small fraction of the plumed seeds may gain sufficient independence from the source to be transported along the expected trajectories.

Among trees, the genus *Pinus* has received much attention. Boyer (1958) trapped seed at 3-m increments from a forest wall of *Pinus palustris* to 27 m beyond the wall. The number of sound seeds collected was halved each 11 m from the forest edge, with few seeds reaching beyond 100 m. The distribution of seed fall also was measured in *Pinus taeda* (Jemison and Korstian, 1944). As in the previous species, most of the seeds were deposited near the source and few were transported beyond 100 m. Bannister (1965) suggests that this pattern also applies to *Pinus radiata,* but also notes that the tail of the distribution may in some years extend as far as 10 miles. In *Eucalyptus regnans,* seeds travel a mean distance of about 40 m from trees over 80 m tall (Gilbert, 1958). A small proportion of seeds move as far as 400 ft. The concentration of 127,070 *Fraxinus* fruits in 10-m-wide concentric rings about a source tree was determined by Kohlermann (cited in Geiger, 1950). The percentage distributions in nine rings between 10 and 100 m were as follows: 26, 22, 14, 12, 10, 6, 4, 3, and 2%. A similar pattern of concentration may be inferred from data of Isaac (1930) on seed dispersion in the Douglas fir (*Pseudotsuga mensiesii*). Most of the seeds fell within 60 m of the source, but the tail of the distribution extended as far as 300 m. Seed dispersal distances of spruce (*Picea engelmannii*) have been measured by Roe (1967) and Ronco (1970). Seed concentration declined abruptly with distance, and dispersal over long distances was infrequent, although some seeds were carried as far as 60 m. In *Picea mariana,* the number of seeds per acre declined from 300,000 within a stand to 84,000 at 3 m, 24,000 at 25 m, 12,000 at 50 m, and none at 100 m (Anonymous, 1939).

The incidence of seedlings as a function of distance from a forest edge is distributed in a manner similar to that for seeds. This is seen in the data of Hofmann (1911) on several species of gymnosperms. In *Pinus monticola,*

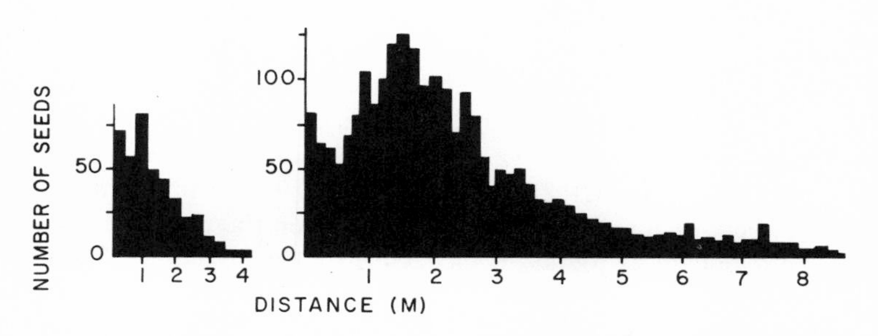

FIG. 10. Unidirectional, axial, seed dispersal distances of *Phlox pilosa* (left) and *Liatris aspera* (right). After Levin and Kerster (1968, 1969*b*).

the number of seedlings per acre declines from 600 at 6 m from the source to 200 at 12 m, 50 at 18 m, and 5 at 24 m. The shape of the distribution is almost identical to that described above for seed concentration in *Pinus palustris*. The seedlings of *Tsuga heterophylla* also occur primarily within 24 m of the source, but they have been observed at distances as great as 72 m. The rate of regression of *Abies alba* seedlings with distance is not as rapid as in the aforementioned species, the number of seedlings between 18 and 48 m declining by only 50%, with a considerable number of seedlings being located beyond 38 m.

Although wings on seeds and fruits are recognized as enhancing diaspore dispersal, the only definitive aerodynamic study of samaras (winged diaspores) has been that by Norberg (1973). On the basis of analytical investigations, he shows how the structure of the samara induces autorotation, which reduces its sinking speed and increases the distance it may be transported by winds. Norberg discusses aerodynamic, mechanical, and structural properties affecting autorotation rates and samara dispersability, and provides formulas for the calculation of performance from data gathered empirically. Data from *Acer platanoides* and *Picea abies* showed the single wing to be a relatively very efficient structure in braking the speed of the falling diaspore when compared to performances by birds, bats, insects, and a parachute.

The shape of propagules is often overlooked when adaptations for seed dispersal are being considered, yet it may have a significant effect. With weight being a constant, the alteration of shape from spherical to ellipsoid will reduce the terminal velocity by 22% (Chamberlain, 1967). Were the seed to become oblong or cylindrical, the terminal velocity would be reduced by about 34% if the length/width ratio were 3:1, and by 57% if the length/width ratio were 4:1.

With the exception of passive dispersal by air currents, quantitative analyses of deposition as a function of distance have been conducted only on species which explosively discharge their seeds. This mode of dispersal, which is achieved by a variety of mechanisms, is found in a large number of families and is represented in a wide range of floras (Ridley, 1930). To our knowledge, the only detailed study of explosive dispersal has been conducted on *Phlox pilosa* (Levin and Kerster, 1968). Seeds (395) were scattered an average of 1.2 m, with a maximum of 3.6 m, from a linear array of plants (Fig. 10). Other species of *Phlox* have similar distribution curves. Numbers of *Crotalaria* seeds thrown when the pods burst were determined in triplicate as found on successive foot-wide bands of soil beginning at the row of seeding plants (Duggar, 1934). The percentages of shed seeds found in these bands were 28 in the first foot, 26 in the second, 17 in the third, 7 in the fourth, 6 in the fifth, 5 in the sixth, 4 in the seventh, and thereafter about 1% in each of the foot-wide bands to a distance of 14 ft. Observations of autodispersal in many genera suggest that seeds are usually scattered only a few meters (Table IX).

Seed and seedling distances from the source suggest that the dispersal of a majority of windborne seeds is rather restricted. It is likely that these data are representative of species whose seeds are medium to large size and windborne or autodispersed.

Animal-Mediated Seed Flow

In addition to serving as a dispersal medium for plants, seeds and fruits also serve as a food source for a broad array of herbivores (van der Pijl, 1969; Janzen, 1971*b*). Seeds may be harvested while on the plant or after they have reached the ground. There are three basic strategies for the escape of seed predation by the plant. First, the plants may produce smaller seeds that are unworthy food sources or egg deposition sites. This strategy probably would reduce the seed dispersability, as the predators which the plant has escaped may have been a dispersal agent as well. Second, the plant may reproduce such large seed crops with each reproductive pulse that the predator is satiated and some seeds remain unharmed. This strategy would have no effect on the dispersability of the seed, nor would the third strategy, chemical protection. It would be most interesting to compare the genetic structure of populations of species which have opted

TABLE IX. Maximum Distances Achieved by Explosive Seed Discharge

Species	Distance (m)
Cardamine hirsuta	1.4
Oxalis acetocella	2.3
Corydalis sibirica	2.2
Geranium columbinum	1.5
Montia fontana	2.0
Impatiens parviflora	3.4
Euphorbia marginata	4.0
Dorstenia contrayerva	5.0
Lupinus digitatus	7.0
Viola elatior	4.6
Acanthus mollis	9.5

After P. Müller (1955).

for small seed size with that of populations of species escaping by other means. Janzen (1969) provides us with a list of legume species using the size strategy and another using the chemical strategy. Janzen (1970, 1971*b*, 1972; Wilson and Janzen, 1972) has shown that the intensity of predation by some insects increases with seed density and the proximity of the seeds to the parent plant owing to the density-responsive foraging behavior of the predators. Thus the probability of a seed giving rise to a reproductive individual is positively correlated with the distance from the parent plant. In this case, there will be a great disparity between potential and actual seed dispersal, the latter being greater. This pattern of dispersal results in hyperdispersion of plants but not in greater outbreeding or interpopulation gene flow.

Although defense against predation is an important goal, compromises are made with those animals which effectively transport a portion of the seed to a favorable location. This involves synchronizing the fruiting period with the time of greatest activity on the part of the vector. This relationship has been shown to exist between some elements of the New World tropics (Snow, 1966; Smythe, 1970; Croat, 1974). Gaining access to effective transport also involves providing a nutritional reward for the vector. In the case of frugivorous animals, the reward may be some of the seeds or a fruit pulp high in oils and proteins as well as carbohydrates and water. Plants serviced by specialized frugivorous birds have few large soft seeds in fruits high in fats and proteins. The seeds either pass through the digestive tract undestroyed or are regurgitated (*cf.* McKey, 1974). Plants serviced by opportunistic birds tend to have succulent fruits rich in sugars and organic acids and to have many small seeds, a large fraction of which may not pass through the digestive tract and be in a germinable condition. Opportunistic birds do not possess a mechanism for seed evacuation which is as efficient and quick as that of specialized birds. Thus the distance over which seeds are transported is apt to be less if the bird is a specialist than if it is an opportunist.

The distances seeds are dispersed by pre- or postdispersal predators are poorly understood, most attention being paid to the tail of the distribution (e.g., Cruden, 1966). Unfortunately, we do not have a body of data on dispersal of inanimate objects which is relevant here. The distance animals move seeds might be a function of animal size or the availability and predictability of the food source. Larger species must collect more energy to supply their requirements than small species. A larger energy demand will necessitate a larger area for food gathering (McNab, 1963; Armstrong, 1965; Schoener, 1968; Turner *et al.*, 1969), so that seeds are apt to be carried over longer distances by larger species. On the other hand, the home ranges of species may decrease in size as a community becomes more productive or a given food source becomes more plentiful (Connell and Orias, 1964). Thus in highly productive regions foragers are likely to move seeds over short distances. This is consistent with our inferences about frugivorous birds.

ACTUAL GENE FLOW

The continued development of new crop cultivars with superior and specific agronomic characteristics demands programs for seed production which insure the retention of genetic purity and prevention of interbreeding. Isolation by distance is a prime method of safeguarding the genetic purity of strains. In entomophilous crops, isolation requirements also are intimately connected with the foraging behavior of pollinators, since all heterovarietal pollen has to be brought by them. In anemophilous crops, isolation requirements also must take into account source height, the terminal velocity of the pollen, and presence and height of barriers.

Indications of the isolation distances that are necessary between seed crops of different strains may be obtained by either (1) growing a cluster of genetically marked plants of one strain in the center of a population of another and recording the amount of interbreeding at different distances from the donor, (2) finding the amount of contamination that occurs at different distances from the line where two varieties are juxtaposed, or (3) determining the amount of interbreeding between two varieties whose populations are separated by different distances. From information obtained in part from these procedures, seed grower associations across the world have specified minimal isolation distances for the maintenance of varietal purity. These isolation distances are summarized for various groups of crop plants in Table X. In most instances, total isolation may be achieved by distances of a few hundred meters in both wind- and insect-pollinated plants, and in plants which outcross to various degrees. The mean isolation distance for primarily self-fertilizing species is 295 m $\pm$ 151 vs. 804 m $\pm$ 243 in primarily or exclusively cross-fertilizing species; in species where cross- and self-fertilization are of similar importance, the mean isolation distance is 349 m $\pm$ 83. The difference between the outbreeders and the other two groups is statistically significant ($P < 0.05$).

Investigations of contamination as a function of distance have been conducted on only a few entomophilous species, and these are pollinated almost exclusively by bees. Afzal and Khan (1950) found that in cotton with a genetically marked block in the center of a field most gene flow occurred within 3.8 m of the source and was very sporadic beyond this distance, with none occurring beyond 30.5 m. Another crop with narrow gene dispersal is

TABLE X. Isolation Requirements for Seed Crops as a Function of Breeding System

Species	Breeding system[a]	Pollination agent[b]	Isolation requirement (m)
Gossypium spp.	S	I	200
Linum usitatissimum	S	I	100–300
Lupinus spp.	S	I	500
Vigna spp.	S	I	130
Camellia sinensis	S	I	800–3000
Lactuca sativa	S	I	30–60
Avena sativa	S	W	180
Hordeum vulgare	S	W	180
Oryza sativa	S	W	15–30
Panicum spp.	S	W	180
Echinochloa spp.	S	W	180
Pennisetum spp.	S	W	180
Sorghum vulgare	S	W	190–270
Triticum aestivum	S	W	1.5–3.0
Cajanus cajan	SC	I	180–360
Dolichos spp.	SC	I	180–360
Capsicum spp.	SC	I	360
Citrullus vulgaris	SC	I	900
Cucumis spp.	SC	I	900
Apium graveolens	SC	I	1100
Carthamus tinctorius	SC	I	180–270
Papaver somniferum	SC	I	360
Phaseolus spp.	SC	I	45
Vicia faba	SC	I	90–180
Pastinaca sativa	SC	I	500
Hibiscus spp.	SC	I	100–300
Voandzeia subterranea	SC	I	180–360
Nicotiana tabacum	SC	I	400
Poa, Bromus, Festuca, Phleum spp.	SC	W	540–1000
Coffea arabica	SC	IW	500
Hevea brasiliensis	C	I	2000
Helianthus annuus	C	I	800
Brassica campestris	C	I	900
Daucus carota	C	I	900
Lycopersicon esculentum	C	I	30–60
Anethum graveolens	C	I	300
Cassia spp.	C	I	720
Medicago, Trifolium, Lotus spp.	C	I	720–1600
Brassica oleracea	C	I	600

TABLE X. (Continued)

Species	Breeding system[a]	Pollination agent[b]	Isolation requirement (m)
Allium cepa	C	I	900
Raphanus sativus	C	I	270–300
Brassica oleracea	C	I	970
Fagopyrum spp.	C	I	180–360
Carica papaya	C	I	100–1600
Cucurbita spp.	C	I	400
Zea mays	C	W	180
Chenopodium ambrosoides	C	W	180–360
Agropyron spp.	C	W	
Cannabis sativa	C	W	500
Secale cereale	C	W	180
Beta vulgaris	C	IW	3200

From Kernick (1961).

[a] S, Species which are predominantly self-fertilizing; SC, species in which self- and cross-fertilization are of similar importance; C, species in which there is predominant or exclusive cross-fertilization.

[b] I, Insect pollination; W, wind pollination.

tomato. Currence and Jenkins (1942) found that the maximum amount of gene flow in tomato was adjacent to the source and rapidly declined with distance, less than 0.1% hybridization occurring at 22 m and beyond. In lima beans, the percentage gene flow declined from 5% at a distance of 1 m to 1% at a distance of 9 m (Barrons, 1939). Pedersen *et al.* (1969) found that in alfalfa cross-breeding between adjacent plots averaged 15% as compared to 11% at 50 m and 11% at 75 m. Bradner *et al.* (1965) found that hybridization in alfalfa declined with distance from 32% at 15 m to 6% at 200 m, but then declined slowly to 3% hybridization at 1500 m. Intervarietal crossing in turnip and radish decreased rapidly with increasing distance, but there was a progressive reduction in the rate of decline brought about by repeated increases in isolation distance (Bateman, 1947*a*). In the experiment most conducive to gene flow, the amount of hybridization declined from 57% at 7 m to 1% at 20 m and remained at this level as the isolation distance was increased to 200 m. Crane and Mather (1943) had previously obtained similar results with radish. A significant aspect of their study was the rapid decline of hybridization from 30–40% at the proximal edges of two varieties to 1% at 5 m.

The effect of the strain arrangement on the incidence of hybridization within populations further attests to the spatial restriction of gene flow in bee-pollinated plants. Nieuwhof (1963) established a series of plots containing cabbage and kale intermixed and clustered to various degrees. The experimental design and the percentage of cabbage progeny which were hybrids are presented in Fig. 11. Where the cabbage had been planted in single-row groups, the level of hybridization varied from 43 to 47%. In two-row groups, the level of hybridization varied from 39 to 42%. However, when planted in four-row groups, the level of hybridization was reduced to about 20%. Rick (1947) tested the amount of natural cross-fertilization in different planting arrangements of male-fertile and male-sterile tomatoes. The production of fruits and seeds by the latter was a positive function of the mean spatial proximity of the two strains. Gene exchange in *Lupinus luteus* was about 20% when plants of two varieties were intermingled as opposed to 1–4% when the varieties were grown in discrete rows (Free, 1970).

Similar phenomena have been observed in orchard-planted fruit trees.

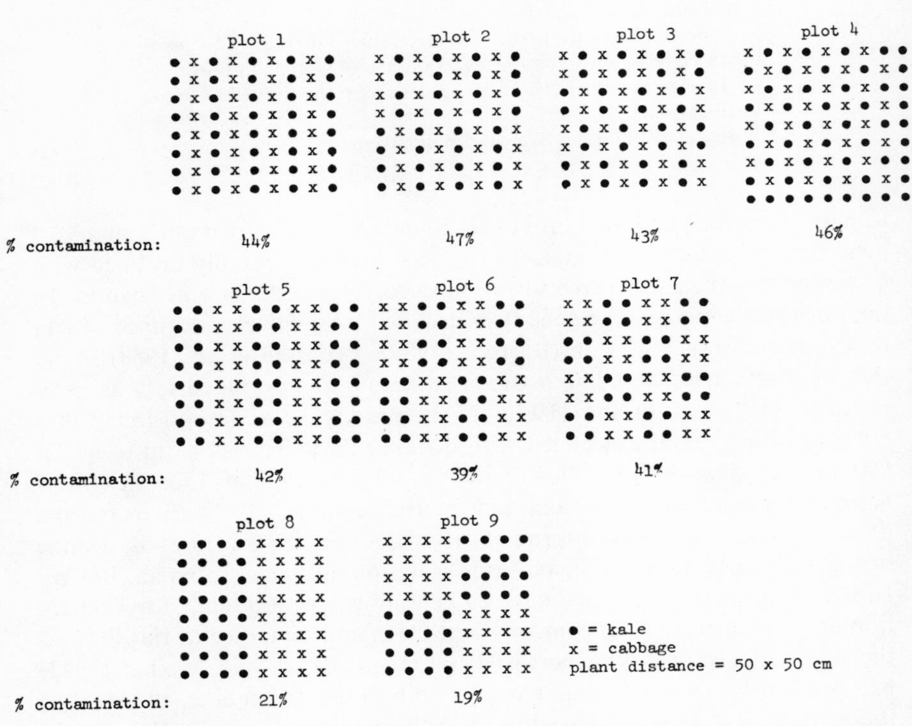

FIG. 11. Layout of seed fields containing kale and cabbage, and the percentage hybridization of cabbage seed. After Nieuwhof (1963).

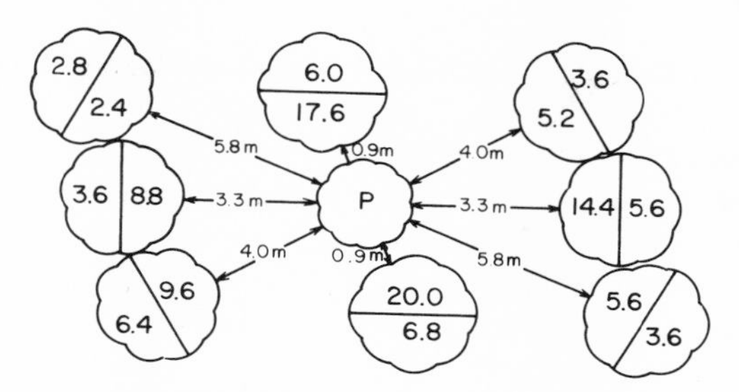

FIG. 12. Mean percentages of "Cox" apple flowers that set fruit, with "Worcester" pollinizers (P). After Free (1962).

Stephen (1958) found that cross-breeding in apples was greater, the greater the intermixture of varieties throughout an orchard. The proximity to the pollen source is so important that the fruit set of trees may differ significantly from one side to another. For example, in plum trees adjacent to the pollinizer 10.8% of the flowers on the side facing the pollinizer set fruit as compared to 4.3% on the far side (Free, 1962). The differential between the facing and far side, however, is dependent on the distance between the pollen donor and recipient, as Free (1962) demonstrated in apples (Fig. 12).

Little information on gene flow is available for entomophilous species which are not primarily bee-pollinated. Of the two studies known to us, one involves lepidopteran pollination, the other bee and fly pollination. Contamination of a series of recessive populations (300 plants) of *Phlox drummondii* by a donor population (400 plants) bearing a dominant corolla marker was analyzed in the experimental garden by the authors. The percentages of hybrids in the progeny of the recessive populations and their distances were as follows: 3.34% at 4 m, 0.52% at 9 m, 0.22% at 15 m, 0.04% at 22 m, and 0.08% at 50 m. Populations separated from the donor by 62, 80, and 95 m were free of contamination. *Allium cepa* is pollinated by flies and bees. Van Der Meer and Van Bennekom (1968) planted a strain carrying a dominant marker gene in the center of a population of homozygotes recessive for that gene and scored the fraction of progeny at various distances which were hybrids. Gene receipt at 15 cm from the donor was 14% and declined to 2% at 180 cm and 1% at 300 cm.

The amount of hybridization between two populations at a given isolation distance is significantly affected by their size and density. Bateman (1947*a*) established a square plot of Icicles turnips and stringers of Scarlet Globe comprising 1, 3, 6, and 12 rows. Increasing the number of arms decreased hybridization. With one-row stringers, hybridization of plants at distances 1, 2, 3, 5, and 10 ft averaged 35% as compared to 25% with three-row stringers, 23% with four-row stringers, and 15% with 12-row stringers. The addition of rows did not alter the relative rate of decrease in gene receipt with distance, only the absolute rate. Size and density proved similar to isolation distance in mode of action, constant increases in mass producing progressively smaller decreases in contamination. Using two radish varieties, Crane and Mather (1943) had previously demonstrated the effect of varietal mass on the extent of intervarietal hybridization. In one experiment, two varieties were juxtaposed and laid out in square plots. Hybridization declined from 30–40% at the proximal edges of the plots to 1% 5 m away. In another experiment in which a one-row stringer extended from a square plot of another variety, contamination fell to 1% only at a distance of 50 m. The researchers concluded that a mass of plants and a profusion of flowers hold bees to a small area, and that mass plays a considerable role in isolation. Williams and Evans (1935) found that the purity of a red clover seed crop was more dependent on the abundance of flowers within a plot than isolation distance, provided the isolated plot was relatively large and the plants therein were readily cross-compatible.

In species which are self-fertilizing, the level of gene exchange between populations also is dependent on the pollinator service per flower. For example, the extent of natural crossing in cotton apparently is closely related to the frequency and timeliness of the visits of pollinators (Simpson, 1954). There is a regional pattern for hybridization which is somewhat coincident with the cotton acreage distribution. Where there is intensive cotton cultivation, the ratio of cotton flowers to pollinators is high and cross-fertilization is infrequent. Conversely, in areas where cotton is less abundant, the ratio of flowers to pollinators is low, and cross-fertilization is frequent. Thus gene exchange between cotton populations is density dependent.

The findings on herbs and trees are compatible with data on bee foraging behavior presented above. Pollinators have narrow foraging ranges and tend to move between neighboring plants when flowers are profuse. Presumably, the extraneous pollen that a pollinator is carrying soon becomes unavailable because it is either packed into pollen baskets, diluted with local pollen, or deposited on a part of the flower other than the stigma. As a consequence, insect-mediated gene flow in large and dense populations is apt to be over very small distances. This conclusion is supported by bumblebee foraging observation and population genetic studies in *Mimulus guttatus* (Kiang, 1972).

Gene flow estimates are available for only a few anemophilous species, and of these species *Lolium perenne* (perennial ryegrass) and *Zea mays* (corn) have been the most intensively studied. Griffiths (1950) conducted a series of experiments on the crossing patterns from red pigmented *Lolium* into spaced nonred plants. A rapid decrease in intervarietal gene exchange

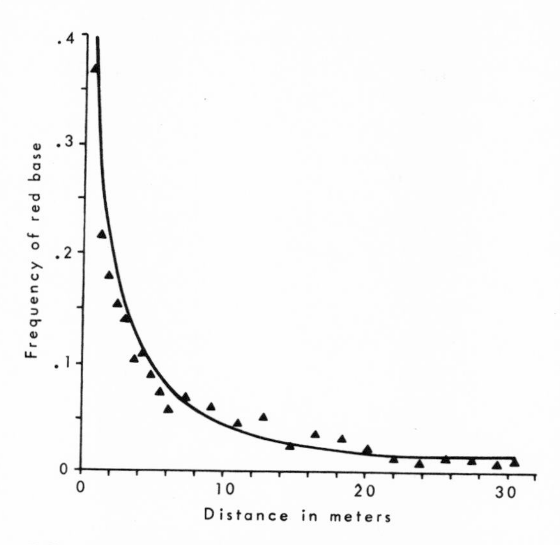

FIG. 13. Data points (average of east and west arms) from Griffiths' (1950) experiment 3, and a curve fitted by $f(x) = 1/x^{1.8347}$. After Gleaves (1973).

was obtained within very short distances. In one experiment, which involved distances up to 1000 m, contamination was reduced to 5.2% of that found in the marginal plants within the first 300 m, but beyond this distance the level of contamination failed to decrease significantly. Increasing the density of the recipient reduced the level of contamination per isolation distance. In another experiment, the proportion of outcrossing was reduced from 40% at the intervarietal margins to about 1% within a distance of less than 35 m. Gleaves (1973) noted that the gene flow in this experiment could be well described by an inverse power function where $f(x) = 1/x^k$, with $k = 1.8347$ (Fig. 13). Griffiths attributes this rapid decline to the increasing quantities of local pollen becoming available with each increment of distance from the red variety. Accepting the validity of this view, the effect of intravarietal pollen in reducing out-pollination is similar to that of distance in that an increase in density will produce a more rapid reduction in contamination. On the basis of varied field designs, Griffiths suggested that gene exchange within *Lolium* occurs primarily between neighboring units. Conformation for this impression was forthcoming from breeding of clonal plantations in polycross fields. Wit (1952) observed that tester clones set out for polycross tests were fertilized at an average rate of 40% by the two adjacent clones and at 74% by the three neighboring clones on each side. This was accomplished in spite of differences in the commencement of clone flowering extending to 1 week.

Gene flow in corn is over shorter distances than in most wind-pollinated plants because of its excessively large pollen grains. Salamov (cited in Jones and Brooks, 1950) determined the rate of crossbreeding as a function of distance in the Northern Caucasus of the USSR. Hybrid seed was reduced from 3.3% at 12 m to 0.3% at 50 m, and then gradually declined, hybrids still being formed at 800 m from the source of extraneous pollen. Similar results recently were reported by Paterniani and Short (1974). Gene flow from a central pollen source (yellow endosperm) declined sharply with distance near the source of extraneous pollen and then remained more or less constant at greater distances. It was also reported that the distance from the central plant in numbers of plants is of greater importance than physical distance measured in meters. If the contribution by all plants to effective pollination is on the average the same as with the central plant, it is possible to estimate for each plant the number of kernels resulting from pollination by plants standing at distances of up to 1 m, 2 m, etc. Paterniani and Short judged that nearly 50% of the kernels of any individual plant resulted from pollen within a radius of 12 m. This involved 800–2500 plants, depending on the size of the population in different experiments. This study indicates that maize populations are highly outbred, with a great number of parents contributing to seed set on a given plant.

Knowles and Ghosh (1968) determined the isolation requirements of smooth bromegrass, *Bromus inermis*. They note that the average hybridization of green by yellow-leaved plots was 9.6%, 1.0%, and 0.2% for isolation distances of 1, 61, and 183 m, respectively. Additional evidence of proximity effects was apparent in the nonrandom pollination observed in tester clones comprising polycross plots. Knowles (1969) found that the greatest crossing (15.4%) occurred when two yellow-leaved plants were among the ring of eight immediate green-leaved neighbors. Almost as much interbreeding (14.0%) occurred when only one yellow plant appeared in the immediate circle of eight. However, green plants which were once removed from yellow plants with green intervening experienced lower gene receipt (9.4%).

Having treated the more definitive information on interbreeding in wind-pollinated crops, we may now consider the relationship between pollen flow and gene flow in wind-pollinated plants. In one *Lolium perenne* experiment discussed above (Griffiths, 1950), it was noted that contamination was greatly reduced in the first 300 m, but failed to decline substantially with distances up to 1000 m. This pattern is similar in essentials to that obtained by Jensen and Bogh (1941) for numbers of *L. perenne* pollen grains captured on slides. However, as density of the recessive strain increased, the rate of contamination declined and contamination was more confined to the border of the plot facing the alternative strain. The level of pollen immigration remained constant during these changes. This serves to illustrate an important and general feature. The departure of the curves representing contamination and alien pollen receipt as a function of distance is due to the effect of intravarietal pollen, and will increase as the density of the recipient increases. Accordingly, pollen flow data offer estimates of maximal levels of gene flow, rather than that which actually occurs in most situations. The disparity between actual and potential gene flow would be even greater if a species were self-compatible because self pollen would be included in the local pollen pool.

Bateman (1947*b*) has proposed a general mathematical expression for the effect of distance on contamination in entomophilous and anemophilous crops which describes the aforementioned curves rather well. If contamination (F) is small,

$$F = \frac{ye^{-kD}}{D} \tag{4}$$

where y is the contamination at zero isolation distance, k is the rate of decrease of contamination with distance, and D is the distance from the source. If this formula is expressed as a graph in which the log of the proportion of contamination is plotted against distance, a curve of negative slope is obtained, its steepness decreasing as distance increases. In other words, successive increases in isolation distance become less and less effective in retarding gene exchange. The formula is particularly useful in that it affords a means of predicting the level of contamination at any isolation distance provided contamination at two distances is known.

In the absence of evidence to the contrary, population geneticists have assumed that gene dispersal follows a normal distribution. Bateman (1951) assembled a body of evidence, in part from plants, to support the general rule that diffusionary processes in populations of living organisms give rise to spatial distributions which are leptokurtic. We concur with Bateman's view that pollen-mediated gene dispersion in plants is normally leptokurtic. Windborne seeds may also be distributed in a similar fashion.

Bateman (1951) suggests that for insect-mediated gene flow leptokurtosis may be traced to the heterogeneity in pollinated flight distance due to the superimposition of localized foraging flights and long-distance exploratory flights. The basic assumption is that movement of pollinators is nonrandom. There is ample support for this premise, as a review of the pollinator data revealed. In addition, the leptokurtosis of pollinator-mediated gene flow may be accentuated over that observed for the pollinator if the pollen deposition schedule itself is leptokurtic. An explanation for the leptokurtosis of wind-mediated gene flow is sought in the general eddy diffusion formulas of Gregory (1973).

Thus far, we have considered gene dispersion between populations only with regard to the average level of gene receipt. We may now ask whether gene receipt is concentrated in particular zones of discontinuous populations or whether it is random. The pattern of hybridization is distinctly nonrandom in both wind-pollinated and insect-pollinated plants. The pattern of contamination in corn as described by Jones and Brooks (1950) is summarized in Table XI. From reading across the columns it is apparent that the mean level of cross-breeding decreased as the distance between the recipient fields increased. Reading down the column, we find that cross-breeding was progressively less from the margin of the population nearest the pollen source toward the most distant side. The side of a field nearest a pollen source received approximately 5 times more extraneous genes than the back side. The back rows of the 25-m plot, which extended up to 55 m away from the source field, experienced less cross-breeding than did the proximal rows of the 125-m plot. Jones and Brooks calculated that a 10-acre population extending for 100 m along a hybridization population and 400 m deep should have the same average gene receipt as a 40-acre population extending 400 m along a hybridizing population and 400 m deep. Even more striking evidence of hybridization being associated with border rows is seen in data on smooth bromegrass (Knowles and Ghosh, 1968). In a plot isolated 61 m from an alternative strain, the first or facing row averaged 11.1% contamination over a 3-year period. The average hybridization in the second row was only 3.5%, and was down to 1.5% in the third row. From rows 4 through 45, the average contamination was about 1%.

The relatively severe hybridization of the proximal portion of a popu-

TABLE XI. Average Percentage of Hybrid Corn in Groups of Five Rows at Eight Distances of Isolation

	Distance north of pollen source (m)							
Rows	0	25	75	125	200	300	400	500
1–5	51.67	31.90	16.00	7.32	4.40	1.78	0.54	0.46
6–10	28.12	12.49	5.60	3.13	1.10	0.55	0.26	0.15
11–15	19.02	8.73	3.54	1.48	0.99	0.32	0.23	0.21
16–20	16.82	6.41	2.53	1.77	0.74	0.46	0.07	0.12
21–25	9.53	5.64	2.95	1.60	0.67	0.23	0.19	0.12
Average	25.38	13.08	6.12	3.08	1.57	0.65	0.27	0.22

After Jones and Brooks (1950).

lation of anemophilous plants ostensibly is the consequence of local deficiencies in pollen. As noted above, Pedersen *et al.* (1961*b*) reported that the intravarietal pollen concentration above the windward side of a rye field was much below that closer to the center of the field, and that the ratio of extraneous to local pollen also was greater on the windward side of the field. Additional support for the pollen deficiency hypothesis comes from Knowles and Baenziger (1962), who obtained lower fertility of grasses on borders of fields.

The localization of hybridization also has been observed in bee-pollinated plants. Green and Jones (1953) determined the incidence of hybridization on a row-by-row basis in cotton. A rapid decline in hybridization from the proximal to the distal end was observed in the population adjacent to the alternative strain. Over short intervarietal distances a gradual decline was noted from the proximal to distal end, while over greater distances the trend was absent. A row-by-row analysis in alfalfa revealed a gradual decline in gene receipt from the proximal to the distal end of populations, but the slope was independent of isolation distances, which varied from 25 to 75 m (Pedersen *et al.,* 1969). Pollinator behavior responsible for the hybridization gradients within populations has not been observed. We may infer, however, that on flights between populations pollinators will tend to commence feeding soon after they reach a population.

The concept of utilizing border rows as an alternative means (to distance) of achieving satisfactory isolation has been adopted for maintaining crop purity. For fields of hybrid corn, the International Crop Improvement Association (1963) requires a minimum isolation distance of 220 m regardless of field size. Each addition of a border row, which is discarded after pollination, reduces the isolation requirement by 12.5 m. In sorghum, modification from the standard isolation distance of 200 m down to a minimum of 100 m is permitted when two additional border rows are planted for each 5.5 m of reduction in isolation distance.

In most isolation experiments, interpopulation gene exchange was not hampered by the presence of alternative vegetation which might capture windborne pollen or compete with the crop for pollinator service. Since monoculture with space between plots is a highly artificial situation, it is of interest to consider the effect which vegetational heterogeneity may have on the level of gene flow. A few experiments conducted by plant breeders offer some enlightenment in this regard. Pope *et al.* (1944) have studied the effect of three-, six-, and nine-row corn barriers on natural crossing in cotton. The corn barrier was 8–10 ft tall prior to the flowering of cotton. This barrier reduced the amount of hybridization, and the reduction tended toward linearity for the different barrier widths. The average percentage hybrids for four replicates was as follows: no barrier, 27%; three rows, 18%; six rows, 15%; nine rows, 13%. In contrast to these results, Afzal and Khan (1950) found that sorghum was no more effective a barrier than open space. The fact that corn afforded a taller screen than sorghum may account for the disparate results of the cotton experiments. Archimowitsch (1949) tested the value of growing hemp and sorghum between blocks of beets. When plant screens were 6, 10, and 12 m wide, the contamination between adjacent blocks was 17.1, 5.4, and 0.7%, respectively. In this case, a 12-m-wide plant screen was equivalent to 200 m of open space. In each of the experiments described, the barrier to cross-fertilization was an anemophilous plant. Although a discontinuity in species composition may reduce the incidence of interpopulation pollinator flights, the effect ostensibly would have been greater if the contrasting species could be utilized as a food source by the pollinator. Goplen *et al.* (1972) report that presence of a competitor, *Brassica napus,* for bee service between populations of *Melilotus alba* affords a very effective isolation barrier. Interpopulation gene flow is 6% in the absence of *Brassica,* but only 0.2% in its presence.

Jones and Brooks (1952) conducted an experiment to determine the effect of an American elm barrier on intervarietal hybridization in corn. They found that the tree barrier was effective in reducing the amount of hybridization by 50% immediately behind the tree line when compared to the hybridization obtained at the same distance without a barrier, but at greater distance was less effective. The elm barrier reduced hybridization to amounts similar to 75-m distance without a barrier. Tracy (1910) conducted a similar experiment with hedge barriers of osage orange, and found that hybridization was limited within 50 m of the hedge, but occurred with normal frequency beyond that distance. These results are compatible with what is known about pollen deposition beyond a barrier. Jensen and Bogh (1941) found that hedgerows and plantations protected ryegrass, cocksfoot, and mangolds in proportion to the height of the barrier. Protection corresponding to an isolation distance of about 200 m of open ground was obtained behind hedges, even as far downwind as 5–10 times their height.

Wind velocities to the lee of a vegetation barrier are known from field and wind-tunnel experiments on the effect of shelterbelts. If the barrier is impermeable to the wind, large eddies and vortices may be formed to the lee, and the zone of protection from alien genes will be relatively shallow because rapid downmixing of air takes place. However, if the barrier is penetrable, reduced wind velocity and turbulence will reach to distances 20–30 times the height of the barrier (Gloyne, 1954; Carborn, 1957). The wind shear to the lee of penetrable vegetation is insufficient for the development of large vortices and rapid downward mixing of the air. When the pollen cloud is carried toward a band of trees, it is split into two parts, one being forced above the crown by the windward edge, the other passing through the trunk space. The pollen passing through the latter will be transported by winds of lower velocity and much of it may be lost by deposition or impaction. The effects will be large for heavy pollen grains and small for light ones. The pollen carried above the canopy will reach the ground some distance beyond the edge of the trees because it is slowly mixed down to lower levels.

Although barriers of space or heterospecific plants may reduce interpopulation gene flow, a population of the same type as the recipient is much more efficacious in this regard. This has been shown in cotton (Simpson and Duncan, 1956; Afzal and Khan, 1950) and *Lolium* (Griffiths, 1950). For insect-pollinated plants, intervening populations act as "pollinator traps," since these organisms tend to stop at the first population similar to that in which they had foraged. For wind-pollinated plants, intervening populations of the same variety as the recipient provide additional "local" pollen which serves to dilute the extraneous pollen. Accordingly, the effectiveness of an isolation distance between distant populations is tempered by the number of intervening populations.

POPULATION STRUCTURE

The breeding structure of plant species is poorly understood, although it is clear that a population is not a panmictic unit and that gene exchange among populations occurs at very low levels. Since it is of heuristic value to deal with this problem, we will consider gene distribution schedules in relation to models of population structure. This procedure is useful in providing insight into the effect of migration patterns and migration–selection and migration–drift interactions on population and species structure.

Neighborhood Analyses

The first important contributions to the mathematical treatment of subdivided populations and species in relation to local differentiation of gene frequencies were those of Wright (1940, 1943*a*, 1951). His neighborhood or isolation-by-distance model has proved valuable in thinking about population structure. A neighborhood is the area of a colony within which mating is assumed to be random. In his models, the reference colony for panmixia is one composed of dioecious individuals with equal numbers of both sexes, random mating, and a Poisson distribution of number of offspring per parent. A subject colony is said to have a genetically effective size, N_e, if it undergoes the same rate of decay of gene frequency variance as a reference colony of size N. The effective size of a neighborhood is equivalent to the number of reproducing individuals in a circle whose radius is equivalent to twice the standard deviation of the gene dispersal distance. A circle of this type will include 86.5% of the parents of the individuals at its center. The standing crop may be used as an estimator of effective density if the colony is stable (in age structure) and stationary (in numbers) and if the distribution per parent of offspring reaching maturity is Poisson (Kimura and Crow, 1963). Deviations from these conditions undoubtedly exist, and accordingly the effective density usually will be less than the standing crop (flowering plants) in the prescribed circle (Falconer, 1960). The neighborhood area is N_e/d, where d is the genetically effective density. Genetically effective density is approximately the density of flowering plants.

Wright (1946) applies $N_e = 12.6\ \sigma^2 d$ as the neighborhood size in a population of hermaphrodites where male and female gametes show the same amount of axial dispersion. Expanding Wright's equation so that gene dispersal is effected by pollen (p) and seeds (s) yields

$$N_e = 12.6\ d[(\sigma_p^2 + \sigma_s^2)/2] = 6.3\ d\ (\sigma_p^2 + \sigma_s^2) \tag{5}$$

Wright's equation is based on the assumption that populations do not move in space. Accordingly, there must be no net movement of pollen or seeds in a given direction. By substituting calculated variances that have zero means in equation (5), we obtain

$$\sigma_p^2 = \Sigma\ (p_i - p)^2/N_p = \Sigma\ (p_i - 0)^2/N_p = \Sigma\ p_i^2/N \tag{6}$$

$$\sigma_p^2 = \Sigma\ (s_i - s)^2/N_s = \Sigma\ (s_i - 0)^2/N_s = \Sigma\ s_i/N_s \tag{7}$$

To bring pollen (haploid) and seed (diploid) dispersal into accord, we use one-half the absolute pollen dispersal. Combining the two dispersal components in the same equation gives

$$N_e = 6.3d(p_i^2/N_p + s_i^2/N_s) \tag{8}$$

By incorporating the proportion of outcross progeny (r) into equation (8), we arrive at equation (9):

$$N_e = 6.3\ dr(\Sigma\ p_i^2/2N_p + \Sigma\ s_i^2/N_s) \tag{9}$$

The neighborhood size may also be estimated for clone-forming facultative apomicts. The reader is referred to Levin and Kerster (1971) for the equations and attendant rationale.

By employing the neighborhood size as an indicator of breeding structure, we are relating gene flow to the decay of genetic variance. In amphimictic plants, the narrower the area from which parents are drawn and the stronger the correlation of parental genes by descent, the smaller is the neighborhood size. Self-fertilization greatly restricts neighborhood size, because it represents zero gene dispersal by pollen. In facultative apomicts, the rate at which variance decays is dependent not only on the movement of pollen and seed but also on the level of sexual reproduction. Apomixis is a unique reproductive method in that no genetic variance is lost from one generation to another regardless of population size. With apomixis alone, there is no loss of variance because there is no sexuality. Facultative apomicts will have larger neighborhoods than amphimictic plants having identical pollen and seed dispersal statistics by virtue of this mechanism. The formation of spatially discrete and nonoverlapping clones will further enlarge the neighborhood size, because long-distance pollen dispersal will be at a premium.

Neighborhood size has been estimated from the dispersion of gene disseminules in only a few wild species. There are three cases where both pollen and seed dispersal have been taken into account. *Phlox pilosa* is pollinated by lepidopterans and disperses its seeds via an explosive capsule. The neighborhood size varied from 75 to 282 and the area from 11 to 21 m^2 (Levin and Kerster, 1968). *Liatris aspera* is pollinated primarily by bees, and its seeds are dispersed by the wind. Neighborhood size varied from 30 to 191 and area from 4.5 to 29 m^2 (Levin and Kerster, 1969*b*). Research in progress suggests that *L. cylindracea* has neighborhood statistics similar to those of *L. aspera.* They have almost identical seed dispersal profiles and are serviced by the same pollinators. Both species are obligate outbreeders. *Lithospermum caroliniense* is largely bee-pollinated and has a dual breeding system; i.e., it produces chasmogamic flowers which are outcrossed and cleistogamic flowers which are selfed. The small nutlets are very hard and relatively heavy and have no special mechanism for dispersal. Assuming no gene dispersal by seeds, the neighborhood size varied from 2.2 to 5.4 and the area from 4 to 21 m^2 (Kerster and Levin, 1968). From pollen dispersion data, Wright (1953) estimated neighborhood size to be nine individuals in white ash, 11 in pinyon pine, 26 in Douglas fir, and 207 in Atlas cedar. Neighborhood areas were 0.9, 0.44, and 0.5 acre, respectively.

The neighborhood statistics on wild plants are based on pollen and seed dispersion rather than gene flow *per se.* As a consequence, the values obtained for anemophilous plants are overestimates because local pollen is ignored. A similar argument can be made for gene flow via seed being less than seed flow in both anemophilous and entomophilous plants where populations are under density regulation.

The calculation of neighborhood size from gene dispersal data has certain weaknesses. It tends to *underestimate* total inbreeding because it utilizes standing crop density as the genetically effective density, and thus fails to take into account catastrophic events, prolonged periods of environmental stress, or annual fluctuations in the percentage of potentially breeding individuals. These phenomena which may depress the effective density for one or more generations greatly affect neighborhood size. The average genetically effective size over a period of generations is the harmonic mean of the genetically effective densities of each generation, and harmonic means are dominated by low values (Wright, 1938). Neighborhood size also may be estimated from the spatial variance of gene frequencies. Wright (1943*b*) has judged the neighborhood size in *Linanthus parryae* to be 14–27 from an analysis of the spatial variance in the occurrence of two flower-color morphs. This rationale is circular, as it uses the very effects that are supposed to be indiced by finite size to measure that size (Lewontin, 1967). Considerable error may be introduced by selection. Epling *et al.* (1960) have provided evidence that moderate selection may be in force in *Linanthus*.

In *Liatris cylindracea*, we may estimate the extent to which neighborhood size calculated from pollen and seed deviates from that calculated from the spatial variance of gene frequency. The ecological measurements were made by Levin and the genetic measurements by Schaal (1974) on a population located near Zion, Illinois. Maruyama (1972) has provided formulas for determining the variance of gene dispersion distance from the value F_{ST}, which is the correlation between random gametes within subpopulations relative to the gametic array of the total population (Wright, 1965). The assumptions made in Maruyama's calculations are that the population is in a steady state in a torus-like habitat, population density is uniform, and there are no mutations. The variance of gene dispersal from ecological measurements is about 2.50, adjusted for zero mean distribution. The mean F_{ST} for polymorphic allozyme loci is about 0.065. This value is equivalent to a density–variance product of 10 in Maruyama's calculations. With a mean density of 12 flowering plants per meter, the variance of dispersal (adjusted for zero mean distribution) is approximately 0.83. Incorporating this value into equation (9) above, the genetically determined neighborhood size is 52, vs. 189 for the ecologically determined size. Thus there is about a 3.5-fold difference in the two measures.

Using the isolation-by-distance model, Rohlf and Schnell (1971) simulated the distributional pattern of gene frequencies within a population and the manner in which a population as a whole underwent random differentiation with neighborhood sizes of 9, 25, and 10,000. They report that the variance in gene frequencies among areas in a population is increased by decreasing the neighborhood size, and that this variance is greater near the periphery of the population. A key contribution lies in the graphic presentation of gene frequencies and the information derived from such.

The results of simulation for 100 generations are shown in Fig. 14 for three populations, each of which contained 10,000 individuals but differed in neighborhood size. With a neighborhood size of 10,000, gene frequencies varied little over the population (between 0.4 and 0.6). The population with a neighborhood size of 25 displayed an uneven texture, with gene frequencies varying from 0 to 1. The roughness in surface was even more pronounced in the population with a neighborhood size of 9. Considering gene frequency variance among areas in time, Rohlf and Schnell found that with the large neighborhood size the highs and lows in the gene frequency topography were independent of location from one generation to the next. When the neighborhood sizes were smaller, a prominent surface feature tended to persist for many generations, although their exact placement was due to chance. The correlations between gene frequencies in successive generations in one area were about 0.5 where $N_e = 10{,}000$ and about 0.75 where $N_e = 25$ or 9.

The neighborhood structure of a species is not a static property, but subject to variation in time and space. In wind-pollinated plants the distance which pollen is carried is dependent on properties of the source (e.g., terminal velocity of pollen and height). If these properties remain more or less constant as the population becomes less dense, the number of potential mates within a given distance will decline. Bannister (1965) has calculated neighborhood size for *Pinus radiata* under different densities assuming that the pollen dispersal distance has a standard deviation of 225 ft as reported by Wang *et al.* (1960) and seed dispersal was negligible. The neighborhood size declined from 379 when trees were 10 ft apart to 795 at 20 ft, 353 at 30 ft, 73 at 66 ft, 18 at 131 ft, and 8 at 198 ft. At the same time the neighborhood area remained at about 7 acres. There is no reason to expect this statistic to change unless the plants are self-compatible. In this case, the amount of self-pollen relative to outcross pollen would increase with decreasing density, selfing would be more frequent, and the neighborhood area would decline. The neighborhood size would fall faster than projected above, because with greater rates of selfing the standard deviation of the "effective" pollen dispersal distance would decline.

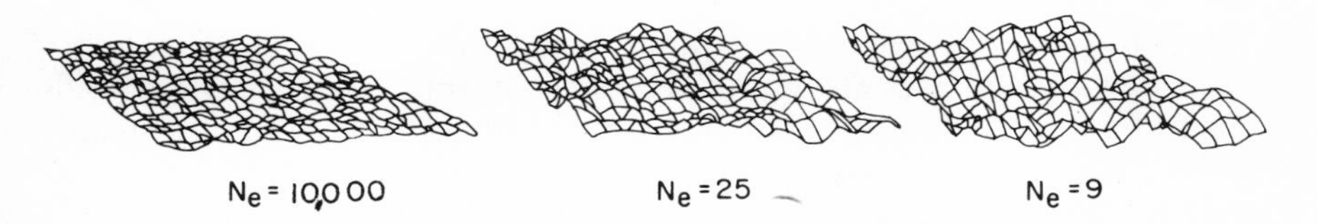

FIG. 14. A perspective view of surfaces of gene frequencies after 100 generations over a 100 × 100 array of individuals with neighborhood sizes of 9, 25, and 10,000. After Rohlf and Schnell (1971).

With density-independent neighborhood area but density-dependent neighborhood size, the potential for random differentiation is density dependent. As the neighborhood size declines, inbreeding increases and with it the likelihood of local differentiation. Thus we might expect sectors of sparse stands of an anemophilous species to show greater variance in gene frequencies than would sectors of the same size in a dense stand, stand dimensions remaining constant. Correlatively, a sparse stand is less able to respond to selection because of genetic drift among population subunits.

In animal-pollinated plants, the distance which pollen is transported is dependent on the foraging behavior of the vector. As noted earlier, pollinators typically move from a plant to one of its nearest neighbors. Levin and Kerster (1969*a*) have shown that bee feeding-flight distances are controlled almost exclusively by the plant spacing pattern and thus are highly deterministic, plant species notwithstanding. As plant density declines, the variance of pollinator flight distance and thus pollen dispersal distance increase. The neighborhood size, however, remains relatively constant, the reduction in effective density being counterbalanced by an increase in the variance of pollen dispersal distances. For example, in *Liatris aspera* colonies with densities of 1, 3, 5, and 11 plants/m^2, neighborhood sizes calculated on the basis of pollen movement were 29, 51, 42, and 60, respectively. The (pollen) neighborhood area decreased from 29 m^2 in the sparse population to 16, 8.4, and 4.5 m^2 in the denser colonies. When the seed component of the neighborhood equation was included, the neighborhood size rose with increased plant density, but the area still declined by almost 50%. We have suggested that in wind-pollinated plants capable of self-fertilization, greater interplant distances would be accompanied by a greater incidence of self-fertilization. This would also be so for animal-pollinated plants as long as the amount of pollinator service per flower was influenced by plant density. Bateman (1956) has shown that the wider the spacing, the greater the incidence of self-fertilization in *Cheiranthus cheiri*. The percentage of self-fertilization varied from 24.2 in a plot with 6-inch spacing to 26.3 in a plot with 1-ft spacing and 31.3 in a plot with 2-ft spacing. Apparently, the ratio of pollinations between flowers on the same plant to pollination between plants increases as density declines. The relationship between self-fertilization and density has been studied in *Lithospermum caroliniense,* which produces chasmogamic flowers yielding seeds only by selfing (Levin, 1972*b*). Interpopulation differences in cleistogamy and selfing were positively correlated with plant density. In this species, the density-dependent dispersal of pollen was reinforced by density-dependent selfing via cleistogamy.

Neighborhood diameter is a useful parameter for measuring the degree to which two points in a population are isolated by distance. The neighborhood concept was so formulated that gene migration over one neighborhood diameter occurs at a given rate per generation regardless of other neighborhood parameters. Therefore, the greater the number of neighborhoods separating two points, the greater is the number of generations required for genes to flow from one point to the other. It follows that in zoophilous plants gene migration between two points will be much slower in high-density arrays than in low-density arrays, for neighborhood diameters would be smaller in arrays where high density and close spacing prevailed. Density and spacing, through their effects on gene migration, will affect the ability of a population to undergo selective differentiation and subdivision. Should population dimensions remain constant, differentiation in response to a heterogeneous environment is most likely to occur in arrays of high density and close spacing, because under such circumstances gene migration is most restricted and swamping by migration is minimized.

F-Statistics Analyses

The genetic structure of a subdivided population can be analyzed by means of Wright's *F*-statistics (Wright, 1943*a*, 1951, 1965), a heterogeneous group of statistics which afford a means of partitioning an observed pattern of genetic variation or genetic relationships. They were developed to analyze subdivisions in hierarchical populations so that one may obtain the correlation between uniting gametes within subdivisions, between subdivisions, and throughout the population as a whole. Three *F*-statistics commonly are used to describe a population. They are (1) F_{IS}, the average correlation between uniting gametes relative to the gametic arrays of their own subdivision, which gives the average deviation of the subpopulation genotype proportions from Hardy–Weinberg expectations; (2) F_{ST}, the correlation between random gametes within subpopulations relative to the gametic array of the total population; and (3) F_{IT}, the correlation between gametes that unite to produce individuals relative to the gametes of the population as a whole. *F*-statistics have been applied extensively to analyzing population structure in man (e.g., Nei and Imaizumi, 1966*a,b;* Workman and Niswander, 1970; Neel and Ward, 1972). This method has been used only once to describe population structure in plants, this being by Schaal (1974) on a population of *Liatris cylindracea* (Compositae).

Liatris cylindracea is a perennial herb of climax prairies and is self-incompatible. Pollination is performed by an array of bees, butterflies, and beetles. The perennating organ is a corm which forms annual rings. The population analyzed occupies a hillside near Lake Michigan on the east side of Zion, Lake County, Illinois. The population was divided into 66-m^2

quadrats. Sixty corms were collected from each quadrat where plant density permitted. Genetic data were obtained from enzyme electrophoresis. Allozyme studies were performed on 12 enzyme systems encoded by 27 loci; 15 loci were polymorphic. The results of the *F*-statistics analysis for each polymorphic locus are presented in Table XII. F_{IS} varied from 0.10 to 0.51, with an average of 0.41. This value is surprisingly high for an obligate outbreeder, and most likely is due to near-neighbor and consanguineous pollination. F_{ST} values fell in the range of 0.01–0.22, with an average of 0.07. We see that the gene frequency heterogeneity between quadrats is small, although the population has experienced considerable inbreeding. The total fixation index, F_{IT}, varied from 0.11 to 0.61, with an average of 0.43. This value is dominated by the high F_{IS}. The Shannon information measure was used to ascertain the genetic diversity within the population and to apportion the diversity to the quadrat, row and column. Almost all diversity observed in the population is accounted for within quadrats, 0.93 being the mean for the 15 polymorphic loci. This finding is consistent with the F_{ST} interpretation. Nei's (1972) genetic distance was calculated for 27 quadrats in all pairwise combinations in order to determine whether genetic distance was correlated with spatial distance.

TABLE XII. F Statistics of a *Liatris cylindracea* Population

Locus	F_{IS}	F_{ST}	F_{IT}
GOT	0.3773	0.0184	0.3885
MDH	0.4853	0.0903	0.5318
ADH	0.4508	0.0452	0.4755
AP_1	0.4669	0.2240	0.5863
AP_2	0.5050	0.0438	0.5267
Est_1	0.5020	0.0464	0.5249
Est_2	0.5059	0.2190	0.6141
Pep	0.3025	0.0256	0.3203
G-6PDH	0.4013	0.0677	0.4419
6-PGDH	0.3629	0.0756	0.4110
Per^-	0.1004	0.0139	0.1121
Per	0.2579	0.0395	0.1004
PGI	0.4401	0.0767	0.4830
AlkP	0.4148	0.0361	0.4358
Est_3	0.4289	0.0091	0.4344
$\bar{x}$	0.4070	0.0687	0.4257

After Schaal (1974).

Genetic distance is plotted as a function of spatial distance in Fig. 15. The genetic distance designated is the average for all genetic distance measurements between sample quadrats of a given spatial distance. The linear relationship between these distance parameters is striking and suggests that isolation by distance in the sense of Wright (1943*a*) is the cause of genetic differentiation between quadrats.

Stepping-Stone Model

Species often are distributed more or less discontinuously to form numerous discrete populations. Should the populations be hundreds of meters apart, gene exchange in plants will be almost exclusively between adjacent populations. The effect of this breeding structure on the genetic structure of a population system may be considered in terms of Kimura's (1953) stepping-stone model. In this model, a population system consists of a rectangular array of populations each located at a grid point in the two-dimensional lattice. A population exchanges migrants with four surrounding populations, but its genetically effective size remains constant. If the migration rates are equal in both directions, the correlation coefficient $r(k_1 k_2)$

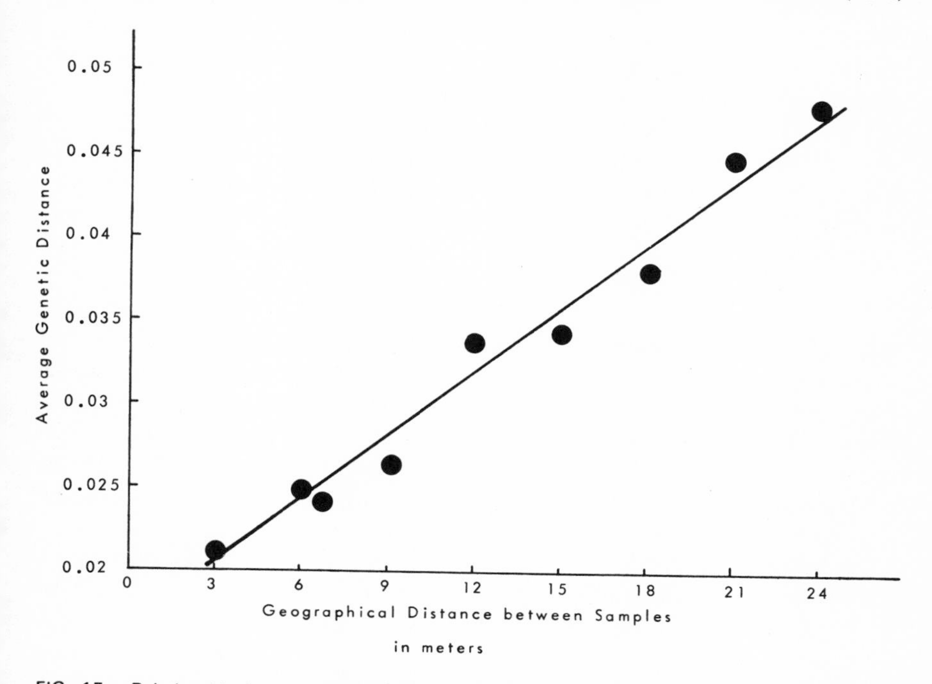

FIG. 15. Relationship between genetic distance and spatial distance in a population of *Liatris cylindracea*. After Schaal (1974).

of gene frequencies between populations which are k_1 steps apart in the vertical direction and k_2 steps apart in the horizontal direction becomes

$$r(\rho) \propto \frac{\exp[-\rho(4\bar{m}_\infty/m_1)^{1/2}]}{(\rho)^{1/2}} \tag{10}$$

where $\rho = (k_1^2 + k_2^2)^{1/2}$ is the distance between two populations, $\bar{m}_\infty$ is long-range migration, and m_1 is the migration rate per generation in one direction (Kimura and Weiss, 1964). If gene A is unconditionally deleterious, the selection coefficient against it may be included in $\bar{m}_\infty$, and with no long-range migration $m_\infty = s$. In this model, the correlation of gene frequencies does not depend on the number of individuals or the density of the populations. Genetic correlation in the stepping-stone model with nonsymmetrical migration rates has been considered by Maruyama (1969).

The stepping-stone model is applicable where the spatial dimensions of a population are very small relative to the distance between populations. For the sake of discussion, assume that small populations are located 50 m apart on a rectangular grid, that migration between adjacent populations in one direction is 0.1, and that long-range migration is 0.00004. The genetic correlation between adjacent populations is about 0.65, but it declines to about 0.25 at 500 m and about 0.15 at 1000 m. This is in the absence of selection, but with very liberal migration. An increase in N_e or m_∞ would decrease the correlation between gene frequency and distance.

In addition to the correlation of gene frequencies as a function of distance, it is also of interest to estimate the probability that two homologous genes randomly extracted from the population system are identical by descent. If gene neutrality is assumed, the aforementioned probability ($\bar{f}$) is approximately

$$\bar{f} = (1 - f_0)(4N_e u) \tag{11}$$

where f_0 is the probability that two homologous genes extracted from the same population are identical by descent, N_e is the effective number of the population system, and u is the rate of mutation per generation to new alleles. This relationship was first established by Maruyama (1970*c*) for the stepping-stone model and shown by Crow and Maruyama (1971) to be generally applicable for any subdivided structure and patterns of migration provided the total population size and breeding structure remain the same.

Maruyama (1970*a,b,c,* 1971) has performed a series of mathematical analyses of the stepping-stone model of finite size, being concerned in part with the specific relationship between local differentiation of selectively neutral gene frequencies by drift and the amount of migration. He has shown that in the two-dimensional stepping-stone models marked interpopulation differentiation is possible when Nm is smaller than unity, where N is the effective size of each population and m is the rate at which each colony exchanges migrants with four surrounding colonies per generation. This is a very severe restriction for gene flow between neighboring colonies, because the number of individuals which each colony exchanges with its neighbors on the average must be less than one per generation regardless of the size of each colony. On the other hand, if Nm is greater than 4, the whole population behaves as if it were panmictic and the gene frequencies become uniform over the entire range (Kimura and Maruyama, 1971).

Recalling that migration between discrete populations of plants may be very low when populations are separated by a few hundred meters and that the effective size is controlled largely by the phase of small numbers, it is likely that many species of plants would undergo local differentiation for neutral genes were they grown in the spatial pattern required by the model. Correlatively, we surmise that random differentiation has occurred in natural population systems where local units undergo wide fluctuations in size and are isolated by hundreds of meters. Kimura and Maruyama (1971) have run Monte Carlo experiments illustrating the pattern of random differentiation expected if $Nm = 4$ and 0.25 (Fig. 16). Note that in the latter case, in which local differentiation is evident, a cline is formed between regions where mutant frequencies are high and those where they are low. The resulting pattern is similar to a gene frequency cline due to selection, even though the genes in question are neutral.

Before examining other models useful to the student of plant gene

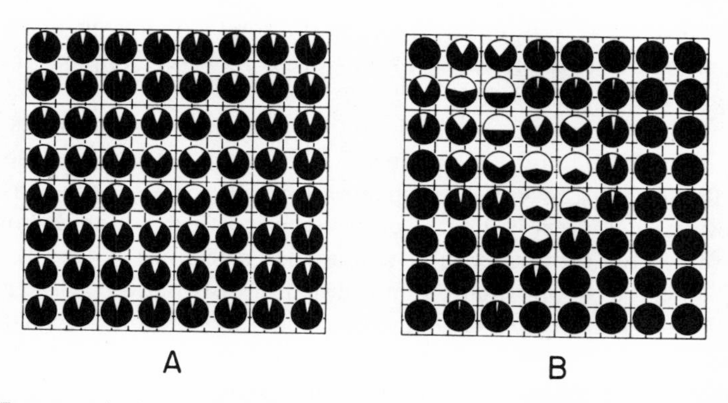

FIG. 16. Two examples from Monte Carlo experiments simulating the genetic change in a two-dimensional stepping-stone model with 20×20 populations arranged on a torus. In diagram A, the product of the population size (N) and the migration rate (m) is 4.0; in diagram B, the product is 0.25. The average frequency of one of the two alleles is 0.1. The light sector in each circle represents the average frequency of this allele in 6.25 neighboring colonies. After Kimura and Maruyama (1971).

flow, it is desirable to compare and contrast those treated thus far. Breeding structure models of Wright and of Kimura are devices used with data on gene flow to infer genetic properties of populations and their subdivisions and clusters. They focus on population structure at equilibrium and are based on *accumulated* inbreeding history. Each model includes such simplifying assumptions as breeding units of the same size and uniform density and immigrants having gene frequencies which are the regional or local average. Although different paths were followed in the construction of the models, it is important to note that their major results are in accord. Thus the models of relationships between demes and the larger arrays in which they are embedded are robust in the sense of Levins (1968).

Wright's neighborhood model was developed with path correlation techniques from genealogical considerations and from the relationships among the hierarchy of levels within a continuously and uniformly occupied region. Wright is interested in the identity of genes by descent in a continuum. Kimura's stepping-stone model, elaborated from consideration of statistical expectation, is designed to describe the relationships among islands (local panmictic breeding clusters), although the model can be adapted to treat continuous arrays. Unlike Wright, Kimura is interested in the identity of genes regardless of their ancestry.

The models differ in data used to infer relationships and in weight given data. Wright's model requires no estimate of systematic effects, although such could be added (Wright, 1969, p. 309). This is a fortunate circumstance in the absence of such measurements, but a circumstance leading one to question the model's validity. Kimura's model requires estimates of systematic effects and local migration. Adaptation of the stepping-stone model to continuous arrays does, however, require the same local information plus density.

The data on gene flow and population structure in seed plants do not permit the precision of the models to be tested. What is needed are parent–offspring distance measurements for many species, as well as an accurate estimation of very low level systematic effects (long-distance migration rates, mutation rates, and region-wide selection). These data must be complemented by information on the spatial distribution of like genes at the local and regional levels, and on local and regional selective differentials. Scraps of information are available for several species, but we lack a complete body of knowledge on any given species.

Models Employing Computer Simulation

The facility for local differentiation in plants has clear implications for the ecological adaptation to environments that are spatial mosaics (for discussion of theoretical treatments and references, see Maynard Smith,

1970*a,b*, and Dickinson and Antonovics, 1973*a,b*). Models which have been designed to provide insight into patterns of genetic differentiation observed in plant populations have been presented by Jain and Bradshaw (1966), Antonovics (1968*a*), and Gleaves (1973).

Jain and Bradshaw simulated a wide variety of model situations, with particular emphasis on gene distribution patterns. Their model assumes that (1) a population is continuously distributed along a linear habitat which is divided into two different environments, (2) the effective breeding units are subpopulations which are partially isolated by distance but of sufficient size to preclude random differentiation, (3) selection favors allele a_1 in one environment and a_2 in the other, (4) gene flow occurs only through pollen, and (5) random differentiation of subpopulations is absent. Four nodes of pollen dispersal were treated: stepping-stone model, island model, normal curve, and leptokurtic curve. These are shown in Fig. 17. Subpopulations with initial gene frequencies of q = 0.95 or 0.99 were allowed to respond to gene flow and selection until a stable equilibrium was reached. The distribution of gene frequencies over the linear array gives clines of varying slopes. The slope of the cline is dependent on the selection coefficients, but also is greatly influenced by the pollen distribution schedule. At similar levels of gene flow (approximately 50%), the island model confers the greatest genetic load on each subpopulation, followed by the leptokurtic distribution, normal distribution, and stepping-stone model in that order (Fig. 18). Of particular importance is the fact that moderate to strong selection can maintain local differences at short distances beyond the junction

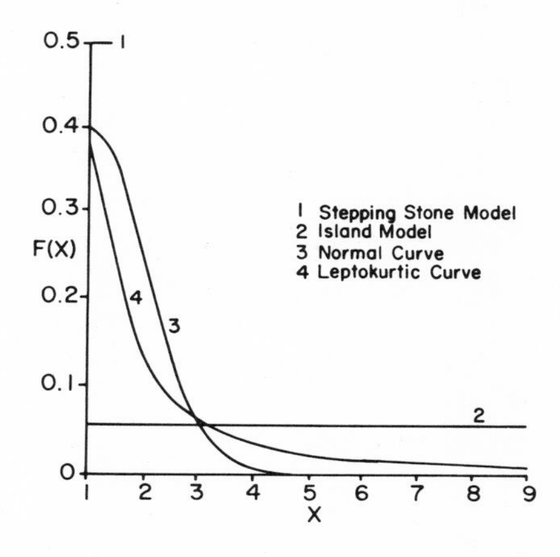

FIG. 17. Mode of pollen dispersal as given by $f(x)$, a function of distance (x). 1, Stepping stone: $f(x)$ = 0.50. 2, Island: $f(x) = e^{-x}/x$. After Jain and Bradshaw (1966).

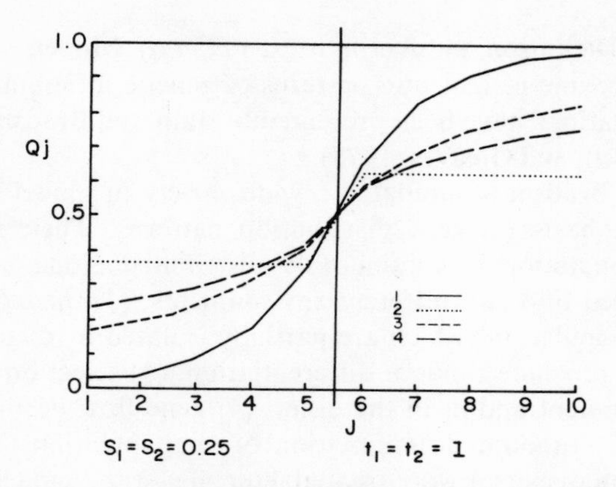

FIG. 18. Distribution curves of allelic frequencies (*Qj*) at equilibrium with symmetrical selection and generation time under four modes of pollen dispersal. The numbered curves refer to the modes of the same number in Fig. 17. After Jain and Bradshaw (1966).

of the two environments in spite of substantial gene flow. Polymorphism also can be maintained in the presence of unequal rates of gene flow in reciprocal directions if selection is strong; weak gene flow results in the fixation of the allele which "migrates" most frequently.

Whereas Jain and Bradshaw (1966) emphasized the importance of the interplay between gene flow and selection in determining patterns of differentiation between adjacent populations, Antonovics (1968*a*) considered the effects of continued gene flow on the genetic structure of a single population. His model assumes that (1) a recipient population is infinitely large and random breeding and consists of three genotypes (*AA, Aa, aa*), (2) selection operates against genotype *aa,* and (3) gene flow by pollen is opposed after the incoming genotypes have mated whereas gene flow via seed is opposed in the preexisting population and in the incoming genotypes before they mate. A population with an initial gene frequency of $A = 1.0$ was allowed to respond to gene flow and selection until a stable equilibrium was reached. The results of the simulation may be summarized as follows. Gene flow via pollen or seed is remarkably effective in maintaining a gene in a population in the face of selection against it (Fig. 19). If it is sufficiently strong and selection is weak, the favored gene can be obliterated. On the other hand, if gene flow between populations separated by a few hundred meters occurs at the low levels that are suggested in the literature, its effect in the face of weak selection would be minor. Antonovics also observed

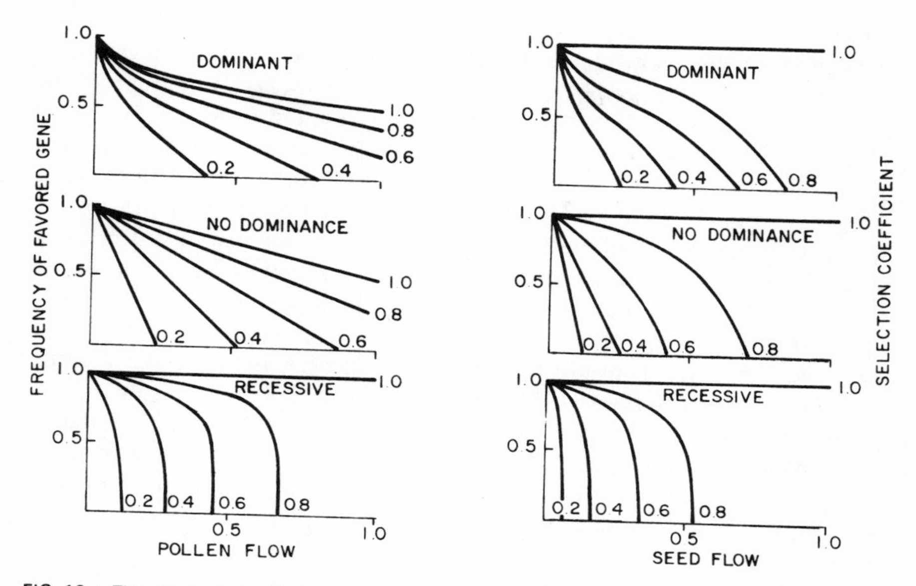

FIG. 19. The effect of selection and pollen flow (left) or seed flow (right) on gene frequency at equilibrium. The selection coefficients and degree of dominance of favored gene are shown on the graphs. After Antonovics (1968*a*).

that small populations would suffer more under a given immigration pressure than large populations because the balance of local to extraneous pollen grains would be less favorable in the former. Including the density-dependent effect in the model led to simulations wherein initial colonizers were nearly all heterozygotes when strong selection was in force. As colony size increased, the frequency of the favored gene did likewise. Again we are reminded that factors other than selection determine the consequence of a given input of extraneous pollen.

Gleaves (1973) considered the effect of plant aggregation within a population on the level of gene incorporation from a uniform rain of exogenous pollen. The model populations took the form of square areas containing plants of one genotype, with and without aggregation. The source of exogenous pollen was assumed to be very distant so that each plant received the same amount. The amount of endogenous pollen received by each plant corresponded to the distances between each plant and every other plant in the population. The amount of pollen as a function of distance was expressed as an inverse power function so that near neighbors contributed more endogenous pollen than distant plants within the population. Employing computer simulation, Gleaves found that more endogenous pollen is available to a plant at the center of an aggregate than to one near or at the periphery. In addition, the more highly aggregated the plants within a population, the less gene flow resulted from the same level of exogenous pollen receipt. Correlatively, the greater the density of a population, number of plants remaining constant, the less the effect of the exogenous pollen. In general, gene flow into a population is density dependent; it decreases as the average nearest-neighbor distance decreases, whether this be the product of an increase in aggregation or in population density. These conclusions are consistent with those obtained empirically (e.g., Griffiths, 1950).

The stochastic theory of gene frequency change as first elaborated by Fisher (1930) and Wright (1931 and later) has played a prominent role in population genetics theory. Although much has been written on the subject, most treatments deal with random-mating populations. We have attempted to determine the effect of alternate gene dispersal patterns on the tempo of gene substitution in order to gain a more realistic picture of gene frequency dynamics in natural populations. We have established a model based on a population of annual plants composed of 225 organisms in a 15 by 15-point square of "safe sites." By virtue of safe-site locations, plant spacing is uniform. Each plant produces pollen and seeds which are distributed in one of four ways: (1) to one of the four nearest neighboring locations, (2) to plants dictated by a leptokurtic distribution, (3) to locations chosen at random, or (4) dispersal may be zero. For the sake of brevity, the modes of pollen and seed dispersal will be referred to as "zero," "stepping-stone," "leptokurtic," and "random." Each plant is fertilized by pollen whose origin depends on the model, the genotype of the seeds being a probabilistic determinant of pollen and ovule genotypes. Each plant produces 18 seeds. Selection occurs at the seedling stage among seedlings at each of the 225 localities. The simulation begins with three homozygotes (*aa*) distributed at random (but at the same locations in all runs) in a population of homozygous dominant plants. The relative fitnesses are 0.2 for the dominant homozygote (*AA*), 0.2 for the heterozygote (*Aa*), and 1 for the recessive homozygote (*aa*).

The mean numbers of generations until the *a* allele reaches 0.5 and until fixation under all combinations of pollen and seed dispersal modes are summarized in Tables XIII and XIV. The values and their standard errors are from ten replicates. It is evident that the modes of pollen and seed dispersal have a pronounced effect on the rate of gene substitution. With zero pollen dispersal and random seed dispersal, the *a* allele reaches 0.5 in an average of 4.8 generations vs. 37.9 with a stepping-stone mode of pollen dispersal and no seed dispersal. If we consider the time to near-fixation (*a*

TABLE XIII. Average Number of Generations Until *a* Reaches 0.5 with Gene Dispersal as a Variable (Levin–Kerster Model)

Pollen dispersal mode	Seed dispersal mode			
	Zero	Stepping-stone	Leptokurtic	Random
Zero		11.7 ± 3.94	5.5 ± 1.84	4.8 ± 1.61
Stepping-stone	37.9 ± 12.97	20.8 ± 7.22	17.6 ± 5.99	21.0 ± 7.33
Leptokurtic	32.2 ± 11.24	26.1 ± 8.82	22.6 ± 7.64	22.2 ± 7.83
Random	28.0 ± 9.50	20.9 ± 7.10	24.2 ± 8.40	17.1 ± 5.78

≥0.95) (Table XIV), the effect of the dispersal modes also is dramatic. Random dispersal of pollen and seeds results in the least time to fixation, an average of 22.1 generations in contrast to an average of 78.4 for zero pollen dispersal and random seed dispersal. The most rapid increases early in the substitution process are engendered by zero pollen dispersal (or self-fertilization), whereas the most rapid substitution times are achieved with the widest broadcasting of pollen and seeds. Self-fertilization is associated with early increase in the favored allele because it eliminates the possibility of heterozygotes (recall that we start with three favored homozygotes).

A comparison of Tables XIII and XIV reveals that the time to a gene frequency exceeding 0.5 is not well correlated with the time to fixation. In fact, the modes of dispersal which yielded the most rapid shift to 0.5 (zero pollen dispersal and random seed dispersal) also yielded the least rapid substitution time. Thus the shapes of the gene substitution curves through time as well as the time to attaining a specific gene frequency are dependent on the dispersal modes.

TABLE XIV. Average Number of Generations to Near-Fixation with Gene Dispersal as a Variable (Levin–Kerster Model)

Pollen dispersal mode	Seed dispersal mode			
	Zero	Stepping-stone	Leptokurtic	Random
Zero		51.3 ± 17.28	72.6 ± 24.5	78.4 ± 26.21
Stepping-stone	64.1 ± 20.81	40.7 ± 13.94	24.4 ± 8.28	25.7 ± 8.82
Leptokurtic	42.0 ± 14.13	33.4 ± 11.18	28.1 ± 9.46	27.6 ± 9.52
Random	37.1 ± 12.49	25.7 ± 8.66	28.8 ± 9.79	22.1 ± 7.44

Another feature evident from Tables XIII and XIV is that the tempo of *a* allele increase to 0.5 and to fixation is similar whether pollen dispersal is leptokurtic or random, or seed dispersal is leptokurtic or random. The values for random dispersal (pollen and seeds) provide a fair (albeit high) approximation for leptokurtic schedules. Thus stochastic gene frequency models assuming random mating are not without value for species in which pollen and seed dispersal is leptokurtic.

In running our simulations, it was apparent that the favored gene often was lost from the population. Thus it was of interest to determine the probability of fixation with gene dispersal as a variable. Three dispersal modes were considered: stepping-stone for pollen and seeds, leptokurtic for pollen and seeds, and random for pollen and seeds. On the basis of 100 replicates of the general model discussed above, the probability of fixation of the *a* gene with the double stepping-stone mode is 0.86, in contrast to 0.57 for the double leptokurtic mode and 0.49 for the double random mode. Thus restricted gene flow by virtue of inbreeding and exposing favored homozygotes for selection enhances the probability that a rare gene will not be lost from the population, although gene flow restriction retards the rate of gene substitution.

CONCLUSIONS

Prevailing evolutionary and ecological theories are based on the premise that breeding units are extensive and that species are assemblages of individuals which maintain a common gene pool through bonds of mating. This interpretation is inconsistent with information on potential and actual gene flow in plants. The foraging behavior of animals, pollen and seed dispersal by wind or animals, and the isolation distances necessary for the maintenance of crop purity indicate that most gene flow in plants is restricted in space. Granted that pollen and seeds are collected tens or hundreds of miles from the source, this fact in itself is not decisive. The effect of such dispersal on the breeding structure of a species is decided by the relation between the quantity of pollen or seeds produced within a population or subdivision thereof and that which comes from greater distances. Pollen and seed dispersal are either exclusively local or highly leptokurtic. Thus local pollen and seeds undoubtedly are at a decided numerical advantage unless the recipient population is very small. In species which are self-pollinating and self-fertilizing, local (including self) pollen must be at a very great numerical advantage, although the imbalance might be adjusted somewhat by gametophytic selection against self-pollen tubes. The level of gene flow in most species must average less than 0.01 between populations or population subdivisions a few hundred meters apart, and at least two orders of magnitude less between populations a mile or more apart. In predominantly self-fertilizing species, without special mechanisms for long-distance dispersal, effective gene exchange between populations may ordinarily be at levels approaching mutation rates. This interpretation is consistent with that recently made by Bradshaw (1972). He states that "whatever indications of gene flow between populations we have, they are likely to be overestimates, and not underestimates of the true situation. Since values for gene flow where there are no special restrictions are already low, in natural situations the gene flow must be very low indeed. . . . Effective population size in plants is to be measured in meters and not kilometers." We would add that the numbers within panmictic units are to be measured in tens and not hundreds. Whatever the actual movement of pollen and seeds in natural and artificial populations, it is sufficiently restricted as to be overriden by natural selection. This is reflected in the fact that the pattern of differentiation is in accord with the pattern of the environment even when the environment varies over distances of less than 100 m (e.g., *Galium pumilum,* Ehrendorfer, 1953; *Agrostis tenuis,* Bradshaw, 1959, McNeilly, 1968; *Eschscholtzia californica,* Cook, 1962; *Agrostis stolonifera,* Aston and Bradshaw, 1966; *Potentilla erecta,* Watson, 1970; *Anthoxanthum odoratum,* Snaydon, 1970, Davies and Snaydon, 1973*a,b*). In fact, the discontinuities among neighboring populations within these and many other species, especially those that are self-compatible, may be as great as those reported between populations isolated by distance in the Aegean (Snogerup, 1967; Strid, 1970) and Caribbean (Morley, 1972) archipelagos. Accordingly, it seems likely that most species of seed plants are composed of multiple isolated or semi-isolated breeding units of various sizes and areas, each of which may adapt to local environmental conditions. This form of species structure is especially advantageous for the storage of genetic variability and evolutionary transformation in time and space by better permitting localized adaptation to different habitats (Wright, 1931, 1951, 1960).

The level of gene exchange between populations is determined by a multitude of factors including the size, density, and shape of the donor and recipient populations, plant height and breeding system, characters of the surrounding vegetation, terminal velocity of pollen and seeds, pollen and seed production, and the foraging behavior of pollen and seed vectors, as well as the distance between populations. To further confound matters, many of these factors vary in both time and space. For example, the degree of self-fertilization and/or apomixis within many species varies over a broad spectrum as a function of genetic influences or environmental heterogeneity (Stebbins, 1950; Clausen, 1954; Fryxell, 1957; Jain and

Marshall, 1968; Jain, 1973). The joint effect of documented parameters will differ for essentially every pair of populations within a given species as well as vary among species. Thus a simple predictor of gene flow over a given distance either within or between populations is out of the question. Experiments such as those described above, and hopefully more imaginative ones, will have to be conducted before any meaningful statement about gene flow in a given situation can be made.

A review of the models used by population geneticists brings to attention the predominant role of a single parameter, N_e, which may represent the effective size of a neighborhood, local population, or entire population system, depending on the model in question. Correlations of gene frequencies as a function of distance, the probability of genes being related by descent, the effective number of alleles in populations or population systems, and the rate of decay of variability and attendent loss of fitness all are functions of effective size and are inferred from it. The effective size of a breeding unit is calculated from the rate of decay of gene frequency variance, the latter value not being readily amenable to investigation in natural plant populations. In the absence of data, effective size has been estimated from dispersal statistics couched in terms of a number of assumptions relating to age structure and population size in time, and the distribution per parent of offspring. From these assumptions and data accrue effective sizes which ostensibly are gross overestimates of those existing in natural populations (*viz*. the *Liatris* example). If we are to understand the breeding biology and population structure of plant species, it is imperative that the assumptions employed in estimating effective size be tested and that the variation patterns of populations (and their subdivisions and aggregates) be studied through time.

The level of gene flow via pollen and seeds within species seems to be related to the general features of the communities in which species are integrated. Communities having relativity great longevity and stability are characterized by species in which reproductive rates are low, the defense budget is high, interspecific competition, predation, and parasitism are intense, and mortality is very high and in part genotype dependent. Such species may be regarded as K-selected (MacArthur and Wilson, 1967). The most strongly K-selected assemblages are found in the moist tropics (MacArthur and Wilson, 1967; MacArthur, 1972). On the other hand, some communities occupy transient or unstable habitats and are composed of species in which net reproductive rates are high, the defense budget is low, intraspecific competition is more intense than interspecific competition, predation and parasite pressure are not intense, and mortality is primarily genotype independent. Such species may be regarded as r-selected. Assemblages of these species are best developed in the temperate regions.

Gene flow via pollen and outbreeding in tropical floras is promoted by a high incidence of self-incompatibility, asynchronous flowering, and long flowering periods (Corner, 1954; Medway, 1972; Bawa, 1974). Moreover, the breeding systems of tropical floras exhibit an excess of dioecy relative to temperate floras or seed plants as a whole. Twenty-six percent of the tree species in a Malayan rainforest (Ashton, 1969) and 11% of the trees in an Ivory Coast flora (Aubreville, 1938) are dioecious. Twenty-two percent of the tree species in Costa Rica forest are dioecious (Bawa, 1974). This is in contrast to 3% of the flora of southern California (Gilmartin, 1968), 4% of the Australian flora (Parsons, 1958), 2% of the British flora (Lewis, 1942), and 5% of the seed plants of the world (Yampolsky and Yampolsky, 1922). Thus in species of the moist tropics, zero gene dispersal via pollen is infrequent. In contrast, the most intensively r-selected species, i.e., those living in the most transient or open habitats, are either predominantly self-pollinating and self-fertilizing or apomictic (Stebbins, 1950, 1957, 1965; Baker, 1959, 1965, 1972; Allard, 1965, Allard *et al.* 1968). Thus gene dispersal via pollen gene in transient or open habitats is likely to be low.

Gene flow via seeds in tropical floras tends to be restricted. Dispersal is primarily by animals which tend to have small home ranges and establish a seed cache close to the seed source (Connell and Orias, 1964; van der Pijl, 1969; Janzen, 1970, 1971*b*; Harper *et al.*, 1970). Long-distance seed dispersal methods are poorly developed. On the other hand, gene flow via seeds in transient floras is relatively great. Dispersal by wind is important, and the most elaborate and effective mechanisms for dispersal by this vehicle are evident in these floras (Stebbins, 1950; van der Pijl, 1969). In addition, the seeds of weedy species are very long-lived, whereas those of tropics may survive only a few days after dispersal (Crocker, 1938; Barton, 1961). Thus the long-distance dispersal of an r-selected species is more likely to result in effective gene flow than that of a K-selected species.

The breeding structure of populations also is determined by mean population size and variance in progeny numbers. Populations whose size tends to be constant through generations will have a greater effective size than populations which fluctuate in size even though the mean population sizes of the former and latter are the same, because genetically effective size is the harmonic mean of the genetically effective densities of each generation, and harmonic means are dominated by low values (Wright, 1938). Populations in which individuals tend to have the same number of offspring will have a greater effective size than those in which there is a high variance in offspring number (Wright, 1938). In general, K-selected species have more uniform population sizes and lower variance of offspring per generation than do r-selected species (MacArthur and Wilson, 1967; Pianka, 1970). These features contribute to K-selected species being more outbred than r-selected species. Another feature by which r- and K-selected species might differ is the extent to which generations overlap. K-selected species probably have greater overlap by virtue of great individual longevity and long (perhaps discontinuous) reproductive span, which more than compensate for the seed storage pool of r-selected species, especially if sites are occupied for only a few generations. However, which species type has greater overlap seems unimportant in terms of the breeding structure of populations, because the gain in N_e with overlapping generations over the N_e with discrete generations is very small (Felsenstein, 1971; Hill, 1972; Crow and Kimura, 1972).

The general relationship between relative levels of gene flow and life style of species is portrayed in Fig. 20. Based on the work referred to in this chapter, we judge gene flow by both vehicles jointly to be greater within and between populations of K-selected species than populations of r-selected species. This view is vindicated in part by the variation patterns of these species types, although factors other than gene flow obviously are important here. Given that the amount of gene flow between populations of selfing species is less than in outbreeders, we would expect greater differences between populations of the former than between populations of the latter. This pattern has been demonstrated in *Bromus mollis* (Knowles, 1943), *Collinsia heterophylla* (Weil and Allard, 1964), *Avena fatua* and *Avena barbata* (Imam and Allard, 1965; Jain and Marshall, 1967; Clegg and Allard, 1972); *Agrostis tenuis* and *Anthoxanthum odoratum* (An-

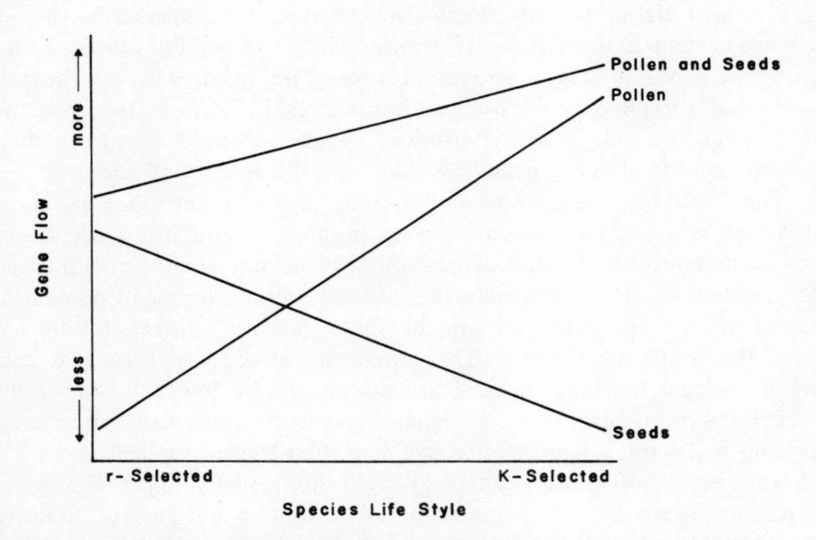

FIG. 20. Putative relationship between the level of gene flow (by pollen, seeds, or both) within a species and the life style of that species.

tonovics, 1968*a,b*), *Arabidopsis thaliana* (Jones, 1971), *Triticum longissimum* and *T. speltoides* (Hillel *et al.*, 1973), and *Phlox drummondii* and *P. cuspidata* (Levin, 1975). Correlatively, we might expect and we find greater gene frequency heterogeneity in space within populations of selfing species than within populations of outcrossing species. This pattern has been demonstrated in *Clarkia exilis* (Vasek, 1967), *Trillium* spp. (Fukuda, 1967; Haga, 1969). *Avena barbata* (Hamrick and Allard, 1972), *Picea abies* (Tigerstedt, 1973), *Liatris cylindracea* (Schaal, 1974), and *Phlox pilosa* (Kerster and Levin, unpublished).

The numbers of generations in which a specific pattern of gene frequency heterogeneity persists within and among populations is a function of the breeding structure and demography of species, as is the persistence of a pattern is space. Demographic conditions most conducive to the maintenance of a pattern are overlapping generations, which may take the form of long-lived seed pools or long-lived adult plants. Neighborhood size also influences the persistence of a specific pattern. Rohlf and Schnell (1971) have shown that the correlation between gene frequencies of neighborhoods in successive generations is greatest when neighborhoods are small. Thus the most persistent patterns of gene frequencies in space are likely to be found in self-fertilizing species with narrow seed dispersal but a long-lived seed pool. The work of Epling *et al.* (1960) on *Linanthus parryae* and that of Gottlieb (1974) on *Stephanomeria exigua* lend some credence to this hypothesis.

Having considered some effects of the breeding system and demographic features on the genetic structure of populations and species, we may deal with the consequences of adaptive radiation for different pollen and seed vectors. Adaptive shifts from one pollen vector to another has occurred within genera and among genera of the same family (Stebbins, 1970*b*). Adaptive radiation for different pollinators is an ecological strategy which in and of itself is not likely to have a profound effect on the breeding structure of populations. All groups of pollinators ostensibly forage in the most economical manner and thus will tend to work small areas within a population and move from a plant to one of its near neighbors. Species serviced by strong fliers (e.g., bees or hummingbirds) may experience more long-distance dispersal than species serviced by weak fliers (e.g., beetles) and thus may display a greater correlation of gene frequencies between populations and subpopulations than other species. Unfortunately, definitive empirical data for comparative purposes are lacking. Shifts from insect to wind pollination and *vice versa* also are known in several families (Stebbins, 1970*b*). This shift is associated with the occupation of more open and arid habitats, and ostensibly results in greater gene flow within and among populations than with insect pollination, if the species is self-incompatible.

Adaptive shifts from one seed vector to another as well as from animal to wind dispersal have been documented in many genera and families (Stebbins, 1971). Again, these shifts are ecological adaptations which only secondarily may have important genetic consequences. As we suggested earlier, seed dispersal by large animals generally will be more widespread than dispersal by very small ones. Dispersal by an opportunistic species may be more widespread than that by a specialist. Ingestion of fruits and seeds by animals, especially birds, is one of the most common methods of zoophilous seed dispersal. There is no body of evidence suggesting that this mode of transport will carry seeds farther than active transport and cache formation, or transport by adhesion. However, shifts to internal transport occur in K-selected taxa, so that it is tempting to surmise that this mode of transport reduces seed dispersal. The consequences of shifting from animal to wind dispersal depend in part on the effectiveness of the dispersal mechanism. An effective wind dispersal mechanism could yield greater seed broadcasting than might be accomplished by animals. This is suggested by the high incidence of wind dispersal in weedy species.

From consideration of migration statistics, we conclude that the local gene pool is little dependent on migration for the maintenance of genetic diversity. Two factors prompted this judgment. First, migration rates are very low and are almost exclusively between neighboring populations, which tend to have correlated gene pools. Second, populations retain a high level of genetic polymorphism in spite of vanishingly small inputs of extraneous genes. The best evidence for the latter proposition comes from the work of Allard and associates. Allard *et al.* (1968) reviewed a series of investigations which demonstrated that crop lines and isolated local populations are highly variable at many exomorphic loci, and retain this variability, breeding system notwithstanding. Individuals within populations often differ in single- and multiple-unit polymorphisms, and for continuously varying characters as well. Moreover, individuals are frequently heterozygous at numerous loci, as seen in the segregational variability which exceeds predictions based on the mating system alone. Recently, analyses have been extended to include electrophoretic variation at loci encoding for structural proteins (summarized in Allard and Kahler, 1971, 1972). Again, both outbreeding and inbreeding species and crop lines exhibited extensive heterogeneity within and between populations. The main factors involved in the morphological and allozymic polymorphisms are seen as (1) balancing selection where heterozygotes have an advantage, (2) epistatic interactions which render the genotype or larger fractions thereof the unit of selection, and (3) frequency-dependent selection associated with intergenotypic interactions.

If the population is little dependent on migration for the maintenance of genetic variability, it may also receive little novel germplasm from neighboring populations. What then are the sources of alleles for populations well isolated by distance? They are mutation and, more importantly, introgressive hybridization. Related species could serve as valuable sources of new and tested alleles, and indeed have served in this capacity in numerous genera (Heiser, 1949, 1973; Anderson, 1953; Stebbins, 1959). Harlan and deWet (1963) have proposed the concept of the compilospecies, which is a species which "plunders" genes from various relatives and assumes characteristics of each of these congeners from region to region, in the process increasing its adaptive range. In many genera, we surmise that interspecific gene exchange occurs at a greater level than interpopulation hybridization when conspecific populations are widely spaced and related species are biotically sympatric.

The role of introgression in the enrichment of local gene pools has been difficult to determine because the characters analyzed often are under polygenic control and in sone instances the underlying genes display dominance, epistatic interactions, and pleiotropy. On the other hand, the development of electrophoretic separation of enzyme variants greatly enhances the possibility of detecting and quantifying introgression. We favor the view that the absence of absolute barriers to gene exchange between closely related species is in some measure an adaptation to permit periodic infusions of new germ plasm. The burden imposed by such infusions could be handled by the reproductive excess of plants, especially perennials. Even if there is selection for isolation, it will not necessarily lead to the complete blocking of hybridization so long as the population is safeguarded from the net deleterious effects of hybridization.

ACKNOWLEDGMENTS

The authors are indebted to V. Grant, J. Antonovics, and Y. Linhart for their critical reading of the manuscript and valuable suggestions for improving it. Heretofore unpublished data by the authors cited herein were obtained with the assistance of National Science Foundation grant GB-35967X to DAL.

REFERENCES

Afzal, M., and Khan, H., 1950, Natural crossing in cotton in western Punjab, *Agron J.* **42:**14–19, 89–93, 202–205, 236–238.

Allard, R. W., 1965, Genetic systems associated with colonizing ability in predominantly self-pollinated species, in: *The Genetics of Colonizing Species* (H. G. Baker and G. L. Stebbins, eds.), pp. 50–75, Academic Press, New York.

Allard, R. W., and Kahler, A. L., 1971, Allozyme polymorphisms in plant populations, *Stadler Symp. Vol.* **3**:9–24.

Allard, R. W., and Kahler, A. L., 1972, Patterns of molecular variation in plant populations, in: *Proceedings of the 6th Berkeley Symposium on Mathematical Probability and Statistics,* Vol. 5, pp. 237–254. University of California Press, Berkeley.

Allard, R. W., Jain, S. K., and Workman, P. L., 1968, The genetics of inbreeding species, *Advan. Genet.* **14**:55–131.

Anderson, E., 1953, Introgressive hybridization, *Biol. Rev.* **28**:280–307.

Anonymous, 1939, Black Spruce Is a Limited Air Traveler, USDA Forest Service, Lake States Forest Experiment Station Technical Bulletin No. 147.

Antonovics, J., 1968*a*, Evolution in closely adjacent plant populations. VI. Manifold effects of gene flow, *Heredity* **23**:507–524.

Antonovics, J., 1968*b*, Evolution in closely adjacent plant populations. V. Evolution of self-fertility, *Heredity* **23**:219–238.

Archimowitsch, A., 1949, Control of pollination in sugar beets, *Bot. Rev.* **15**:613–628.

Armstrong, J. T., 1965, Breeding home range in the nighthawk and other birds: Its evolutionary and ecological signifcance, *Ecology* **53**:350–361.

Ashton, P. S., 1969, Speciation among tropical forest trees: Some deductions in light of recent evidence, *Biol. J. Linn. Soc. Lond.* **1**:155–196.

Aston, J. L., and Bradshaw, A. D., 1966, Evolution in closely adjacent populations. II. *Agrostis stolonifera* in marine habitats. *Heredity* **21**:649–664.

Aubreville, A., 1938, La forêt coloniale: Les forets de l'Afrique occidentale francaise, *Ann. Acad. Sci. Paris* **9**:1–245.

Baker, H. G., 1959, Reproductive methods as factors in speciation in flowering plants, *Cold Spring Harbor Symp. Quant. Biol.* **24**:177–190.

Baker, H. G., 1963, Evolutionary mechanisms in pollination biology, *Science* **139**:877–883.

Baker, H. G., 1965, Characteristics and modes of origins of weeds, in: *The Genetics of Colonizing Species* (H. G. Baker and G. L. Stebbins, eds.), pp. 147–168, Academic Press, New York.

Baker, H. G., 1970, Evolution in the tropics, *Biotropica* **2**:101–111.

Baker, H. G., 1972, Seed weight in relation to environmental conditions in California, *Ecology* **53**:997–1010.

Baker, H. G., 1973, Evolutionary relationships between flowering plants and animals in American and African tropical forests, p. 145–159. *In* B. J. Meggers, E. S. Ayensu, and W. D. Duckworth (eds.), *Tropical Forest Ecosystems in Africa and South America: A Comparative Review.* Smithsonian Institution Press, Washington, D.C.

Bannister, M. H., 1965, Variation in the breeding system of *Pinus radiata,* in: *The Genetics of Colonizing Species* (H. G. Baker and G. L. Stebbins, eds.), pp. 353–372, Academic Press, New York.

Barrons, K. C., 1939, Natural crossing in beans at different degrees of isolation, *Proc. Am. Soc. Hort. Sci.* **36**:637–640.

Barton, L. V., 1961, *Seed Preservation and Longevity,* Hill, London.

Bateman, A. J., 1947*a*, Contamination of seed crops. I. Insect pollination, *J. Genet.* **48**:257–275.

Bateman, A. J., 1947*b*, Contamination of seed crops. III. Relation with isolation distance, *Heredity* **1**:303–336.

Bateman, A. J., 1951, Is gene dispersion normal? *Heredity* **4**:253–263.

Bateman, A. J., 1956, Cryptic self-incompatibility in the wallflower: *Cheiranthus cheiri* L., *Heredity* **10**:257–261.

Bawa, K. S., 1974, Breeding systems of tree species of a lowland tropical community, *Evolution* **28**:85–92.

Bene, F., 1946, The feeding and related behavior of hummingbirds, *Mem. Boston Soc. Nat. Hist.* **9**:403–478.

Bonnier, G., 1906, Sur la division du travail chez les abeilles, *Comp. Rend. Acad. Sci. Paris* **143**:941–946.

Boyer, W. D., 1958, Longleaf pine seed dispersal in south Alabama, *J. Forestry* **56**:265–268.

Bradner, N. R., Frakes, R. V., and Stephen, W. P., 1965, Effects of bee species and isolation distances on possible varietal contamination in alfalfa, *Agron. J.* **57**:247–248.

Bradshaw, A. D., 1959, Population differentiation in *Agrostis tenuis*, I. Morphological differentiation. *New Phytol.* **58**:208–227.

Bradshaw, A. D., 1972, Some of the evolutionary consequences of being a plant, *Evol. Biol.* **5**:25–44.

Brittain, W. H., and Newton, D. E., 1933, A study in the relative constancy of hive bees and wild bees in pollen gathering, *Can. J. Res.* **9**:334–349.

Brower, L. P., 1961, Studies on the migration of the monarch butterfly. I. Breeding populations of *Danaus plexippus* and *D. gilippus berenice* in south central Florida, *Ecology* **42**:76–83.

Brownlee, J., 1911, The mathematical theory of random migration and epidemic distribution, *Proc. Roy. Soc. Edinburgh* **31**:262–289.

Brussard, P. F., and Ehrlich, P. R., 1970, The population structure of *Erebia epipsodea* (Lepidoptera: Satyrinae), *Ecology* **51**:119–129.

Burrows, F. M., 1973, Calculation of the primary trajectories of plumed seeds in steady winds with variable convection, *New Phytol.* **72**:647–664.

Butler, C. G., Jeffree, E. P., and Kalmus, H., 1943, The behavior of a population of honeybees on an artificial and on a natural crop, *J. Exp. Biol.* **20**:65–73.

Buzzard, C. N., 1936, De l'organisation du travail chez les abeilles, *Bull. Soc. Apic. Alpes-Marit.* **15**:65–70.

Calde, R. D., 1965, *Particle Size*, Reinhold, New York.

Carborn, J. M., 1957, Shelterbelts and Microclimate, Forestry Commission Bulletin No. 29, Edinburgh.

Carlquist, S., 1966, The biota of long-distance dispersal. II. Loss of dispersability in Pacific Compositae, *Evolution* **20**:30–48.

Chamberlain, A. C., 1967, Deposition of particles to natural surfaces, in: *Airborne Microbes* (17th Symposium of the Society for General Microbiology, *Cambridge*), pp. 138–164. Cambridge University Press, Cambridge.

Clausen, J., 1954, Partial apomixis as an equilibrium system in evolution, *Caryologia* **6**:469–479, (suppl.).

Clegg, M. T., and Allard, R. W., 1972, Patterns of genetic differentiation in the slender wild oat species *Avena barbata*, *Proc. Natl. Acad. Sci.* **69**:1820–1824.

Connell, J. H., and Orias, E., 1964, The ecological regulation of species diversity, *Am. Naturalist* **98**:399–413.

Cook, S. A., 1962, Genetic system, variation and adaptation in *Eschscholzia californica*, *Evolution* **16**:278–299.

Corner, E. J. H., 1954, The evolution of the tropical forest, in: *Evolution as a Process* (J. Huxley, A. C. Hardy, and E. B. Ford, eds.), Allen and Unwin, London.

Crane, M. B., and Mather, K., 1943, The natural cross-pollination of crop plants with particular reference to the radish, *J. Appl. Biol.* **30**:301–308.

Croat, T. B., 1974, A case for selection for delayed fruit maturation in Spondias (Anacardiaceae), *Biotropica* **6**:135–137.

Crocker, W., 1938, Life-span of seeds, *Bot. Rev.* **4**:235–274.

Crow, J. F., and Kimura, M., 1972, The effective number of a population with overlapping generations: A correction and further discussion, *Am. J. Hum. Genet.* **24**:1–10.

Crow, J. F., and Maruyama, T., 1971, The number of neutral alleles maintained in a finite, geographically structured population, *Theoret. Pop. Biol.* **2**:437–453.
Cruden, R. W., 1966, Birds as agents of dispersal for disjunct plant groups of the temperate western hemisphere, *Evolution* **20**:516–532.
Currence, T. M., and Jenkins, J. M., 1942, Natural crossing in tomatoes as related to distance and direction, *Proc. Am. Soc. Hort. Sci.* **41**:273–276.
Dansereau, P., and Lems, K., 1957, The grading of dispersal types, *Contrib. Inst. Bot. Montréal,* No. 71.
Davies, M. S., and Snaydon, R. W., 1973*a*, Physiological differences among populations of *Anthoxanthum odoratum* L. collected from the Park Grass experiment, Rothamsted. I. Response to calcium, *J. Appl. Ecol.* **10**:33–45.
Davies, M. S., and Snaydon, R. W., 1973*b*, Physiological differences among populations of *Anthoxanthum odoratum* collected from the Park Grass experiment, Rothamsted, II. Response to aluminum, *J. Appl. Ecol.* **10**:47–55.
den Hartog, C., 1964, Over de oecologie van bloeiende *Lemna trisulca, Gorteria* **2**:68–72.
Dethier, V. G., and MacArthur, R. H., 1964, A field's capacity to support a butterfly population, *Nature* **201**:728–729.
Dickinson, H., and Antonovics, J., 1973*a*, The effects of environmental heterogeneity on the genetics of finite populations, *Genetics* **73**:713–735.
Dickinson, H., and Antonovics, J., 1973*b*, Theoretical considerations of sympatric divergence, *Am. Naturalist* **107**:256–274.
Dingle, A. N., Gill, G. C., Wagner, W. H., Jr., and Hewson, E. W., 1959, The emission, dispersion, and deposition of ragweed pollen, *Advan. Geophys.* **6**:367–386.
Dowdeswell, W. H., Fisher, R. A., and Ford, E. B., 1940, The quantitative study of populations in the Lepidoptera. I. *Polyommatus icarus* Rott., *Ann. Eugen.* **10**:123–136.
Dowdeswell, W. H., Fisher, R. A., and Ford, E. B., 1949, The quantitative study of populations in the Lepidoptera. II. *Maniola jurtina* L., *Heredity* **3**:67–84.
Dressler, R. L., 1968, Pollination by euglossine bees, *Evolution* **22**:202–210.
Duggar, J. F., 1934, The distance to which *Crotalaria* seed are thrown by pods, *Ann. Rep. Alabama Agr. Exp. Sta.* **35**:27.
Ehrendorfer, F., 1953, Okologisch-geographische Mikro-Differenzierung einer Population von *Galium pumilum* Murr. s. str., *Oesterr. Bot. Z.* **100**:616–638.
Ehrendorfer, F., 1965, Dispersal mechanisms, genetic systems, and colonizing abilities in some flowering plant families, in: *Genetics of Colonizing Species* (H. G. Baker and G. L. Stebbins, ed.), Academic Press, New York.
Ehrlich, P. R., 1965, The population biology of the butterfly *Euphydryas editha.* II. The structure of th Jasper Ridge colony, *Evolution* **19**:327–336.
Ehrlich, P. R., and Gilbert, L. E., 1973, Population structure and dynamics of the tropical butterfly *Heliconius ethilla, Biotropica* **5**:69–82.
Ehrlich, P. R., and Raven, P. H., 1969, Differentiation of populations, *Science* **165**:1228–1232.
Ehrman, L., 1962, Hybrid sterility as an isolating mechanism in the genus *Drosophila, Quart. Rev. Biol.* **37**:279–302.
Emmel, T. C., 1964, The ecology and distribution of butterflies in a montane community near Florissant, Colorado. *Amer. Midl. Natur.* **72**:358–373.
Emmel, T. C., 1968, The population biology of the neotropical satyrid butterfly *Euptychia herma., J. Lipid. Res.* **7**:153–165.
Emmel, T. C., 1972, Dispersal in a cosmopolitan butterfly species (*Pieris raphae*) having open population structure, *J. Lipid. Res.* **11**:95–98.
Epling, C., Lewis, H., and Ball, F. M., 1960, The breeding group and seed storage: A study in population dynamics, *Evolution* **14**:238–255.

Faegri, K., and van der Pijl, L., 1966, *Principles of Pollination Ecology*, Pergamon Press, New York.

Falconer, D. S., 1960, *Introduction to Quantitative Genetics*, Ronald Press, New York.

Fales, J. H., 1959, A field study of the flight behavior of the tiger swallowtail butterfly, *Ann. Entomol. Soc. Am.* **52**:486–487.

Felsenstein, J., 1971, The effective size of a population with overlapping generations, *Genetics* **68**:581–597.

Fisher, R. A., 1930, *The Genetical Theory of Natural Selection*, Clarendon Press, Oxford.

Fosdick, M. K., 1972, A population study of the neotropical nymphalid butterfly, *Anartia amalthea*, in Ecuador, *J. Res. Lepid.* **11**:65–80.

Free, J. B., 1960, The behavior of honeybees visiting the flowers of fruit trees, *J. Anim. Ecol.* **29**:385–395.

Free, J. B., 1962, The effect of distance from pollinizer varieties on the fruit set on trees in plum and apple orchards, *J. Hort. Sci.* **37**:262–271.

Free, J. B., 1964, The behavior of honeybees on sunflowers (*Helianthus annuus* L.), *J. Appl. Ecol.* **1**:19–27.

Free, J. B., 1968, The foraging behavior of honeybees (*Apis mellifera*) and bumblebees (*Bombus* spp.) on blackcurrant (*Ribes nigrum*), raspberry (*Rubus idaeus*) and strawberry (*Fragaria* × *Ananassa*) flowers, *J. Appl. Ecol.* **5**:157–168.

Free, J. B., 1970, *Insect Pollination of Crops*, Academic Press, New York.

Fryxell, P. A., 1956, Effect of varietal mass on the percentage of outcrossing in *Gossypium hirsutum* in New Mexico, J. Hered. **57**:299–301.

Fryxell, P. A., 1957, Mode of reproduction in higher plants, *Bot. Rev.* **23**:135–233.

Fukuda, I., 1967, The formation of subgroups by the development of inbreeding systems in a *Trillium* population, *Evolution* **21**:141–147.

Gadgil, M., 1970, Dispersal: Population consequences and evolution, *Ecology* **52**:253–261.

Gardner, E. J., 1946, Wind pollination in guayule, *Parthenium argentatum* Gray, *J. Am. Soc. Agron*, **38**:264–272.

Geiger, R., 1950, *The Climate Near the Ground*, Harvard University Press, Cambridge, Mass.

Gilbert, J. M., 1958, Forest succession in the Florentine Valley, Tasmania, *Proc. Roy. Soc. Tasmania* **93**:129–151.

Gilbert, L. E., and Singer, M. C., 1973, Dispersal and gene flow in a butterfly species, *Am. Naturalist* **107**:58–72.

Gilmartin, A. J., 1968, Baker's law and dioecism in the Hawaiian flora: An apparent contradiction, *Pacific Sci.* **22**:285–292.

Giltay, E., 1904, Über die Bedeutung der Krone bei den Blüten und über das Farbenunterscheidungsvermögen der Insekten, *Jahrbuch Wiss. Bot.* **40**:368–402.

Gleaves, J. T., 1973, Gene flow mediated by wind-borne pollen, *Heredity* **31**:355–366.

Gloyne, R. W., 1954, Some effects of shelterbelts upon local and microclimate, *J. Forestry* **27**:85–95.

Goplen, B. P., Cooke, D. A., and Pankiw, P., 1972, Effects of isolation distance on contamination in sweetclover, *Canad. J. Plant Sci.* **52**:517–524.

Gottlieb, L. D., 1974, Genetic stability in a peripheral isolate of *Stephanomeria exigua* ssp. *coronaria* that fluctuates in population size, *Genetics* **76**:551–556.

Grant, K., and Grant, V., 1968, *Hummingbirds and Their Flowers*, Columbia University Press, New York.

Grant, V., 1958, The regulation of recombination in plants, *Cold Spring Harbor Symp. Quant. Biol.* **23**:337–363.

Grant, V., 1971, *Plant Speciation*, Columbia University Press, New York.

Green, H. L., and Lane, W. R., 1957, *Particulate Clouds: Dusts, Smokes, and Mists*, Span, London.

Green, J. M., and Jones, M. D., 1953, Isolation of cotton for seed increase. *Agron J.* **45**:366–368.

Gregory, P. H., 1973, *The Microbiology of the Atmosphere,* 2nd ed., Wiley, New York.

Griffiths, D. J., 1950, The liability of seed crops of perennial ryegrass (*Lolium perenne*) to contamination by wind-borne pollen, *J. Agr. Sci.* **40**:19–38.

Haberlandt, G., 1914, *Physiological Plant Anatomy,* Macmillan, London.

Haga, T., 1969, Structure and dynamics of natural populations of a diploid *Trillium,* in: *Chromosomes Today,* Vol. 2 (C. D. Darlington and K. R. Lewis, eds.), pp. 207–217, Oliver and Boyd, London.

Hamrick, J. L., and Allard, R. W., 1972, Microgeographical variation in allozyme frequencies *Avena barbata, Proc. Natl. Acad. Sci.* **69**:2100–2104.

Harlan, J. R., and deWet, J. M. J., 1963, The compilospecies concept, *Evolution* **17**:497–501.

Harper, J. L., Lovell, P. H., and Moore, K. G., 1970, The shapes and sizes of seeds, *Ann. Rev. Ecol. Syst.* **1**:327–356.

Heinrich, B., and Raven, P. H., 1972, Energetics and pollination ecology, *Science* **176**:597–602.

Heiser, C. B., 1949, Natural hybridization with particular reference to introgression, *Bot. Rev.* **15**:645–687.

Heiser, C. B., 1973, Introgression reexamined, *Bot. Rev.* **39**:347–366.

Hill, W. G., 1972, Effective size of populations with overlapping generations, *Theoret. Pop. Biol.* **3**:278–289.

Hillel, J., Feldman, M. W., and Simchen, G., 1973, Mating systems and population structure in two closely related species of the wheat group. I. Variation between and within populations, *Heredity* **30**:141–167.

Hodgson, H. J., 1949, Flowering habits and pollen dispersal in Pensacola Bahia grass, *Paspalum notatum* Flugge, *J. Am. Soc. Agron.* **41**:337–343.

Hofmann, J. V., 1911, Natural reproduction from seed stored in the forest floor, *J. Agr. Res.* **11**:1–26.

International Crop Improvement Association, 1963, Minimum Seed Certification Standards, Publication 20.

Imam, A. G., and Allard, R. W., 1965, Population studies in predominantly self-pollinated species. VI. Genetic variability between and within natural populations of wild oats from differing habitats in California, *Genetics* **51**:49–62.

Isaac, L. A., 1930, Seed flight in the Douglas-fir region, *J. Forestry* **28**:492–499.

Jain, S. K., 1973, Population structure and the effects of the breeding system, in: *Plant Genetic Resources: Today and Tommorow* (FAO/IBP Conference, O. H. Frankel and J. G. Hawkes, eds.).

Jain, S. K., and Bradshaw, A. D., 1966, Evolutionary divergence among adjacent plant populations. I. Evidence and its theoretical analysis, *Heredity* **21**:407–441.

Jain, S. K., and Marshall, D. R., 1967, Population studies in predominantly self-pollinating species. X. Variation in natural populations of *Avena fatua* and *A. barbata, Am. Naturalist* **101**:19–33.

Jain, S. K., and Marshall, D. R., 1968, Simulation of models involving mixed selfing and random mating. I. Stochastic variation in outcrossing and selection parameters, *Heredity* **23**:411–432.

Janzen, D. H., 1969, Seed eater versus seed size, number, toxicity and dispersal, *Evolution* **32**:1–27.

Janzen, D. H., 1970, Herbivores and the number of tree species in tropical forests, *Am. Naturalist* **104**:501–528.

Janzen, D. H., 1971*a*, Euglossine bees as long-distance pollinators of tropical plants, *Science* **171**:203–205.

Janzen, D. H., 1971*b*, Seed predation by animals, *Ann. Rev. Ecol. Syst.* **2**:465–492.

Janzen, D. H., 1972, Escape in space by *Sterculia apatala* seeds from the bug *Dysdercus fasciatus* in a Costa Rican deciduous forest, *Ecology* **53**:360–361.

Jemison, G. M., and Korstian, C. F., 1944, Loblolly pine seed production and dispersal, *J. Forestry* **42**:734–741.

Jensen, I., and Bogh, H., 1941, On conditions influencing the danger of crossing in the case of wind pollinated cultivated plants, *Tidsskr. Planteavl* **46**:238–266.

Johnson, C. G., 1969, *Migration and Dispersal of Insects by Flight*, 763 pp., Methuen, London.

Jones, E. W., 1956, Ecological studies on the rain-forest of southern Nigeria. II, *J. Ecol.* **44**:83–117.

Jones, M. D., and Brooks, J. S., 1950, Effectiveness of Distance and Border Rows in Preventing Outcrossing in Corn, Oklahoma Agricultural Experiment Station Technical Bulletin No. T-38.

Jones, M. D., and Brooks, J. S., 1952, Effect of Tree Barriers on Outcrossing in Corn, Oklahoma Agricultural Experiment Station Bulletin No. T-45.

Jones, M. D., and Newell, L. C., 1946, Pollination Cycles and Pollen Dispersal in Relation to Grass Improvement, Nebraska Agricultural Research Bulletin No. 148.

Jones, M. E., 1971, The population genetics of *Arabidopsis thaliana*. I. Breeding system, *Heredity* **27**:39–50.

Kaplan, S. M., 1970, Aspects of the reproductive biology of *Thalictrum*, Ph.D. thesis, University of Massachusetts, Amherst.

Keay, R. W., J., 1957, Wind-dispersed species in a Nigerian forest, *J. Ecol.* **45**:471–478.

Keller, E. C., Mattoni, R. H. T., and Seiger, M. S. B., 1966, Preferential return of artificially displaced butterflies, *Anim. Behav.* **14**:197–200.

Kerner, M. A., 1898, *Pflanzenleben*, Vol. II, 2nd ed., Leipzig, Wien.

Kernick, M. D., 1961, Seed production of specific crops, in: *Agricultural and Horticultural Seeds*, pp. 181–547, FAO Agricultural Studies No. 55.

Kerster, H. W., and Levin, D. A., 1968, Neighborhood size in *Lithospermum caroliniense*, *Genetics* **60**:577–587.

Kiang, Y. T., 1972, Pollination study in a natural population of *Mimulus guttatus*, *Evolution* **26**:308–320.

Kimura, M., 1953, Stepping stone model of population, *Ann. Rep. Natl. Inst. Genet.* **3**:63–65.

Kimura, M., and Crow, J. F., 1963, The measurement of effective population number, *Evolution* **17**:279–288.

Kimura, M., and Maruyama, T., 1971, Pattern of neutral polymorphism in a geographically structured population, *Genet. Res.* **18**:125–133.

Kimura, M., and Weiss, G. H., 1964, The stepping stone model of population structure and the decrease of genetic correlation with distance, *Genetics* **49**:561–576.

Knowles, P. F., 1943, Improving annual bromegrass, *Bromus mollis* L., for range purposes, *J. Am. Soc. Agron.* **35**:584–594.

Knowles, R. P., 1969, Nonrandom pollination in polycrosses of smooth bromegrass, *Bromus inermis* Leyss., *Crop Sci.* **9**:58–61.

Knowles, R. P., and Baenziger, H., 1962, Fertility indices in cross-pollinated grasses, *Canad. J. Plant Sci.* **42**:460–471.

Knowles, R. P., and Ghosh, A. W., 1968, Isolation requirements for smooth bromegrass, *Bromus inermis*, as determined by a genetic marker, *Crop Sci.* **3**:371–374.

Levin, D. A., 1968, The breeding system of *Lithospermum caroliniense:* Adaptation and counteradaptation, *Am. Naturalist* **102**:427–441.

Levin, D. A., 1972*a*, Interspecific pollen exchange as a function of species proximity in *Phlox*, *Evolution* **26**:251–258.

Levin, D. A., 1972*b*, Plant density, cleistogamy, and self-fertilization in natural populations of *Lithospermum caroliniense, Am. J. Bot.* **59**:71–77.

Levin, D. A., 1974, The oil content of seeds: An ecological perspective, *Am. Naturalist* **108**:193–206.

Levin, D. A., and Berube, D., 1972, *Phlox* and *Colias:* The efficiency of a pollination system, *Evolution* **26**:242–250.

Levin, D. A., and Kerster, H. W., 1967, An analysis of interspecific pollen exchange in *Phlox, Am. Naturalist* **101**:387–400.

Levin, D. A., and Kerster, H. W., 1968, Local gene dispersal in *Phlox pilosa, Evolution* **22**:130–139.

Levin, D. A., and Kerster, H. W., 1969*a*, The dependence of bee-mediated pollen dispersal on plant density, *Evolution* **23**:560–571.

Levin, D. A., and Kerster, H. W., 1969*b*, Density-dependent gene dispersal in *Liatris, Am. Naturalist* **103**:61–74.

Levin, D. A., and Kerster, H. W., 1971, Neighborhood structure under diverse reproductive methods in plants, *Am. Naturalist* **104**:345–354.

Levins, R., 1968, *Evolution in Changing Environments,* Princeton University Press, Princeton, N.J.

Lewis, D., 1942, The evolution of sex in flowering plants, *Biol. Rev.* **17**:46–67.

Lewontin, R. C., 1967, Population genetics, *Ann. Rev. Genet.* **1**:37–70.

Linhart, Y. B., 1973, Ecological and behavioral determinants of pollen dispersal in hummingbird-pollinated *Heliconia, Am. Naturalist* **107**:511–523.

MacArthur, R. H., 1972, *Geographical Ecology,* Harper and Row, New York.

MacArthur, R. H., and Wilson, E. O., 1967, *The Theory of Island Biogeography,* Princeton University Press, Princeton, N.J.

MacDaniels, L. H., 1931, Further experience with the pollination problem, *Proc. N.Y. State Hort. Soc.* **76**:32–37.

Maruyama, T., 1969, Genetic correlation in the stepping stone model with non-symmetrical migration rates, *J. Appl. Prob.* **6**:463–477.

Maruyama, T., 1970*a*, On the rate of decrease of heterogeneity in circular stepping stone models of populations, *Theoret. Pop. Biol.* **1**:101–119.

Maruyama, T., 1970*b*, Rate of decrease of genetic variability in a subdivided population, *Biometrika* **57**:299–311.

Maruyama, T., 1970*c*, Effective number of alleles in a subdivided population, *Theoret. Pop. Biol.* **1**:273–306.

Maruyama, T., 1971, Analysis of population structure. II. Two-dimensional stepping stone models of finite length and other geographically structured populations, *Ann. Hum. Genet.* **35**:179–196.

Maruyama, T., 1972, Rate of decrease of genetic variability in a two-dimensional continuous population of finite size, *Genetics* **70**:639–651.

Mattoni, R. H. T., and Seiger, M. S. B., 1963, Techniques in the study of population structure in *Philotes sonorensis, J. Res. Lepid.* **1**:237–244.

Maynard Smith, J., 1970*a*, Population size, polymorphism, and the rate of non-Darwinian evolution, *Am. Naturalist* **104**:231–237.

Maynard Smith, J., 1970*b*, Genetic polymorphism in a varied environment, *Am. Naturalist* **104**:487–490.

McCubbin, W. A., 1944, Relation of spore dimensions to their rate of fall, *Phytopathology* **34**:230–234.

McKey, D., 1974, The ecology of coevolved seed dispersal systems, in: *Plant–Animal Coevolution* (P. H. Raven and L. E. Gilbert, eds.), University of Texas Press, Austin.

McNab, B. K., 1963, Bioenergetics and the determination of home range size, *Am. Naturalist* **93**:133–140.

McNeilly, T., 1968, Evolution in closely adjacent plant populations. III. *Agrostis tenuis* on a small copper mine, *Heredity* **23**:99–108.
Medway, L., 1972, Phenology of a tropical rain forest in Malaya, *Biol. J. Linn. Soc.* **4**:117–146.
Minderhoud, A., 1931, Untersuchungen über das Betragen der Honigbiene als Blütenbestauberin, *Gartenbauwissenshaft* **4**:342–362.
Morley, B. D., 1972, The distribution and variation of some gesneriads on Caribbean Islands, in: *Taxonomy, Phytogeography and Evolution* (D. H. Valentine, ed.), pp. 239–257, Academic Press, New York.
Müller, H., 1882, Versuche über die Farbenliebhabeire der Honigbiene, *Kosmos* **12**:273–299.
Müller, P., 1955, Verbreitungs, biologie der Blutenpflanzen, *Verh. Geobot. Inst. Zurich*, Publication 30.
Neel, J., and Ward, R. H., 1972, The genetic structure of a tribal population, the Yanomama Indians, *Ann. Hum. Genet.* **36**:255–279.
Nei, M., 1972, Genetic distance between populations, *Am. Naturalist* **106**:283–292.
Nei, M., and Imaizumi, Y., 1966*a*, Genetic structure of human populations. I. Local differentiation of blood group gene frequencies in Japan, *Heredity* **21**:9–35.
Nei, M., and Imaizumi, Y., 1966*b*, Genetic structure of human populations. II. Differentiation of blood group gene frequencies among isolated populations, *Heredity* **21**:183–190.
Nieuwhof, M., 1963, Pollination and contamination of *Brassica oleracea* L., *Euphytica* **12**:17–26.
Norberg, R. A., 1973, Autorotation, self-stability, and structure of single-winged fruits and seeds (samaras) with comparative remarks on animal flight, *Biol. Rev.* **48**:561–596.
Parsons, P. A., 1958, Evolution of sex in the flowering plants of Australia, *Nature* **181**:1673–1674.
Paterniani, E., and Short, A. C., 1974, Effective maize pollen dispersal in the field, *Euphytica* **23**:129–134.
Pedersen, N. W., Hurst, R. L., Levin, M. D., and Stoker, G. L., 1969, Computer analysis of the genetic contamination of alfalfa seed, *Crop. Sci.* **9**:1–4.
Pedersen, P. N., Johansen, H. B., and Jorgensen, J., 1961*a*, Pollen spreading in diploid and tetraploid rye. I. Importance of pollen quantity and pollen distribution for the percentage of seed setting in the ears, *Royal Vet. Agr. Coll. Ann. Yearbook*, pp. 54–67.
Pedersen, P. N., Johansen, H. B., and Jorgensen, J., 1961*b*, Pollen spreading in diploid and tetraploid rye. II. Distance of pollen spreading and risk of intercrossing, *Royal Vet. Agr. Coll. Ann. Yearbook*, pp. 68–86.
Pianka, E. R., 1970, On r- and K-selection, *Am. Naturalist* **104**:592:597.
Pitelka, F. A., 1942, Territoriality and related problems in North American hummingbirds, *Condor* **44**:189–204.
Plummer, G. L., and Keever, C., 1963, Autumnal daylight weather and camphor-weed dispersal in the Georgia Piedmont region, *Bot. Gaz.* **124**:283.
Pope, O. A., Simpson, D. M., and Duncan, E. N., 1944, Effect of corn barriers on natural crossing in cotton, *J. Agr. Res.* **68**:347–361.
Raynor, G. S., 1967, Effects of a forest on particulate dispersion, in: *USAEC Meteorological Information Meeting Proceedings* (C. A. Mawson, ed.), Chalk River Nuclear Laboraties, Ontario.
Raynor, G. S., and Ogden, E. C., 1965, Twenty-four Hour Dispersion of Ragweed Pollen from a Known Source, Brookhaven National Laboratory Bulletin BNL 957 (T398).
Raynor, G. S., Cohen, L. A., Hayes, J. V., and Ogden, E. C., 1966, Dyed pollen grains and spores as tracers in dispersion and deposition studies, *J. Appl. Meteorol.* **5**:728–729.
Raynor, G. S., Ogden, E. C., and Hayes, J. V., 1970*a*, Dispersion and deposition of ragweed pollen from experimental sources, *J. Appl. Meteorol.* **9**:885–895.
Raynor, G. S., Hayes, J. V., and Ogden, E. C., 1970*b*, Experimental Data on Dispersion and

Deposition of Timothy and Corn Pollen from Known Sources, Brookhaven National Laboratory Bulletin BNL 50266.
Ribbands, C. R., 1949, The foraging method of individual honeybees, *J. Anim. Ecol.* **18**:47–66.
Ribbands, C. R., 1955, The scent perception of the honeybee, *Proc. Roy. Soc. Lond. (B)* **143**:367–379.
Rick, C. M., 1947, The effect of planting design upon the amount of seed produced by male sterile tomato plants as a result of natural cross-pollination, *Proc. Am. Soc. Hort. Sci.* **50**:273–284.
Ridley, H. N., 1930, *The Dispersal and Plants Throughout the World,* Reeve, Ashford.
Roberts, D., 1956, Sugar sprays aid fertilization of plums by bees, *N.Z. J. Agr.* **93**:206–211.
Roe, A. L., 1967, Seed dispersal in a bumper spruce seed year, U.S. Forest Service Intermountain Forest and Range Experiment Station Research Paper INT-39.
Rohlf, F. J., and Schnell, G. D., 1971, An investigation of the isolation by distance model, *Am. Naturalist* **105**:295–324.
Ronco, F., 1970, Englemann spruce seed dispersal and seedling establishment in clearcut forest openings in Colorado, USDA Forest Service Research Note RM-168.
Salisbury, E. J., 1942, *The Reproductive Capacity of Plants,* Bell, London.
Salisbury, E. J., 1961, *Weeds and Aliens,* Collins, London.
Schaal, B. A., 1974, Population structure and balancing selection in *Liatris cylindracea,* Doctoral dissertation, Yale University, New Haven.
Schlising, R. A., and Turpin, R. A., 1971, Hummingbird dispersal of *Delphinium cardinale* pollen treated with radioactive iodine, *Am. J. Bot.* **58**:401–406.
Schmidt, W., 1918, Die Verbreitung von Samen and Blutenstaub durch die Luftbewegung, *Oesterr. Bot. Z.* **67**:313–328.
Schoener, T. W., 1968, Sizes of feeding territories among birds, *Ecology* **49**:123–141.
Scott, R. K., and Longden, P. C., 1970, Pollen release by diploid and tetraploid sugar-beet plants, *Ann. Appl. Biol.* **66**:129–135.
Shapiro, A. M., 1970, The role of sexual behavior in density-related dispersal of pierid butterflies, *Am. Naturalist* **104**:367–372.
Sheldon, J. C., and Burrows, F. M., 1973, The dispersal effectiveness of the achene pappus units of selected Compositae in steady winds with convection, *New Phytol.* **72**:665–675.
Simpson, D. M., 1954, Natural cross-pollination in cotton, U.S.D.A. Technical Bulletin No. 1094.
Simpson, D. M., and Duncan, E. N., 1956, Cotton pollen dispersal by insects, *Agron. J.* **48**:305–308.
Sindu, A. S., and Singh, S., 1961, Studies on the agents of cross pollination of cotton, *Indian Cotton Grow. Rev.* **15**:341–353.
Singh, S., 1950, Behavior studies on honeybees in gathering nectar and pollen, Memoirs of the Cornell Agricultural Experiment Station, No. 288.
Smythe, N., 1970, Relationships between fruiting seasons and seed dispersal methods in a neotropical forest, *Amer. Natur.* **104**:25–35.
Snaydon, R. W., 1970, Rapid population differentiation in a mosiac environment. I. The response of *Anthoxanthum odoratum* populations to soils, *Evolution* **24**:257–269.
Snogerup, S., 1967, Studies in the Aegean flora. IX. *Erysimum* sect. Cheiranthus. B. Variation and evolution in small population systems, *Opera Bot.,* No. 14.
Snow, D. W., 1965, A possible selective factor in the evolution of fruiting seasons in tropical forests, *Oikos* **15**:274–281.
Southwood, T. R. E., 1962, Migration of terrestrial arthropods in relation to habitat, *Biol. Rev.* **37**:171–214.

Stebbins, G. L., 1950, *Variation and Evolution in Plants,* Columbia University Press, New York.

Stebbins, G. L., 1957, Self-fertilization and population variability in the higher plants, *Am. Naturalist* **91**:337–354.

Stebbins, G. L., 1959, The role of hybridization in evolution, *Proc. Am. Phil. Soc.* **103**:231–251.

Stebbins, G. L., 1970*a*, Variation and evolution in plants: *Progress during the past twenty years, in: Essays in Evolutionary Genetics* (M. K. Hecht and W. C. Steere, eds.), pp. 173–208, Appleton-Century-Crofts, New York.

Stebbins, G. L., 1970*b*, Adaptive radiation of reproductive characteristics in angiosperms. I. Pollination mechanisms, *Ann. Rev. Ecol. Syst.* **1**:307–326.

Stebbins, G. L., 1971, Adaptive radiation of reproductive characteristics in angiosperms. II. Seeds and seedlings, *Ann. Rev. Ecol. Syst.* **2**:237–260.

Stephen, W. P., 1958, Pear Pollination Studies in Oregon, Technical Bulletin of the Oregon Agricultural Experiment Station No. 43.

Stephens, S. G., and Finkner, M. D., 1953, Natural crossing in cotton, *Econ. Bot.* **7**:257–269.

Stiles, F. G., 1972, Time, energy, and territoriality in the Anna hummingbird (*Calypte anna*), *Science* **173**:818–820.

Strand, L., 1957, Pollen dispersal, *Silvae Genet.* **6**:129–136.

Strid, A., 1970, Studies in the Aegean flora. XVI. Biosystematics of the *Nigella arvensis* complex, *Opera Bot.,* No. 28.

Sutton, O. G., 1932, A theory of eddy diffusion in the atmosphere, *Proc. Roy. Soc. Lond. Ser. A* **135**:143–165.

Sutton, O. G., 1947, The theoretical distribution of airborne pollution from factory chimneys, *Quart. J. Roy. Meteor. Soc.* **73**:426–436.

Tauber, H., 1965, Differential pollen dispersion and the interpretation of pollen diagrams, *Geol. Surv. Denmark Ser. II,* No. 89.

Tauber, H., 1967, Differential pollen dispersion and filtration, in: *Quaternary Paleoecology* (E. J. Cushing and H. E. Wright, eds.), pp. 131–141, Yale University Press, New Haven.

Thies, S. A., 1953, Agents concerned with natural crossing of cotton, *Agron. J.* **45**:481–484.

Tigerstedt, P. M. A., 1973, Studies on isozyme variation in marginal and central populations of *Picea abies, Hereditas* **75**:47–60.

Tracy, W. W., 1910, The Production of Vegetable Seeds; Sweet Corn and Garden Peas and Beans, U.S.D.A. Bulletin 184.

Turner, F. B., Jennrich, R. I., and Weintraub, J. D., 1969, Home ranges and body sizes of lizard, *Ecology* **50**:1076–1081.

Turner, J. R. G., 1971, Experiments on the demography of tropical butterflies. II. Longevity and home-range behavior in *Heliconius erato, Biotropica* **3**:21–31.

van Der Meer, Q. P., and Van Bennekom, J. L., 1968, Research on pollen distribution in onion seed fields, *Euphytica* **17**:216–219.

van der Pijl, L., 1969, *Principles of Dispersal in Higher Plants,* Springer-Verlag, New York.

Vasek, F. C., 1967, Outcrossing in natural populations. III. The Deer Creek population of *Clarkia exilis, Evolution* **21**:241–248.

Wang, C. W., Perry, T. O., and Johnson, A. G., 1960, Pollen dispersion of slash pine (*Pinus elliottii* Engelm.) with special reference to seed orchard management, *Silvae Genet.* **9**:78–86.

Watson, P. J., 1970, Evolution in closely adjacent plant populations. VI. An entomophilous species, *Potentilla erecta,* in two contrasting habitats, *Heredity* **24**:407–422.

Weaver, N., 1957, The foraging behavior of honeybees on hairy vetch. II. The foraging area and foraging speed, *Insectes Sociaux* **4**:43–57.

Weil, J., and Allard, R. W., 1964, The mating system and genetic variability in natural populations of *Collinsia heterophylla, Evolution* **18:**515–525.

Whitehead, D. R., 1969, Wind pollination in the angiosperms: Evolutionary and environmental considerations, *Evolution* **23:**28–35.

Williams, N. H., and Dodson, C. H., 1972, Selective attraction of male euglossine bees to orchid floral fragrances and its importance in long distance pollen flow, *Evolution* **26:**84–95.

Williams, R. D., and Evans, G., 1935, The efficiency of spatial isolation in maintaining the purity of red clover, *Welsh J. Agr.* **11:**164–171.

Wilson, D. E., and Janzen, D. H., 1972, Predation on Scheelea palm seeds by brucid beetles: Seed density and distance from the parent palm, *Ecology* **53:**954–959.

Wit, F., 1952, The pollination of perennial ryegrass (*Lolium perenne* L.) in clonal plantations and polycross fields, *Euphytica* **1:**95–105.

Wolf, L. L., 1969, Female territoriality in a tropical hummingbird, *Auk* **86:**490–504.

Wolfenbarger, D. O., 1946, Dispersion of small organisms, *Am. Midl. Naturalist* **35:**1–152.

Wolfenbarger, D. O., 1959, Dispersion of small organisms, *Lloydia* **22:**1–105.

Workman, P. L., and Niswander, J. D., 1970, Population studies on southwestern Indian tribes, II. Local genetic differentiation in the Papago, *Am. J. Hum. Genet.* **22:**24–49.

Wright, J. W., 1952, Pollen dispersion of some forest trees, Northeast Forest Experiment Station Paper 46.

Wright, J. W., 1953, Pollen dispersion studies: Some practical applications, *J. Forestry* **51:**114–118.

Wright, S., 1931, Evolution in mendelian populations, *Genetics* **16:**97–159.

Wright, S., 1938, Size of population and breeding structure in relation to evolution, *Science* **87:**430–431.

Wright, S., 1940, Breeding structure of populations in relation to speciation, *Am. Naturalist* **74:**232–248.

Wright, S., 1943*a*, Isolation by distance, *Genetics* **28:**114–138.

Wright, S., 1943*b*, An analysis of local variability of flower color in *Linanthus parryae Genetics* **28:**139–156.

Wright, S., 1946, Isolation by distance under diverse systems of mating, *Genetics* **31:**39–59.

Wright, S., 1951, The genetic structure of populations, *Ann. Eugen.* **15:**323–354.

Wright, S., 1960, Physiological genetics, ecology of populations and natural selection, in: *Evolution After Darwin,* Vol. I (S. Tax, ed.), pp. 429–475, University of Chicago Press, Chicago.

Wright. S., 1965, The interpretation of population structure by *F*-statistics with special regard to systems of mating, *Evolution* **19:**395–420.

Wright, S., 1969, *Evolution and the Genetics of Populations,* Vol. II: *The Theory of Gene Frequencies,* University of Chicago Press, Chicago.

Yampolsky, C., and Yampolsky, H., 1922, Distribution of sex forms in the phanerogamic flora, *Bibl. Genet.* **3:**1–62.

Young, A. M., 1972*a*, The ecology and ethology of the tropical nymphaline butterfly, *Victorina epaphus*. I. Life cycle and natural history, *J. Lepid. Soc.* **26:**155–170.

Young, A. M., 1972*b*, Breeding success and survivorship in some tropical butterflies, *Oikos* **23:**318–326.

6

Artemia: A Survey of Its Significance in Genetic Problems

CLAUDIO BARIGOZZI
Institute of Genetics
University of Milan
Milan, Italy

INTRODUCTION

Artemia is a genus belonging to Branchiopoda Anostraca (Crustacea). Until recently this genus was considered to be comprised of a single species, *Artemia salina,* while now at least three specific entities are known. *Artemia,* commonly called the brine shrimp for its natural habitat, which is salt ponds (e.g., in African cases), salt lakes (e.g., the Great Salt Lake), and salterns of Europe, Asia, Africa, and America, is geographically widely spread, and lives gregariously. The brine shrimp can stand drought while in the embryo stage, protected by a shell—i.e., in the cyst stage, commonly referred to as the egg or permanent egg. Growing *Artemia* in the laboratory is now an easy task; it may be fed on yeast or, with much better results, on green unicellular algae, such as *Dunaliella* (Ballardin and Metalli, 1963). The life cycle (from the egg or the larva laid viviparously) includes, on the average, 4–5 weeks for the attainment of sexual maturity. The number of offspring produced by a single female, although varying widely, may reach 200–300. The biology of *Artemia* is characterized by two phenomena: parthenogenesis and polyploidy. The latter is especially interesting because it is rare among animals; both facts were established by the classic investigations of Brauer (1893) and of Artom (1906, 1907, 1911, 1912, 1921, 1924, 1931). The reason for investigating the genetics of *Artemia* is that the invertebrate species studies so far are mainly Insects (*Drosophila* and other Diptera, *Bombyx mori* and other Lepidoptera, *Habrobracon* and other Hymenoptera, *Tribolium* etc.), nearly all of which are holometabolic. *Artemia,* on the other hand, is characterized by a continuous development, from the first larval stage, called the nauplius, to the adult stage, passing through a series of larval stages and several molts without any dormancy interruption; furthermore, the adult grows indefinitely. The adult also goes though several molts; in particular, the female, molts after each egg laying. These features strongly differentiate the *Artemia* life cycle from those of the other invertebrates studied so far. It should also be recalled that the information we have about other Crustacea (e.g., *Asellus, Gammarus, Thysbe*) is restricted to some special genetic aspects, i.e., to cytogenetics or to sex determination or to population genetics. *Artemia,* on the other hand, is a good material from the point of view of nearly all branches of genetics. It is unquestionable that *Artemia* is a suitable material for chromosome studies, although the chromsome number is high (42 in the diploid form); nauplius cells supply very good somatic metaphases, while the oogenesis stage is exceptionally good for the study of meiotic divisions. Another favorable circumstance is that the body of the brine shrimp is always transparent; therefore, many developmental features are easy to follow by mere inspection of the entire individual. Microscopic anatomy, as well as histology and somatic cytology, is also easy to analyze on microtome sections. Since *Artemia* is composed of a small number of organs of relatively simple constitution, and, on the other hand, cell differentiation is manifold as in insects (epithelial, nervous, muscle, and connective cells are clearly classifiable), developmental genetics on an anatomical basis can be studied much better than in any other arthropod investigated so far. The recent findings on spontaneous mutants in *Artemia,* including hemoglobin polymorphism, provide the possibility of studying gene manifestation in a species basically differing from all holometabolic insects and from vertebrates, as far as development is concerned.

The many aspects and possibilities of future work will be described and discussed critically in this chapter.

SPECIATION IN THE GENUS *ARTEMIA*

Artemia salina Leach, described first by Linné as *Cancer salinus,* was early recognized as comprised of bisexual and parthenogenetic individuals. Artom (1906, 1911) demonstrated that the bisexual shrimps are diploid (42 chromsomes) while the parthenogenetic ones are tetraploid (84 chromosomes). The diploid *Artemia* proved to have epithelial nuclei smaller than the tetraploid, hence for the former the name *univalens* was proposed and for the latter *bivalens* (Artom, 1911, 1912). This nomenclature (*A. salina univalens, A. salina bivalens*) was never used afterward, nor were *micropyrenic* and *macropyrenic* (with small nuclei and large nuclei), subsequently introduced by Artom (1921) to distinguish the two main types. The situation became more complicated when Artom (1931) discovered the parthenogenetic diploid form of Sète, later found by Gross (1932) and Barigozzi (1935*a,b*) in Margherita di Savoia (Apulia); Stefani (1960) later found another diploid and parthenogenetic form at Santa Gilla (Cagliari). *Artemia salina* seemed to Artom definitely composed of several entities, although no biological criterion could be applied to justify calling them separate species, since there is no possibility of crossing a parthenogenetic female with a male. Therefore, Artom (1931) spoke of a "*specie collettiva*" and called the single variants "biotypes." This word does not seem particularly adequate, since it is used in biology with different connotations. Barigozzi (1944) used the term "mutants" for the genetic variants to be included under the common name of *A. salina,* which, at that time, was still considered, in its bisexual form, as one single species, at least referring to the shrimps living in Europe. The same point of view (i.e., singleness of the species) was confirmed by Linder (1941), who recognized only one species in the genus, although in 1939 Kuenen had found that Californian *Artemia* cannot be crossed with Sardinian. Small morphological differences were described, and so it was proposed to subdivide the original *A. salina* species into two entities, *A. gracilis* (the Californian form) and *A. salina* (the European form). The name "*gracilis*" was one of the many synonymous terms for *A. salina* and was used as early as 1869 by Verril. Lockhead (1941) did not confirm the differences claimed, and the systematic singleness of the bisexual form was then reinforced. The chromosome number of the Californian *Artemia* was first determined by Barigozzi and Tosi (1959), who found the same number (42) as for the Cagliari form. However, sexual isolation was found by Gilchrist (1960) and by Bowen (1965). Gilchrist found sexual isolation between *Artemia* from California and *Artemia* from North Africa, and Bowen between the Californian and the Sardinian *Artemia.* Later on, a new case of sexual isolation was found regarding a form with 44 chromosomes (Halfer Cervini *et al.,* 1968; Piccinelli *et al.,* 1968; Piccinelli and Prosdocimi, 1968). From all these contributions, it is apparent that there exists a third form, sexually isolated from the other two, found so far in a saltern in Argentina (Salinas Grandes de Hidalgo) and, more rarely, near Cagliari (San Bartolomeo). This new form was studied morphologically and cytologically, and the development time was determined in comparison with the other *Artemia* also living in Cagliari (San Bartolomeo). The differences and the similarities found in the three "forms," which, being sexually fully isolated, deserve the rank of "species," confirm that the general aspect is much the same. The respective names and differences are indicated in Table I and Fig. 1. The genus *Artemia* is comprised of three sibling species, very similar to each other and partially

(*A. salina* and *A. persimilis*) sympatric. The specific name *franciscana* (which has not yet been proposed) seems suitable for the species found primarily at San Francisco; *americana* is inappropriate, since *A. persimilis* is also American. Since no systematic exploration has been made in the many biotopes where *Artemia* might be found, and also populational research has been so far restricted to very few localities, it can be expected that the number of cryptic species of the genus will increase. The parthenogenetic (bi-, tri-, tetra-, and pentaploid) forms require further comment. Definition in taxonomic terms is difficult; a reason for considering the parthenogenetic *Artemia* as belonging to a species different from the bisexual might be the fact that parthenogenetic females mate with males but produce only parthenogenetic offspring. Other cases are known (Coleoptera, Lepidoptera) where offspring between bisexual and parthenogenetic individuals are produced. If these cases are believed to indicate that both the bisexual and the parthenogenetic individuals belong to a single species, *Artemia* might be considered as an example of full specific discrimination between bisexual and parthenogenetic forms. However, it is too early to draw a definite conclusion.

TABLE I. Characteristics of *Artemia* Species

Chromosome number (2*n*)	Morphological characters	Development time (days)
A. salina		
42	1. Furca with two lobes and many setae 2. Ovisac laterally round 3. Second antennae of the male with a subconic knob 4. Penes without spines	14–35
A. persimilis		
44 (smaller than in the two other species)	1. Furca rudimentary with few state setae 2. Ovisac laterally pointed and with spines 3. Second antennae of the male with a subspheric knob 4. Penes with spines	7–22
A. franciscana		
42 (identical to *A. salina*)	Practically identical to *A. salina*	?

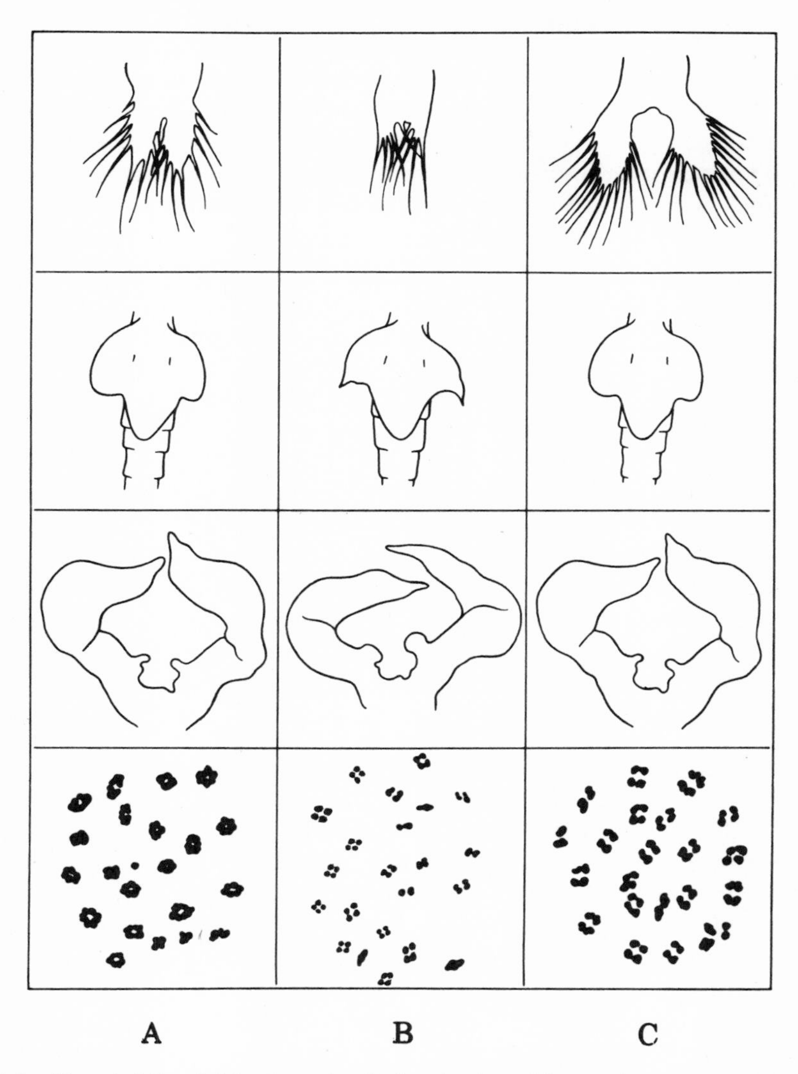

FIG. 1. Characteristics of *Artemia* species. Outline drawings of furca, ovisac antennae, and karyotypes of (A) *A. salina,* (B) *A. persimilis,* and (C) *A. franciscana.*

Mechanisms of Differentiation Between Infraspecific Variants

We have to distinguish between the derivation of parthenogenesis from bisexuality and the derivation of poly- and heteroploidy from diploidy. No case is known of production of parthenogenetic females by bisexual ones. An analysis of the origin of parthenogenesis is particularly difficult, because it is impossible to obtain bisexual offspring by mating a parthenogenetic female with a normal male; thus the origin of parthenogenesis remains obscure. We may better discuss the origin and the relationship between diploidy, polyploidy, and heteroploidy. An important fact is that the differences in meiosis between parthenogenetic and bisexual females are manifold; four are presently known, all of them restoring the basic chromosome number of the individual:

1. Parthenogenetic diploid females of Sète: According to Artom (1931) chromosome pairing, chiasmata formation, and extrusion of the first polar body with 21 dyads generally occur; alternatively, the second polar body is not produced, and the nucleus of the oocytes, still diploid, divides endomitotically, to give rise to nuclei with 42 chromosomes. Stefani (1967) tends to exclude the extrusion of the first polar body and claims that the extruded polar body is the second. However, rare cases of extrusion of the second polar body are known, followed by phases which seem to indicate a nuclear fusion (Artom, 1931).
2. Parthenogenetic diploid females of Sta Gilla: Two different mechanisms are known. After a normal chromosome pairing, the spindles assume a side-by-side and parallel position which gives rise to fusion between them. In this way, the only polar body extruded contains 42 chromosomes, while 42 remain in the germ vescicle. In a second mechanism, the polar body spindles do not fuse, but lead to the production of four haploid nuclei; two of them fuse (automixis), and thus the diploid number is restored (Stefani, 1960). The same mechanism, according to Stefani, is predominant at Sète (1967) (Fig. 2g).
3. Parthenogenetic tri- or tetraploids of various localities (e.g., Pirano, Istria in Yugoslavia): Meiosis is suppressed and only one mitosis takes place, which separates sister chromatids (Brauer, 1893, Artom, 1906; Barigozzi, 1944).
4. Tri- and pentaploid parthenogenetic females of the Dead Sea: A transient chromosome pairing occurs, followed by the extrusion of a single polar body (Goldschmidt, 1952). It is possible that this mechanism (which has not been sufficiently described as far as chromosome pairing is concerned) does not differ from (3) above (Figs. 2 and 3).

The facts indicated reveal a discrepancy in interpretation of the meiotic mechanism predominantly existing at Sète between Artom and Barigozzi on the one hand and Stefani on the other. Two comments seem justified. First, some figures in Artom's paper of 1931 (Figs. 10–14) and

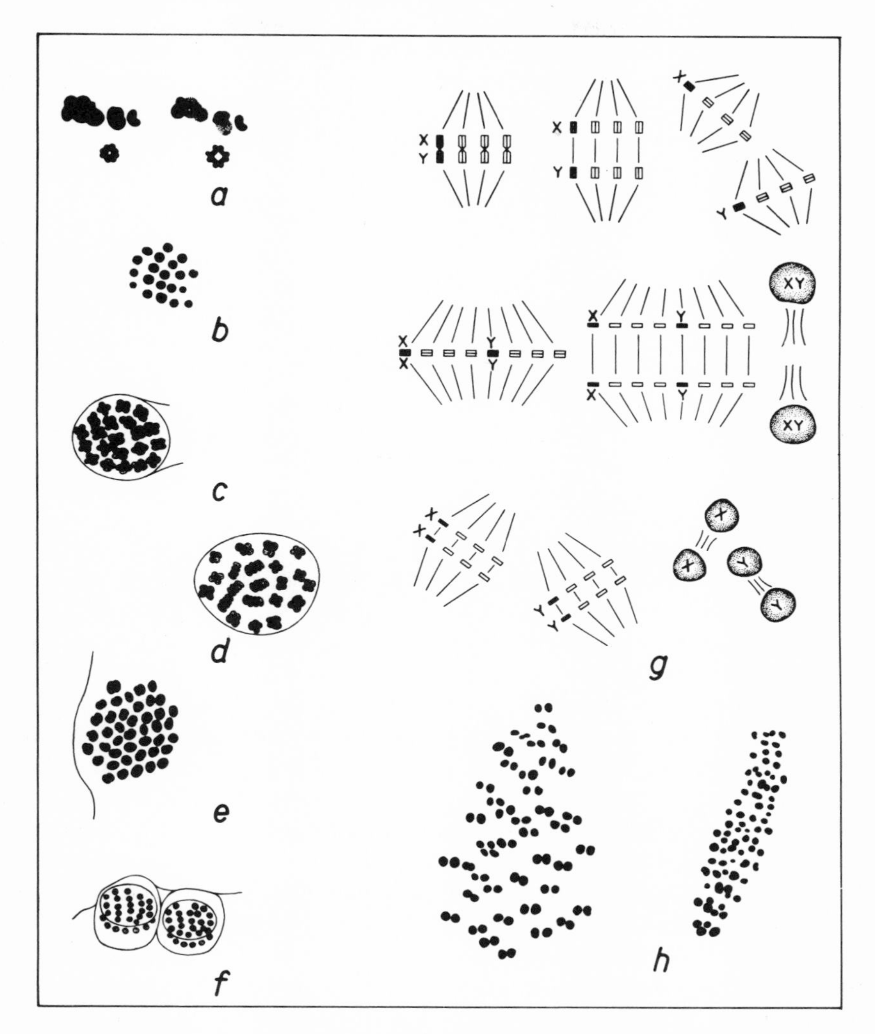

FIG. 2. Mechanisms of meiosis in intraspecific variants of *Artemia*. a: Evidence of the bivalents separating at the first division in *Artemia* from Sète (Barigozzi, 1944). b: Metaphase of the first division in *Artemia* from Sète showing 21 bivalents (Barigozzi, 1944). c: Metaphase of the first division in *Artemia* from Sète showing 21 bivalents (Artom, 1931). d: The same (Artom, 1931). e: Metaphase of oocyte showing 42 bodies, interpreted by Artom (1931) as dyads derived from deconjugation of the bivalents. f: Double metaphase plate, one belonging to the pronucleus and the other to the first polar body in an oocyte, proving the presence of two chromatids in both nuclei (Artom, 1931). g: Diagram to illustrate the two ways of maturation of the egg in the parthenogenetic *Artemia* of Sta Gilla or of Sète. Stefani emphasizes that the bivalent of heterochromosomes at the first division should lead to formation of eggs with X or with Y only (Stefani 1964). Postreduction of the heterochromosomes is a remedy to keep the heterochromosomes XY in each egg, i.e., to determine the feminine sex. h: 42 dyads in two oocytes of Sète (Stefani, 1967).

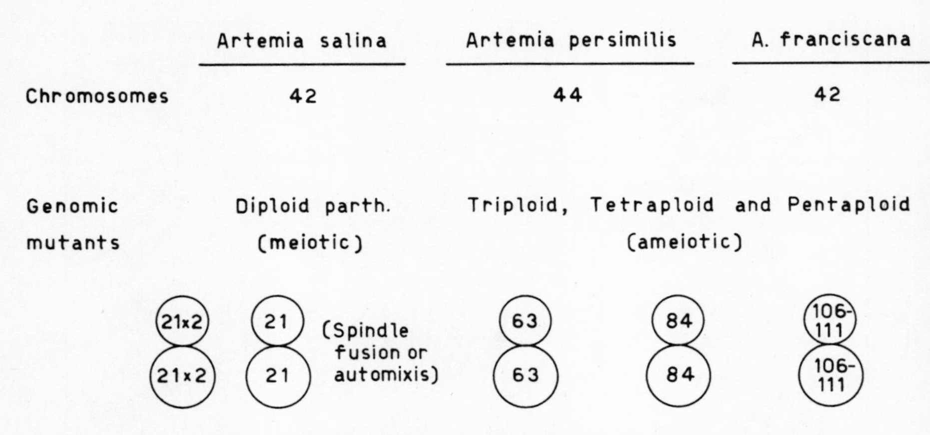

FIG. 3. Scheme of meiosis in parthenogenetic *Artemia*.

in Barigozzi's paper of 1944 (Fig. 2) give strong evidence in favor of Artom's view. Second, Stefani's evidence, although based on an improved technique, is restricted to a few figures, which again may favor one of the mechanisms described by Artom in Figs. 50–52 (see Fig. 2e,f) as a rare occurrence. In conclusion, it seems that between the two points of view there is a different appraisal of the frequency of two phenomena, both certainly occurring. Since 37 years elapsed between the time Artom examined his sample and Stefani examined his, it may be that the population has changed as regards the frequency of the different meiotic mechanisms present in it. In 1943, when Barigozzi studied the *Artemia* of Sète, the conditions were still as described by Artom. A finding related to the phenomenon of parthenogenesis (in general, indefinite and permanent) is the rare appearance of males derived from parthenogenetic females. This phenomenon has been observed several times and has sometimes been described (e.g., Ballardin and Metalli, 1963), but it has been carefully analyzed only by Stefani (1964) with reference to diploid *Artemia* from Sta Gilla. This author advances the highly probable hypothesis that males, which are XX (see p. 233), derive from automixis between two haploid nuclei, each carrying an X chromosome.

Mechanisms of Polyploidization and Heteroploidization

Several authors have tried to explain the polyploidization processes in *Artemia*. The first was Gross (1932), who found in the population of Margherita di Savoia diploid and polyploid parthenogenetic individuals together, some of which showed abnormalities in the meiotic divisions, where the polar body, instead of being extruded, remained in the egg yolk near the pronucleus. If this is considered a preparation for a later nuclear fusion, it might be evidence of polyploidization in progress. Experiments with low temperature (4°C) gave a small number of oocytes showing the same phenomena. Gross's data are not fully convincing as a demonstration of how polyploidization is brought about in nature; however, they may throw light on how the step from diploidy to tetraploidy is performed in individual cases. Similar observations (i.e., automixis) have been made by Artom (1931) on the diploid parthenogenetic *Artemia* of Sète and by Barigozzi (1935*a,b*), again on material from Margherita di Savoia, in both diploid and tetraploid oocytes. To my knowledge, the investigator most recently dealing with the subject (without, however, providing a different point of view) is Goldschmidt (1952), who noticed several abnormalities during the meiosis of tri- and pentaploid *Artemia.* Heteroploidy, first recognized by Barigozzi (1944), was also found by Goldschmidt in her material. The most obvious mechanism seems to be the misdistribution of chromosomes at mitosis, although Goldschmidt finds this explanation unsatisfactory. An alternative mechanism might be the repeated loss of chromosomes in a polyploid set.

Phylogeny of the Polyploid and the Heteroploid Mutants

The first attempt to construct a phylogenetic tree of the *Artemia* forms was made by Artom (1931), who assumed that the diploid bisexual condition was the original one, from which branched off the diploid parthenogenetic form on the one hand and a tetraploid bisexual form on the other, the existence of which was never definitely proved. Additionally, Artom was impressed by some peculiarities revealed by a few males from Odessa, which led him to formulate the hypothesis of their tetraploidy (1924), although the chromosome countings were doubtful. Barigozzi and Tosi (1957, 1959) found some evidence of tetraploidy in the bisexual population of the Great Salt Lake (Utah), but these conclusions too await confirmation. The second evolutionary step, according to Artom, should have been the acquisition of ameiotic tetraploidy from the diploid parthenogenetic form through a hypothetical stage which should have been parthenogenetic and tetraploid, but still characterized by some meiotic stages, as in the diploid *Artemia* of Sète. New data have been provided by Barigozzi (1944), including the heteroploid and the triploid forms and a hypothetical octaploid, without changing the basic ideas expressed by Artom. The rare triploid condition was thought to have been derived from the diploid, leaving aside the fact that the diploid parthenogenetic is meiotic and the triploid is not. The discovery by Goldschmidt (1952) of a large population of tri- and pentaploid shrimps poses the question of the origin of the odd-number ploidy. This is a problem because it is necessary to postulate an independent derivation from the tetraploid ameiotic form. Since this form extrudes only one polar body with 84 chromosomes, no abnormality can be conceived capable of raising the number up to the 106–111 found in the pentaploid strain. It is therefore necessary to postulate a meiotic tetraploid strain, which has never been found. Easier to explain is the origin of the triploid form, which could be an automixis between the pronucleus (42) and a reduced polar body (21) during a meiosis of Sète type. But after this a second step is required to explain the actual conditions of the triploid observed by Barigozzi, i.e., the loss of chromosome pairing. The tetraploid nature of the ameiotic form was, furthermore, a cause of controversy between Artom (1931), Stella (1933), and Barigozzi (1935*a*, 1944) on one side and Hertwig (1931) and Gross (1932) on the other. The discrepancy related to the structure of the 84 chromatic bodies, which for Artom, Stella, and Barigozzi are simple chromosomes but for Hertwig and Gross are double chromosomes. The findings of Barigozzi (1944) are based on a series of observations of all phases of the only mitosis which occurs during the egg maturation of *Artemia* with 84 chromosomes, which failed to show any trace of chromosome pairing (Barigozzi, 1935*a*). On the other hand, the illustrations given by Gross are not entirely clear, and the only fact presented by Hertwig is that the chromosomes of the 84-chromosome form are larger than those of the 42-chromosome diploid form. It is difficult to conceive the existence of double chromosomes on the basis of the observations now available. More recent investigations (unpublished data) on the parthenogenetic shrimps with 84 chromosomes living in Comacchio (Ferrara) confirm the simple structure of the 84 elements. In conclusion, the 84-chromosome *Artemia* must be considered as generally tetraploid, and only exceptionally octaploid. Heteroploidy does not pose any problem, and can be thought of as a branching-off from the bisexual form. However, a presently insoluble difficulty exists: the question of which of the three species of *Artemia* are more likely to give rise to both parthenogenetics and polyploids. No answer can be proposed, because even *A. persimilis* could be taken into consideration; a one-pair difference in chromosome number is not sufficient to exclude its part in producing tetraploid parthenogenetic individuals, with the additional loss of a pair which exceeds diploidy. The final conclusion is that, except for the derivation of hetero- and polyploidy from diploidy, no precise link can be demonstrated between the three species of *Artemia* and the parthenogenetic forms. The parthenogenetic forms could perhaps be grouped under the name *Artemia parthenogenetica,* but the heterogeneity of the chromosome number makes the consistency of this "species" rather illusory.

Geographic Distribution

It is impossible to draw a map of the geographic distribution of the different species and mutants of *Artemia* because of the lack of systematic exploration. Since the accounts given by Stella (1933) and Barigozzi (1957) little progress has been made, except as regards the sibling species. It is noteworthy that *A. salina* and *A. persimilis* are sympatric at Hidalgo (Mexico) and at San Bartolomeo (Cagliari), although in different percentages. Another strange fact is the colonization of the Sta Gilla salterns (which are only a few miles from San Bartolomeo) by a diploid parthenogenetic strain (Stefani, 1960), totally absent at San Bartolomeo, where the shrimps are 100% bisexual. The general impression is that no understandable correlation exists between a given mutant genome or reproduction condition and geographic location. In summary, one may conclude that *A. salina* is found in several European localities and in Mexico in at least one locality (Hidalgo), *A. franciscana* in California, and *A. persimilis* in Italy and in Mexico in at least one locality. Until a few years ago, each population of *Artemia* was considered as homogeneous, with the exception of those of Margherita di Savoia, California (owing to heteroploidy), and, perhaps, Odessa, What was lacking, until recently, was evidence of genic polymorphism. We now have information of the existence of at least one case, that of hemoglobins. Another, still too little known, regards the putative presence of a high frequency of *curved* in the population of Quemado, Mexico (Bowen, 1964). The first data on hemoglobin proteins were provided by Dutrieu (1960), followed by more extensive studies by Bowen *et al.* (1969). Data as yet unpublished have also been collected by Bianchi and Piccinelli. The different hemoglobin types found so far are not fully comparable; thus it is better to keep separate the data on the Californian *Artemia* (Dutrieu), those on the nine populations, both bisexual and parthenogenetic, analyzed by Bowen *et al.* (which include San Bartolomeo, Cagliari), and those on *Artemia* from San Bartolomeo obtained by Bianchi and Piccinelli. However since polymorphism was found in the San Bartolomeo population by Bowen *et al.* and also Bianchi and Piccinelli, their data may be comparable. Since the frequency of the three hemoglobins Hb-1, Hb-2, and Hb-3 (plus a possible fourth) varies from population to population, according to Bowen *et al.*, one may speak of a typical polymorphism. Hb-1 seems to be less common than Hb-2 and Hb-3.

Phenotypic Variation and Salinity

It is a classic notion that *Artemia* reacts to the degree of salinity (water density) in which it lives by morphological transformations of parts of the body, and particularly of the furca, which becomes reduced when salinity is high. The old contributions by Artom (1905, 1906, 1907) definitely proved that salinity exerts purely phenotypic influences only during development. Gross (1932) later demonstrated a correlation between chromosome number and response to salinity. His conclusion can be accepted only up to a point. The alleged presence of a large number of octaploid individuals in his material (as was previously shown) is doubtful; however, the difference of reaction between diploids and polyploids is acceptable. More recently, similar experiments have been carried out with the three species (*salina, persimilis,* and *franciscana*), and the following results have been obtained, as expected: when salinity is increased, the lobes of the furca diminish as well as the mean number of setae in both *A. salina* and *A. persimilis;* however, the specific differences remain recognizable (Halfer Cervini *et al.,* 1967). Salinity (i.e., water density) is the only environmental factor influencing the individual morphology of *Artemia* investigated so far. It would be interesting to investigate other factors, such as temperature and concentration of various ions.

TRANSMISSION MECHANISMS

Generalities on the Chromosomes of *Artemia*

The chromosomes of *Artemia* can be studied in developing eggs, in the still largely undifferentiated dividing cells of the nauplius, and in the maturing germ cells. The techniques used for germ cells (especially oocytes) were first the classic one based on microtome sections and more recently a new one devised by Stefani (1963*a*) based on gentle pressing of fixed and stained oocytes. Whole nauplii can be squashed (Barigozzi, 1942). All the results obtained using the different techniques agree in that they reveal that the chromosomes are small (average 2 μm in the nauplius cells) and very difficult to classify into an idiogram, even using Stefani's technique, which is by far the most useful. Only two pairs are distinguishable during prophase for carrying a nucleolus organizer; only one nucleolus results owing to the fusion of the product of the four elements (Barigozzi, 1942; Stefani, 1963*a*). The problem of the sex chromosomes is discussed later. An important question, as Stefani pointed out (1963*a*), is whether the chromosomes possess a localized or a diffuse centromere. From the cleavage divisions of embryos belonging to the *Artemia* of Sta Gilla (Cagliari), Stefani was able to obtain some convincing pictures, especially from prophases, revealing the outline assumed by chromosomes when the centromere is not localized. However, the data gathered by Barigozzi (1942) from larval cells seem to indicate the existence of a median constriction at least in some chromosomes; even more conflicting with Stefani's evidence is the shape revealed by bivalents during oogenesis seen from side view (Fig. 2a, Barigozzi, 1944). Stefani and Cadeddu (1968) have tried to assess whether fragments of broken chromosomes after X-ray treatment are transmissible. Transmission through mitotic cycles should prove that any piece of broken chromosome is capable of behaving as though it possessed a centromere. The observations seem to be in favor of this view, although is extremely difficult to prove that the transmitted small chromosomes lack the middle portion, which should correspond to the centromere. For the moment, it is impossible to draw any definite conclusion, although the findings by Stefani and Cadeddu carry considerable weight.

Mechanism of Sex Determination

Bowen (1963) and Stefani (1963*b*) have provided evidence that *A. salina* and *A. franciscana* are heterogametic in the feminine sex; there is no reason to exclude *A. persimilis* from the system. Bowen gave a genetic demonstration based on an *ommochromeless* mutant described as *white* (see Table II) in *A. franciscana,* while Stefani found some evidence of the existence of a heteromorphic chromosome pair in the female, which is replaced by a homomorphic one in male *A. salina.* Barigozzi *et al.* (1969) confirmed Bowen's findings with *A. salina,* also using an *ommochromeless* mutant (*apricot,* see Table II). The demonstration of the transmission of *ommochromeless* (*oml*) through the sex chromosomes is clear. By mating a homozygous male *oml/oml* with a homozygous female *Oml/Oml* wild-type individuals of both sexes were obtained. By backcrossing a heterozygous female (*Oml/oml*) receiving *Oml* from the mother with a homozygous male (*oml/oml*), two parental classes (wild-type females and mutant males) and two recombinant classes (wild-type males and mutant females) were obtained. The reciprocal backcross (homozygous female × heterozygous male) gave four classes of equal frequencies, two wild type and two mutant, each composed of a class of males and a class of females (Bowen, 1963, 1965). The conclusions are that (1) the female is heterogametic, thus a pair of heteromorphic chromosomes is predictable, and (2) the marker gene and the sexual factor (or factors) must map at different loci of a pairing segment. The cytological observations by Stefani (1963*b*) substantiate these predictions, but only partially, because the length difference between the putative X and Y chromosomes is not very strong and requires some statistical analysis. It should be pointed out, however, in order to explain a sex-linked inheritance, that the morphological differences of the chromosomes might be small enough to escape microscopic observation. Accepting Stefani's conclusions, we conclude that we may consider the longer element of the heteromorphic pair of the female as the Y and the shorter as the X. The recombination between X and Y, thus between *ommochromeless* and the sex factor (let us consider it as a single locus), requires a special mention. Bowen (1965) found that the Y recombines with the X with a great variation of frequency, between 0.03% and 20%. A given frequency is characteristic of a female, and is transmitted matroclinously. There are at least three types of Y, as far as recombination frequency is concerned, i.e., recombining 0.03%, 1.3%, or 12.0% respectively, with some fluctuations. The control of the recombination frequency (including the suppression typical of stock No. 5) could be of extrachromosomal origin. These conclusions agree with those of Barigozzi *et al.* (1969), using the marker *ommochromeless* (phenotypically *apricot*, see Table II). The authors had no choice of Ys at their disposal to test recombination (the original appearance was in an X, thus it manifested itself in a male), so they could not verify the variability of the recombination frequency. They were able, nonetheless, to find one recombinant (transfer from the X to the Y) in 3312 individuals produced from test crosses, which corresponds to a frequency of 0.03%. This Y seems to correspond to Bowen's first type. While it is established that recombination occurs between X and Y and that, therefore, a pairing segment between them must exist, the estimate of its length in terms of map units is impossible owing to the peculiar variability of the recombination frequency. Another fact to emphasize is the occurrence of recombination in the heterogametic sex. The last question concerning the mechanism of sex determination regards the sex factor itself. It seems preferable to discuss this subject here instead of in the next section, where all factors will be listed and analyzed. Bowen's first interpretation of the sex genotype can be inferred from Fig. 1 of her paper of 1965; she favors the view that the feminine sex factor maps at the junction between the differential and the homologous (pairing) segment of the Y. This interpretation corresponds to the determination model in the mouse, in man, and probably in other mammals. The alternative, later considered also by Bowen (1965), is that postulated in *Drosophila melanogaster*. According to this model, only the X should carry a sex factor (or a cluster of factors) for the feminine sex, the factors for maleness being located in the autosomes. As a consequences, every X0 individual must be male, while, according to the other model, this constitution must correspond to a female. In *Artemia* the situation is reversed, owing to the female heterogamety. Working on *A. salina*, Barigozzi *et al.* (1969) analyzed two findings, which may be useful for an interpretation. The most interesting observations have been carried out on a transversal gynandromorph, anteriorly male and posteriorly female. The second pair of antennae were undoubtedly male, but, at the same time, much smaller than normal. The individual had *ommochromeless* (*apricot*) eyes. Mated to a male, it gave rise to offspring which showed a regular segregation of wild type and *ommochromeless* in the ratio of 1:1; thus the genotype of the female portion was heterozygous (the oocyte chromosomes analyzed were normally 42). The appearance of the mutant eyes together with the male characters could be due either to a loss of the Y or to the presence of two Xs. More than one mechanism could produce both results. The simplest is loss of the Y. In this case, the model assuming that Y carries the feminine factor is confirmed: the anterior part of the body, derived from one of the two first blastomeres, should owe both its male phenotype and the *ommochromleless* eyes to the chromosomal constitution X0. The other blastomere, retaining the 42 chromosomes, should have given rise to heterozygous feminine organs of constitution XY (*ommochromeless*/+) including the germ cells. The smaller antennae seem to be one element in favor of the X carrying the male factor(s), since the presence of only one X could make the antennae of intermediate size between the male and the female condition. Alternatively, the mechanism could be as follows: Bearing in mind that the gynandromorph was *ommochromeless*/+ (X *oml*/Y^+) and the father *oml*/*oml* (X *oml*/X *oml*), the pronucleus of the original egg was certainly Y^+ and the first polar body *oml*/*oml*. If one part of the body derives from the first polar body, this should be the anterior part, male (XX) and *ommochromeless*. The impossibility of verifying the chromosome number of the male organs makes it impossible to rule out the second mechanism. In conclusion, the analysis of the gynandromorph makes the XY mechanism likely but not definitely proved. Another case in point should be mentioned here. An *oml*/+ female revealed one *ommochromeless* eye and one wild type. In this case, the putative loss of the Y^+ should be restricted to the eye-anlage. The same phenotype, however, might have resulted either from mitotic recombination or from loss of the Y, hence no definite conclusion can be drawn (unpublished data).

Transmission in Parthenogenetic Mutants

The three main mechanisms characterizing the Sète-type, the Sta Gilla-type, and the Istrian-type females lead to peculiar features regarding segregation. In the first case, according to Artom (1931) and Barigozzi (1934, 1944), diploidy is preserved through suppression of the second polar body. Since the first meiotic division, in the absence of recombination, keeps in the pronucleus only homozygous genotypes, the Sète type of diploid parthenogenetic females should be in principle homozygotes, with all consequences easily predictable in case of lethality or semilethality. Recombination is a remedy for this condition; thus it is justifiable to suppose (although experimental evidence is lacking) that recombinants should be particularly frequent even in the heterogametic sex. The absence of a localized centromere (if this view is to be accepted) might prevent localization of crossing over, and then be a favorable factor for keeping genetic variability as high as possible. The case of Sta Gilla is more in favor of heterozygosity (Stefani, 1960). Since the main meiotic mechanism is based on the fusion between the spindle which should give rise to the second polar body and that which represents the division of the first polar body (which in many species does not occur), the fusion should permit the permanence of the homologous chromatids, in the same nucleus. Fusion between a polar body and the pronucleus should give the same result. It should be remembered that, according to Stefani (1967), suppression of the second polar body should practically not exist in the Sète *Artemia.* Since there is evidence (Artom; 1931; Barigozzi, 1935*a,b*, 1944), valid at least until recently, that this not the case, both mechanisms must be considered. The ameiotic type (shrimps living in Istrain and Comacchio salterns) does not eliminate heterozygosity and favors the accumulation of semilethals, which, however, can be masked by the presence of four elements instead of two for each chromosome type. Furthermore, it should be emphasized that the heterozygosity of the parthenogenetic form depends on the occurrence of mutations without the help of their dissemination through mating. The conclusion is that the meiotic mechanisms described are capable of preserving a certain degree of genetic variability and therefore that parthenogenetic lines are not all clones *sensu stricto.* The theoretical conclusions are experimentally substantiated by the findings of Ballardin and Metalli (1963), who investigated the transmission of the "fertility" character (i.e., the percentage of eggs which complete embryonic development out of the total number laid) in two parthenogenetic strains, the diploid of Sta Gilla and the tetraploid of Comacchio (Italy). A sort of selection was performed which consisted of using, for the next generation, a small number (one to three or more) of females of known fertility. The experiments, covering ten generations, confirm the predictions. While *Artemia* from Sta Gilla proved to be genetically heterogeneous, and therefore capable of responding to selection, this was not effective in the ameiotic *Artemia* of Comacchio. Needless to say, the environment in which *Artemia* was grown was carefully controlled. The diversity observed among the yields of the different strains thus must be due, at least partially, to the different modes of egg maturation. According to the mechanism described by Stefani, the diploid line of Sta Gilla should be nearly ameiotic save for recombination, predictable as a consequence of pairing.

THE MUTANTS

We know by now some mutants of the genus *Artemia,* nine of which have been found in *A. franciscana,* eight in *A. salina,* and two in *A. persimilis.* They correspond to independent loci. All of them are inherited as simple units, and the majority show a peculiar functional behavior which will be considered later in some detail: change in manifestation during aging. Besides these oligogenic mutants, there are indications of polygenic control of some other characters. With one possible exception (*garnet* in *A. franciscana,* according to Bowen *et al.,* 1966) obtained from X-ray-treated material, all mutants are spontaneous. It is to be remarked that the mutants found in *A. salina* by Barigozzi *et al.* (1969) were obtained from a sample of nearly 20 individuals taken directly from the salterns of San Bartolomeo, Cagliari. A later experiment with a sample of similar size was unsuccessful. Ethyl methanesulfonate was used as mutagen on another sample of *Artemia* from the same origin, but no mutants were obtained. The isolation of the mutants described in this section was obtained from backcrosses or from F_2s. The list presented in Table II is in concise form, since the authors quoted have already published elsewhere a full description of the phenotypes. Three problems will be discussed later: the aging effect, the *ommochromeless* mutant in its food dependence, and the genetic interaction. The sex factor, already discussed, will not be treated again. It is sufficient to mention a large series of mosaics described by Bowen *et al.* (1966), interpretable as intersexes, proving the autonomous differentiation of sex characters. Before describing the eye color mutants, it is necessary to give a short description of the eyes of the wild type. The ocellus of median eye (some authors prefer "ocellus" as the name for each of the "cups" which form the median eye) is visible and pigmented at hatching. Its color is described as red by Vaissière (1961); according to him, the black pigment should appear only in the metanauplius. Bowen seems to agree as regards *A. franciscana,* while in *salina* and *persimilis* the coloration is black at the beginning. The side eyes, or compound eyes, appear later (metanauplius III according to Barigozzi, 1939; third instar according to Vaissière, 1961); they are black or dark brown. The pigmentation is caused by the presence of an ommochrome (Becker, 1942, Bowen *et al.,* 1966) which remains unchanged throughout the whole life cycle. The pigment is contained in the retinular cells (primary neurons), which are five in number, plus an accessory cell, for each elementary eye.

To the mutants, a variant must be added which occurred spontaneously in the tetraploid parthenogenetic line of Comacchio. The phenotype described as *fused* (*fs*) (Ballardin and Metalli, 1967) consists of fusion of the two lobes of the furca, a change in the position of the anus,

TABLE II. List of Mutants

A. franciscana[a]	*A. salina*	*A. persimilis*
1. *Curved* (*Cu*): Antennae of the female curved; *Cu*/+ normal in males, curved in females, although with variable expressivity (semidominant, sex limited); autosomal (Bowen, 1965)	—	—
2. *stump* (*s*): Abdomen twisted dorsally; penetrance incomplete; recessive, autosomal (discovered by Hanson, 1960) (Bowen *et al.*, 1966)	—	—
3. *red* (*r*): Eyes colorless at hatching, later red during larval stages; especially the compound eyes darken in the adult; recessive, autosomal (discovered by Hanson, 1960) (Bowen *et al.*, 1966)	1. *red* (*r*): Ocellus colorless or pink at hatching, then (larval stages) dark red and nearly wild type in the adult; depigmentation occurs in old individuals; compound eyes red during development, then dark brown or nearly wild type in the adult; ocellus loses pigment through maxillary glands; recessive, autosomal (Barigozzi *et al.*, 1969)	—
4. *cyclops* (*cy*): Compound eyes fused into one single anterior mass; viability and penetrance low; transmission irregular; genotype not determined (discovered by Hanson, 1961) (Bowen et al., 1966)	2. *cyclops* (*cy*): Compound eyes fused into an anterior mass; appears erratically; transmission irregular; genotype not determined (Barigozzi *et al.*, unpublished)	—
5. *crinkle* (*c*): Compound eyes mottled due to detachment of retinular	3. *mottled* (*mtd*): Ocellus pink at hatching, then depigmenting; com-	

[a] **Note added in proof:** *A. franciscana*. *Swimmerette* (*sm*): Phyllopedia fused, frequently asymmetrical; penetrance incomplete (30–50%) and expressivity variable; partially sex-linked (Squire and Grösch, 1967).

TABLE II. (Continued)

A. franciscana	*A. salina*	*A. persimilis*
cells after sexual maturity; some patches of retinular cells enter the eyestalk; weakly and irregularly semidominant, autosomal (no indications about the ocellus) (discovered by Hanson, 1960) (Bowen *et al.* 1966)	pound eyes purple, irregularly pigmented, depigmenting during development, colorless in the adult; pigment elimination through the maxillary gland; recessive, autosomal (Barigozzi *et al.*, 1969)	
6. *garnet* (*g*): Compound eye wild type at hatching, then brown or red brown (garnet); in adult stage many cells degenerate; pigment is lost, eliminated in irregular patches through phagocytic cells and maxillary gland; some axons degenerate; old individuals lose orientation to light; ocellus less affected; recessive, autosomal (discovered by Hanson, 1962) (Bowen *et al.* 1966)	4. *vermilion* (*v*): Ocellus permanently vermilion, only rarely depigmenting in the adult; compound eyes vermilion, then in the adult stage first depigmenting in patches, eventually colorless; viability poor, recessive, autosomal (Barigozzi *et al.*, 1969)	—
7. *white* (*w*): All three eyes colorless, sometimes orange or pink, frequently fading; color due probably to storage of some food material; recessive, partially sex linked (Bowen, 1963)	5. *ommochromeless* (*oml*) (*apricot*, *a*): All three eyes lacking ommochrome; carotenoids from food (algae) stored in the eye, producing an apricot color; without carotenoids in the food, colorless (white); recessive, partially sex linked (Barigozzi *et al.*, 1969)	—
8. Hemoglobin gene(s): Inferred through phenotypic differences—three or four hemoglobins found in different populations (Bowen *et al.*, 1969)	6. Hemoglobin gene(s): Inferred through three different hemoglobin patterns, one slow, one fast, and one intermediate (Bianchi and Piccinelli, unpublished)	—

TABLE II. (Continued)

A. franciscana	*A. salina*	*A. persimilis*
	during development, with strong individual variation; penetrance high only in old individuals; recessive, autosomal (Barigozzi *et al.*, 1969)	during development, with strong individual variation; penetrance high only in old individuals; recessive, autosomal, transmission irregular (Barigozzi *et al.*, 1969)
—	8. *brown* (*b*): Ocellus slightly brown at hatching, darker during development, frequently depigmented in the adult; compound eyes light brown in the larval stages, then brown or nearly wild type in the adult; recessive, autosomal (Barigozzi *et al.*, 1969)	—
—	—	2. *curly* (*cr*): Abdomen curled downward; low viability, unable to mate; recessive, autosomal (Cervini, 1965)

which opens dorsally, and the appearance of a small protuberance on the last abdominal segment. Viability is low, with many dying before reaching sexual maturity; penetrance is 4%, with high variability. The character was transmitted through several generations. The mode of inheritance could not be investigated because of the lack of bisexuality. To complete the list of mutants known so far, which mark at least eight chromosomes in *A. franciscana* and in *A. salina* (including the sex chromosomes in both cases), while we do not known whether *cr* and *do* map in the same pair or in two pairs in *A. persimilis,* a very short account must be given of the polygenic effects on fertility (see p. 236), of the morphological variation of the furca depending on environment (see p. 232), and of cell size. The genetic control of fertility can be interpreted as probably polygenic, if one compares the results obtained from the Sta Gilla diploid and the Comacchio tetraploid individuals. Genetic control of the reaction of the furca to salinity can be inferred from Gross's data (1932), although an interpretation in terms of the polygenic system remains very vague.

Another character depending on chromosome number, i.e., the nuclear size of the epithelial cells, gives a clear indication of polygenic determination, which superimposes itself on the basic effect due to the chromosome number (Barigozzi, 1940). Parthenogenetic shrimps of Istrian salterns characterized by 84 chromosomes show a wide variability in the nuclear size of the intestinal cells. A statistical analysis of the progeny obtained from females with different nuclear size, but exhibiting the same chromosome number, proved the existence of a continuous control of the nuclear size; the chromosome number thus is not the only factor controlling cell size.

Mode of Action of the Pigmentation Genes—Aging Effect

Unlike the majority of Metazoa, a number of genes in all three species of *Artemia* have two distinct phases of action: the first works during development, while the second starts after the sexual maturity stage and affects the phenotype during aging. The genes which show an aging effect can be classified into two groups: (1) those revealing a destructive aging effect consisting of a loss of pigment, brought about by cell destruction and elimination of pigment either by means of phagocytic cells or through the maxillary gland (e.g., *crinkle, garnet, mottled*), and (2) those causing changes of pigmentation (*red* and *brown* of *A. salina, red* of *A. franciscana*) without cell destruction and pigment elimination. The two processes are not necessarily identical for both types of eye, the ocellus and the compound one (*red* and *brown* of *A. salina*), nor is the aging effect necessarily observable in both (*depigmented ocellus* in *A. salina* and *A. persimilis, vermilion* in *A. salina*). The situation is summarized in Table III.

The main conclusions which can be drawn from these results are that (1) one single gene is generally polyphenic, although the two phenes refer to organs of similar but not identical structure, and (2) the aging effect is differently expressed as regards the ocellus and the compound eye (e.g., *red* and *brown* in *A. salina*).

Let us now consider the gene action at biochemical level. Becker (1942) was the first to demonstrate that the pigment present in all three eyes of wild-type *Artemia* is a typical ommochrome, while the reddish body color is due to the presence of a carotenoid. The most recent contribution on the subject is by Kiyomoto *et al.* (1968), who confirm Becker's data. These authors have been able to find some evidence that the ommochrome is composed of ommin and ommatin, although the latter compound may be an artifact. Furthermore, they have studied biochemically the pigmentation

TABLE III. Destructive (D) and Nondestructive Aging Effect (ND)

		Ocellus	Compound eye
A. franciscana	*crinkle*	?	D
	red	ND	ND
	garnet	D	D (stronger)
A. salina	*red*	D	ND
	vermilion	—	D
	mottled	D	D
	brown	D (weak)	ND
A. salina and *A. persimilis*	*depigmented ocellus*	D	—

of the mutants *red* and *garnet*. In both cases, the coloration is due again to ommochrome, probably differing from the wild-type form by some substitution in the molecule or by a state of oxidation. Investigations on the chemical nature of pigments (Nasini and Piccinelli, unpublished) based on paper chromatography with collidine–lubidine–water (1:1:2) or with collidine–M/2 phosphate (3:1) after extraction with methanol–HCl (1% HCl) show that the wild type contains mostly xanthommatine as ommochrome, while *red* mutants contain rhodommatin.

Ommochromeless

A case to be considered separately because of the lack of aging effect is that of *ommochromeless*, which can look either white or apricot, and is almost certainly the same in *A. salina* and in *A. franciscana*, at least with regard to its action. In the case studied by Bowen (1963) and by Bowen *et al.* (1966), the phenotype was almost always white because the yeast used as food did not contain carotenoids; *ommochromeless* in *A. salina* could be white or apricot depending on the food used, i.e., yeast or *Dunaliella* (a green alga). The evidence that the apricot color was the result of storage of substances derived from the food is the following: (1) apricot eyes become colorless (white) when the mutant is transferred from algae culture to yeast culture or kept in complete medium for algae but without algae (Fig. 4), and (2) colorless individuals transferred into *Dunaliella* suspension recolor quickly (Barigozzi *et al.*, 1969) (Fig. 5). Microtome sections have revealed a total absence of ommochrome in the eyes of this mutant: thus it seems justified to call it *ommochromeless*. The sections of apricot eyes, after treatment with water or alcohol, are completely colorless, hence it must be concluded that the apricot substance is water and alcohol soluble. An at-

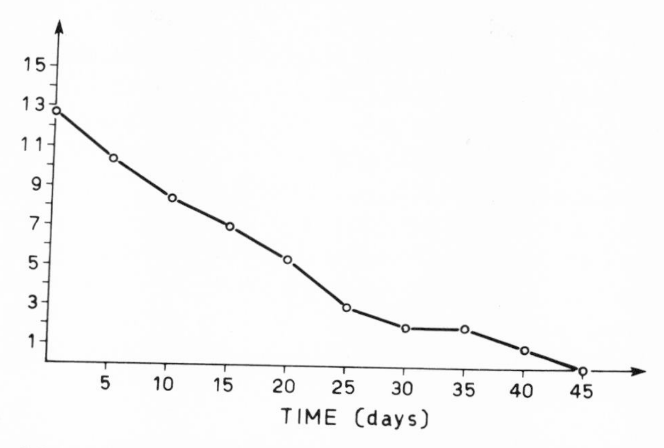

FIG. 4. Time required for apricot eyes to become colorless when *Artemia* specimens are kept in medium without algae. (Ordinate: arbitrary color units.)

tempt to analyze chemically both algae and *Artemia* gave the following results. The extract in ether from *ommochromeless* mutant eyes (nearly 100 individuals) analyzed by means of chromatography on thin sylex layer revealed the presence of two carotenoids: 4-chetocarotene (echinenone) and 4,4′-dichetocarotene (canthaxantine). The apricot color thus is produced by these substances; they are present also in the wild type and in *red*, but ap-

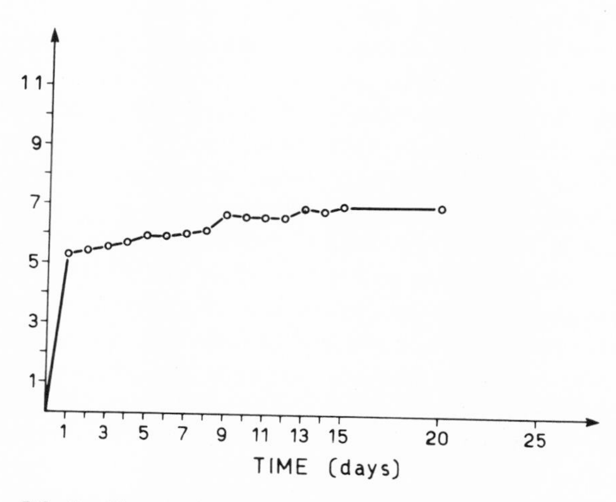

FIG. 5. Time required for colorless eyes to recolor when *Artemia* specimens are transferred into algae suspension. (Ordinate: arbitrary color units.)

parently in lesser quantity. These findings prove that the carotenoids of the shrimp derive from the algae (where they are also present) and are stored in the eye cells irrespective of their pigmentation. In pigmented mutants and in the wild type, the apricot coloration is simply masked (Piccinelli and Nasini, personal communication). Two points are still obscure: the accumulation of carotenoids in *ommochromeless* (since the amount of carotenoids is higher than in pigmented eyes) and the existence of three levels (pale, typical, and orange) of apricot coloration in this mutant. A different capacity for concentration of carotenoids controlled by the genes might be an additional trait of *ommochromeless,* which could exist in two or three alleles. However, this hypothesis remains unproven.

Gene Interactions

Gene interactions have been studied in several cases using the method of double heterozygotes. Both in *A. salina* and in *A. franciscana oml* is epistatic on pigmented mutants. More precisely, in *A. salina oml* is expressed when in double dose together with *red/red, brown/brown,* and *mottled/mottled;* in *A. franciscana, oml* is epistatic on *crinkle/crinkle* and on *garnet/garnet* (Brown *et al.,* 1966). This behavior is perfectly understandable, since all pigmentation mutants have ommochrome, the presence of which, irrespective of its structure, amount, or distribution, is hindered by *oml/oml.* In *A. franciscana* double homozygotes for *crinkle/crinkle;garnet/garnet* show a late manifestation of *crinkle* (detachment of retinular cells) and of *garnet* patches. In *A. salina, mottled/mottled;red/red* gives intermediate conditions of the ocellus; the compound eyes are more similar to *red* during development and to *mottled* in old individuals. These two mutants are thus active in different moments of the life cycle. Throughout the life cycle, *red/red;brown/brown* is pale brown, thus showing a simultaneous action of both loci. A mention should be made of *depigmented ocellus,* since depigmentation of this organ occurs also in *red* and in *brown.* The simultaneous action of one single gene both on the ocellus and on the compound eyes could mean that *r/r* and *br/br* individuals may also carry *do.* However, this can be ruled out, since *do/+;brown/brown* and *do/+;red/red* fail to show any depigmentation of the ocellus. The control of both eye types is thus inherent in *red* and in *brown.* As a general scheme of the gene action on the eye color, the following can be stated: *crinkle, and garnet* (*A. franciscana*) and *red, brown, vermilion. mottled,* and *depigmented ocellus* (*A. salina*) act on the ommochrome itself and/or on its distribution or, additionally, on its maintenance during the life cycle (*crinkle. vermilion, depigmented ocellus, mottled*); *ommochromeless* blocks the ommochrome formation, in a way which is still unknown.

RADIOSENSITIVITY

Artemia is an excellent material for studying radiosensitivity. Early investigators (for references, see Metalli and Ballardin, 1962*a,b*) used X-rays in the attempt to obtain developmental or genetic abnormalities. Two lines of present-day research must be distinguished: (1) analysis of radiosensitivity in different stages of the cell cycle and (2) comparison of the radiosensitivity measured at somatic and cellular levels between different ploidy degrees.

The first type of research takes advantage of the following conditions: the different stages during the first meiotic division can be distinguished macroscopically. When oocytes undergo the first prophase, they are still in the ovary and move toward a cavitarian organ called the ovisac or uterus; when they are in the first metaphase, they group synchronously into two masses on each side of the ovisac; when they perform the first anaphase and telophase of the following meiotic stages, they move toward the median portion of the ovisac where they remain until the end of the maturation. In bisexual individuals, fertilization occurs in the ovisac.

In the parthenogenetic individuals of Sète or Comacchio, where meiosis is frequently reduced to one single mitosis, the telophase of this is followed immediately by embryonic development. It is easy to irradiate at a given stage and score the eggs after delivery, classifying as either living (embryos or larvae) or dead. The dead eggs represent those in which X-rays have induced some damage, interpreted as dominant lethal.

The results obtained by Giavelli and Cervini (1965, 1966) and Cervini and Giavelli (1965*a,b*) from irradiating diploid *Artemia* from Sète with different doses demonstrated that dead eggs were more frequent when X-ray treatment was applied to prophase than to metaphase oocytes. Thus the authors conclude that metaphases is more resistant than prophase. There was no temperature effect (0°C vs. room temperature). Recently, Giavelli (unpublished data) tried to find an explanation of the sensitivity difference between prophase and metaphase on the basis of differential repair processes, which should be more efficient in metaphase than in prophase. Post-treatment with ATP (which, as is well known, interferes with oxidative metabolism) reduced the rate of dead eggs. This occurred after irradiation in both stages of mitosis, but the higher radiosensitivity in metaphase after ATP treatment may partially explain the phenomenon; a more efficient repair in metaphase may be one reason why at this stage the rate of damaged eggs is lower than in prophase. Higher X-ray sensitivity during prophase was also found by Metalli and Ballardin (1962*a,b*) in *Artemia* from Sta Gilla (diploid parthenogenetic) and from Comacchio (tetraploid parthenogenetic). Furthermore, the authors found an effect not only in the treated generation but also in the second generation in the diploid form. The cause remains unexplained. The comparison between diploid and tetraploid *Artemia* at the cellular level was first made using Sta Gilla diploids and Comacchio tetraploids (Metalli and Ballardin, 1962*a,b*): the oocytes of the latter proved more resistant than those of the former. Similar results have been obtained by irradiating developmental stages of diploid bisexual and "tetraploid" bisexual *Artemia* (Metalli *et al.*, 1961). These results require confirmation, since the tetraploidy of the Great Salt Lake shrimps cannot be considered as sufficiently proved. However, it seems premature to consider radiosensitivity as a differential between species. The attempt to compare similar phenotypes in the three species does not imply any similarity or identity with respect to the genotype.

GENERAL REMARKS

The observations reported and discussed critically in the foregoing pages allow some general remarks to be made with reference to four main problems: speciation, sex determination, gene action, and X-ray sensitivity.

Artemia and Speciation in Animals

Artemia is a classic example of polyploidy and parthenogenesis in animals. However, some facts which have arisen recently have changed our original concepts; moveover, some aspects of the evolutionary genetics of *Artemia* remain not fully clear. If we consider the observations made by Suomalainen (1969) on bisexual and parthenogenetic weevils, we may be inclined to add to the three species forming the genus *Artemia* a fourth species consisting of the parthenogenetic forms, distinguished into four ploidy degrees, diploid, triploid, tetraploid, and the pentaploid, leaving aside the uncertain and rare octaploid variant. *Artemia* ranks, in any case, among the few animals (together with *Solenobia, Saga,* and Curculionidae of several genera, to mention some of the best-known examples among Arthropoda) which display repeated phenomena of numerical chromosome changes. A case demonstrating some particular similarities is that of Curculionidae, which is also the one most thoroughly investigated in recent years (Suomalainen, 1969). A further complication in the case of *Artemia* is the existence of more than a single type of egg maturation among the parthenogenetic forms (two to four, according to the different authors) with several genetic consequences. Selection experiments have shown that the genotype controlling metric characters (fertility, nuclear size) is considerably heterozygous not only in the diploid parthenogenetic form (which still exhibits a meiotic process during oogenesis) but also in the ameiotic tetraploid parthenogenetic shrimp. The last observation shows some similarity to what is known about the parthenogenetic curculionid *Otiorrhyncus scaber* (Suomalainen, 1969).

Sex Determination in *Artemia*

The genus *Artemia*, in its two sibling species, *A. salina* and *A. franciscana*, is, together with the isopod *Isotea baltica* (Tinturier-Hamelin, 1959, 1963), the best known among Crustacea regarding sex determination. In her survey on sex determination of Crustacea, Charniaux-Cotton (1960) gives a list of cases (nearly always based only on cytological evidence) where either male or female heterogamety is present, involving either a single chromosome pair (e.g., *Chrirocephalus nankinensis*, male X0; *Eriocher japonica*, male XY; Eucyclopinae, female X0) or a group of three chromsomes (*Heterocypris incongruens*, male XX0; genus *Jaera*, female XY_1Y_2). Physiological evidence exists, moreover, of female heterogamety in *Orchestia gammarella* (Charniaux-Cotton, 1956*a,b*, 1964). The genotype controlling sex is partially known in *Artemia* and in *Isotea baltica* only because of the existence of sex-linked genes, one in *Artemia* (*ommochromeless*) and three in *Isotea* (*flavafusca*, *bilineata*, and *lineata*). In several genera (e.g., *Trichoniscus*, *Asellus*, *Thysbe*), sex is controlled by a complex genotype, as evidenced by a strong fluctuation of the sex ratio, which can reach even the monogenic condition, i.e., the presence of only one sex in the offspring (Montalenti and Vitagliano Tadini, 1963; Battaglia and Malesani, 1959; for a general survey, Bocquet, 1967). The two cases which are interesting for comparison are that of *Artemia* and that of *Isotea*. In the latter species, two of the three sex-linked loci (loci *B* and *L*), should be located at a considerable distance from the "feminizing gene(s)," while one (*F*) should map within the differential segment. Where the masculine gene(s) is located, whether in the X or in the autosomes, remains undecided. It is clear that in *Isotea* and *Artemia* the mechanism is the same. In *Artemia* too, as pointed out on pp. 233–235 definite evidence of the location of the male factor(s) in the X and the female factor(s) in the Y is still lacking: but evidence is probably stronger in *Artemia* than in *Isotea* since mosaics and gynandromorphs in the latter species are apparently still unknown. We may conclude that, in spite of our still incomplete knowledge, some Crustacea unquestionably rank among the few and erratic cases of feminine heterogamety (as in Amphibia) to be added to the classic systematic categories of Lepidoptera, Trichoptera, and birds, where female heterogamety is the general rule.

Gene Action

There is at least one unique aspect of gene action in *Artemia:* a change of the phenotype when the individual grows old. In the mutants *crinkle* and *garnet* in *A. franciscana*, *vermilion, mottled,* and, partially, *brown* in *A. salina,* and *depigmented ocellus* in *A. salina* and in *A. persimilis,* the more or less new phenotype in old individuals consists of a loss of pigment and/or of cells, etc., in all three eyes or in the compound ones only or in the ocellus. It seems of interest to consider the possible cause of this phenomenon, trying to compare it with the very rare similar cases. The first point to clarify is that, in general, animals are considered as old when they have reached sexual maturity and stop growing. In *Artemia*, the life span, after the beginning of sexual activity, is undetermined and includes a variable number of molts (see Introduction). During the interval between two molts, the individual grows; thus sexually mature shrimps can exhibit a considerable size variation depending on the number of molts. Molting, although based on hormonal mechanisms, seems to lack, in the wild type, any effect on eye pigmentation. The local effect of losing pigment in mutants is not explained by any simple model, because the aging effect is not restricted to a block of pigment production and to loss of the retinular cells (*crinkle, garnet*), but also includes the movement of cells (*crinkle*) and phagocytosis and release of the pigment through the maxillary glands (compound eyes and ocellus of *garnet* and *mottled* and the ocellus of *red* of *A. salina*). Without these complications, it might be possible to consider the cessation of pigment production as a halt in mRNA production after the end of development. In fact, two types of mRNA are known: short-lived, occurring in many tissues, and long-lived, characteristic of some cell types lasting life-long (lens cells, feather cells, and reticulocytes in vertebrates, Scott and Bell, 1964). *Artemia* eye mutants, in which there is an aging effect, might represent a third hypothetical case of mRNA living for several weeks or months but not for the duration of the whole life. Why the postulated stop in production and the disappearance of mRNA are followed by cell elimination remains obscure. Other cases of phenotypic changes during aging, except for examples of intensifications (e.g., darkening of color as occurs in several mutants of *Drosophila* such as *brown*), are difficult to find outside the loss of hair pigment in man, the horse, and other mammals. The *ommochromeless* mutant exhibits the peculiarity of revealing a different phenotype according to food. More or less similar examples can be found in the silkworm (Jucci, 1949), where the gene controlling permeability to flavones or to carotenoids ingested with mulberry tree leaves determines the color of the silk. The physiology of *ommochromeless* might be more complicated, since its action consists not only in suppressing the ommochrome but also in storing the carotenes in particularly large quantities. Speculation in this sense seems premature. Another point to emphasize is the different phenotypic complexity controlled by the several mutants. It is remarkable that, frequently, organs which are different structurally and developmentally, such as the compound eye and the ocellus, are controlled by one single gene; it is not impossible that a more complete analysis will reveal that each "gene" is a compound one. Finally, *red* of *A. franciscana* could be the homologue to *d* (*delayed melanin*) described in *Gammarus chevreuxi,* according to Bowen *et al.* (1966); *red* of *A. salina* exhibits, besides some similarities, depigmentation of the ocellus in old individuals which does not fit in with a simple delay in pigmentation, as in the case of *d* of *Gammarus*.

Radiosensitivity

In all other species investigated so far (e.g., *Habrobracon, Mormoniella,* mouse), the most X-ray-sensitive stage of mitosis is metaphase, while in *Artemia* it is prophase. The exceptional behavior requires an explanation, which is still incomplete. One hypothesis is that the different radiosensitivity may be due not simply to the amount of chromosome damage caused by X-rays but also to a difference in repair in the different stages: a stronger repair may simulate a lesser sensitivity. This hypothesis is only partially proved. Irrespective of this problem, *Artemia* may be used for studying other problems within the field of radiobiology. With respect to the different reactions to X-rays in different ploidy degrees, it should not be forgotten that the ploidy degrees of *Artemia* are not comparable to those obtained experimentally, since they are the product of natural selection on a spontaneous mutational basis; many genetic differences thus must exist between strains, irrespective of the chromosome number.

Acknowledgments

The author wishes to express his gratitude to Mrs. S. Bowen Thompson, to Prof. R. Stefani, to Dr. P. Metalli, to Mrs. Dr. M. Lozzi Piccinelli, and to Dr. A. D'Agostino for their valuable criticisms.

REFERENCES

Artom, C. 1905, Osservazioni generali sull'*Artemia salina* Leach delle Saline di Cagliari, *Zool. Anz.* **29**:284–291.

Artom, C., 1906, Ricerche sperimentali sul modo di riprodursi dell'*Artemia salina* Lin. di Cagliari, *Biol. Zentralbl.* **26**:26–32.

Artom, C., 1907, La maturazione, la fecondazione e i primi stadi di sviluppo dell'uovo dell'*Artemia salina* Lin. di Cagliari, *Biologica* (*Racc. Scr. Biol.*) **1**:495–515.

Artom, C., 1911, Analisi comparativa della sostanza cromatica nelle mitosi di maturazione e nelle prime mitosi di segmentazione dell'uovo dell'*Artemia* sessuata di Cagliari (univalens) e dell'uovo dell'*Artemia* partenogenetica di Capodistria (bivalens), *Arch. Zellforsch.* **7**:277–295.

Artom, C., 1912, Le basi citologiche di una nuova sistematica del genere *Artemia:* Sulla dipendenza tra il numero dei cromosomi delle cellule germinative, e la grandezza dei nuclei delle cellule somatiche dell'*Artemia salina univalens* di Cagliari, e dell'*Artemia salina bivalens* di Capo d'Istria, *Arch. Zellforsch.* **9**:87–113.

Artom, C., 1921, Specie micropireniche e macropireniche del genere *Artemia. Ric. Morfol.* **2**:137–155.

Artom, C., 1924, Ancora del tetraploidismo dei maschi dell'*Artemia salina* di Odessa in relazione con alcuni problemi generali di genetica, *Atti Reale Accad. Naz. Lincei Rend. Cl. Sci. Fis. Mat. Nat.* **33**:34–36.

Artom, C., 1931, L'origine e l'evoluzione della partenogenesi attraverso i differenti biotipi di una specie collettiva (*Artemia salina* L.) con speciale riferimento al biotipo diploide partenogenetico di Sète, *Mem. Reale Accad. Ital. Cl. Sci. Fis. Mat. Nat.* **2**:1–57.

Ballardin, E., and Metalli, P., 1963, Osservazioni sulla biologia di *Artemia salina* Leach, *Ist. Lomb. Sci. Lett.* **97**:194–254.

Ballardin, E., and Metalli, P., 1967, Variante ereditaria spontanea di *Artemia salina* portenogenetica tetraploide, *Atti Assoc. Genet. Ital.* **12**:505–508.

Barigozzi, C., 1934, Diploidismo, tetraploidismo e octoploidismo nell'*Artemia salina* partenogenetica di Margherita di Savoia, *Boll. Soc. Ital. Biol. Sper.* **9**:906–908.

Barigozzi, C., 1935*a*, Il legame genetico fra i biotipi partenogenetici di *Artemia salina, Arch. Zool. Ital.* **22**:33–77.

Barigozzi, C., 1935*b*, I caratteri fenotipici del biotipo tetraploide partenogenetico di *Artemia salina* in relazione agli altri biotipi partenogenetici, *Arch. Zool. Ital.* **22**:1–16.

Barigozzi, C., 1939, La biologia di *Artemia salina* Leach studiata in acquario, *Atti Soc. Ital. Sci. Nat.* **78**:137–160.

Barigozzi, C., 1940, Relazione fra numero cromosomico e grandezza nucleare in *Artemia salina* Leach, *Sci. Genet.* **2**:1–25.

Barigozzi, C., 1942, L'azione della colchicina sulla morfologia e sulla struttura dei cromosomi, studiata nelle cellule somatiche di *Artemia salina* Leach, *Chromosoma* **2**:293–307.

Barigozzi, C., 1944, I fenomeni cromosomici delle cellule germinali in *Artemia salina* Leach, *Chromosoma* **2**:549–575.

Barigozzi, C., 1957, Différenciation des génotypes et distribution géographique d'*Artemia salina* Leach: Données et problèmes, *Coll. Int. Biol. Mar. Stat. Biol. Roscoff Ann. Biol.* **33**:241–250.

Barigozzi, C., and Tosi, M., 1957, Nuovi dati sul numero cromosomico di *Artemia salina* anfigonica, *Atti III Riun. Assoc. Genet. Ital. Suppl. Ric. Sci.* **27**:105–107.

Barigozzi, C., and Tosi, M., 1959, New data on tetraploidy of anphigonic *A. salina* Leach and on triploids resulting from crosses between tetraploids and diploids, *Conv. Genet. Suppl. Ric. Sci.* **29**:3–6.

Barigozzi, C., Piccinelli, M., and Prosdocimi, T., 1969, Description of some spontaneous mutants of *Artemia salina, Atti Assoc. Genet. Ital.* **14**:169–181.

Battaglia, B., and Malesani, L., 1959, Ricerche sulla determinazione del sesso del Copepode. *Tysbe gracilis* (T. Scott), *Boll. Zool.* **26**:423–433.

Becker, E., 1942, Über Eigenschaften, Verbreitung und die genetischentwicklungsphysiologische Bedeutung der Pigmente der Ommatin und Ommingruppe (Ommochrome) bei den Arthropoden, *Z. Vererbungsl.* **80**:157–352.

Bocquet, C., 1967, Structure génétique du sexe chez les crustacés, *Ann. Biol.* **6**:225–239.

Bowen, S. T., 1962, The genetics of *A. salina.* I. The reproduction cycle. *Biol. Bull.* **122**:25–32.

Bowen, S. T., 1963, The genetics of *A. salina.* III. Effects of x-irradiation and of freezing upon cysts, *Biol. Bull.* **125**:431–440.

Bowen, S. T., 1963, The genetics of *Artemia salina.* II. White eye, a sex-linked mutation, *Biol. Bull.* **124:**17–23.

Bowen, S. T., 1965, The genetics of *Artemia salina.* V. Crossing over between the X and Y chromosomes, *Genetics* **52:**695–710.

Bowen, S. T., Hanson, J., Dowling, P., and Poon, M., 1966, The genetics of *Artemia salina.* VI. Summary of mutations, *Biol. Bull.* **131:**230–250.

Bowen, S. T., Lebherz, H. G., Poon, M., Chow, V. H. S., and Grigliatti, A., 1969, The hemoglobins of *Artemia salina.* I. Determination of phenotype by genotype and environment, *Comp. Biochem. Physiol.* **31:**733–747.

Brauer, A., 1893, Zur Kenntnis der Reifung des partgen. sich. entw. Eies von *Artemia salina, Arch. Mikrosk. Anat.* **43:**162–222.

Cervini, A. M., 1965, Il mutante spontaneo "Curly" (Cr) del ceppo diploide anfigonico Hildago di *Artemia salina, Atti Assoc. Genet. Ital.* **10:**343–345.

Cervini, A. M., and Giavelli, S., 1965*a*, Radiosensitivity of different meiotic stages of oocytes in parthenogenetic diploid *Artemia salina, Mutation Res.* **2:**452–456.

Cervini, A. M., and Giavelli, S., 1965*b*, Effect of low temperature on radiosensitivity of *Artemia* oocytes, *Progr. Biochem. Pharmacol.* **1:**52–58.

Charniaux-Cotton, H., 1956*a*, Le déterminisme endocrinien des caractères sexuels d'*Orchestia Gammarella* (Crustacé amphipode), *Ann. Sci. Nat. Zool.* **18:**305–310.

Charniaux-Cotton, H., 1956*b*, Déterminatisme hormonal de la différenciation sexuelle chez les Crustacés, *Ann. Biol.* **32:**371–399.

Charniaux-Cotton, H., 1960, Sex determination, in: *The Physiology of Crustacea,* Vol. 1, pp. 411–447.

Charniaux-Cotton, H., 1964, Endocrinologie et génétique du sexe chez les Crustacés supérieurs, *Ann. Endocrinol.* **25:**36–42.

Debaisieux, P., 1944, Les yeux des crustacés, *Cellule* **50:**5–122.

Dutrieu, J., 1960, Quelques observations biochemiques et physiologiques sur le developement d'*Artemia salina* Leach, *Rend. Ist. Sci. Univ. Camerino* **1:**196–224.

Giavelli, S., and Cervini, A. M., 1965, Effetto della bassa temperatura sulla radiosensibilità di ovociti di *Artemia salina* trattati con differenti dosi di raggi X: Risultati preliminari, *Atti Assoc. Genet. Ital.* **10:**241–242.

Giavelli, S., and Cervini, A. M., 1966, La curva di radiosensibilità di ovociti di *Artemia salina, Atti Assoc. Genet. Ital.* **11:**292–300.

Gilchrist, B., 1960, Growth and form of the brine shrimp *Artemia salina* (L), *Proc. Zool. Soc.* **134:**221–235.

Goldschmidt, E., 1952, Fluctuation in chromosome number in *Artemia salina, J. Morphol.* **91:**111–134.

Gross, F., 1932, Untersuchungen über die Polyploidie und die Variabilität bei *Artemia salina, Naturwissenschaften* **20:**962–967.

Halfer Cervini, A. M., Piccinelli, M., and Prosdocimi, T., 1967, Fenomeni di isolamento genetico in *Artemia salina, Atti Assoc. Genet. Ital.* **12:**312–327.

Halfer Cervini, A. M., Piccinelli, M., Prosdocimi, T., and Baratelli Zambruni, L., 1968, Sibling species in *Artemia* (Crustacea Branschiopoda), *Evolution* **22:**373–381.

Hertwig, G., 1931, *Artemia salina,* ein Beispiel für die Entstehung einer gigas-Varietät durch gleichzeitige Verdoppelung der Chromosomenzahl und des Chromosomemenvolumens, *Gegenbaurs Morphol. Jahrbuch.* **67:**371–380.

Jucci, C., 1949, Physiogenetics in silkworms (*Bombyx mori*), in: *Proceedings of the Eighth International Congress of Genetics,* pp. 286–297.

Kiyomoto, R. K., Poon, M., and Bowen, S. T., 1968, Ommochrome pigments of the compound eyes of *Artemia salina, Comp. Biochem. Physiol.* **29:**975–984.

Kuenen, D. J., 1939, Systematical and physiological notes on the brine shrimp, *Artemia, Arch. Neerl. Zool.* **3**:365–449.

Linder, F., 1941, Contributions to the morphology and the taxonomy of the Branchiopoda Anostraca, *Zool. Bid. Uppsala* **20**:101–302.

Lockhead, J., 1941, Studies on the blood and related tissues in *Artemia* (Crustacea Anostraca), *J. Morphol.* **68**:593–632.

Metalli, P., and Ballardin, E., 1962*a*, *Artemia salina* Leach, a new material for radiobiological research, *Strahlenwirk. Milieu* **51**:126–128.

Metalli, P., and Ballardin, E., 1962*b*, First results on X-ray-induced genetic damage in *Artemia salina* Leach, *Atti Assoc. Genet. Ital.* **7**:219–231.

Metalli, P., Ballardin, E., and Barigozzi, C., 1961, Primi risultati del trattamento con raggi X di *Artemia salina* Leach, *Atti Assoc. Genet. Ital.* **6**:409–418.

Montalenti, G., and Vitagliano Tadini, G., 1963, Variability of sex ratio and sex determining mechanism in *Asellus, Genetics Today* (*Proc. XI Int. Congr. Genet.*) **1**:186–187.

Piccinelli, M., and Prosdocimi, T., 1968, Descrizione tassonomica delle due specie *Artemia salina* L. e *Artemia persimilis* n. sp., *Inst. Lomb. Sci. Lett.* **102**:113–118.

Piccinelli, M., Prosdocimi, T., and Baratelli Zambruni, L., 1968, Ulteriori ricerche sull'isolamento genetico nel genere *Artemia, Atti Assoc. Genet. Ital.* **13**:170–179.

Scott, R., and Bell, E., 1964, Protein synthesis during development: Control through messenger RNA, *Science* **145**:711–714.

Squire R. D., and Grösch, D. S., 1967, "Swimmeretti" a new sex-linked recessive mutant in the brine-shrimp, *Artemia salina* Leach, *Biol.* Bull. **133**:487.

Stefani, R., 1960, L'*Artemia salina* partenogenetica a Cagliari, *Riv. Biol.* **52**:463–491.

Stefani, R., 1963*a*, La digametia femminile in *Artemia salina* Leach e la costituzione del corredo cromosomico nei biotipi diploide anfigonico e diploide partenogenetico, *Caryologia* **16**:625–636.

Stefani, R., 1963*b*, Il centromero non localizzato in *Artemia salina* Leach, *Acc. Naz. Lincei Rend. Cl. Sci. Fis. Mat. Nat.* **35**:375–378.

Stefani, R., 1964, L'origine del maschi nelle popolazioni partenogenetiche di *Artemia salina, Riv. Biol.* 147–162.

Stefani, R., 1967, La maturazione dell'uovo nell'*Artemia salina* di Sète, *Riv. Biol.* **60**:599–615.

Stefani, R., and Cadeddu, G., 1967, L'attività centromerica in *Artemia salina* Leach, *Rend. Sem. Fac. Sci. Univ. Cagliari* **37**:287–291.

Stella, E., 1933, Phenotypical characteristics and geographical distribution of several biotypes of *Artemia salina* L., *Z. Induk. Abstammungs-Verebungsl.* **65**:412–446.

Suomalainen, E., 1969, Evolution in parthenogenetic Curculionidae, *Evol. Biol.* **3**:261–293.

Tinturier-Hamelin, E., 1959, La détermination du sexe chez *Idotea balthica basteri* (Audouin) (Isopode Valvifère): Premiers résultats d'une étude génétique, *C. R. Acad. Sci.* **248**:2660–2662.

Tinturier-Hamelin, E., 1963, Définition et analyse génétique du phénotype *pseudolineata* de l'Isopode Valvifère *Idotea balthica, Crustaceana* **5**:133–137.

Vaissière, R., 1961, Morphologie et histologie comparée des yeux des Crustacées Copépodes, *Arch. Zool. Exp. Genet.* **100**:1–125.

Verril, A. E., 1869, Observations on phyllopod crustacea of the family Branchiopidae, with description of some new genera and species from America, *Proc. Am. Assoc. Advan. Sci.* **18**:230–247.

7

Politics and the Evolutionary Process

PETER A. CORNING
Institute of Political Studies
Stanford University
Stanford, California

There are many fearful and wonderful things, but none is more fearful and wonderful than man. He makes his path over the storm-swept sea and he harries old Earth with his plough. He takes the wild beasts captive and turns them into his servants. He has taught himself speech and wind-swift thought, and the habits that pertain to government. Against everything that confronts him he invents some resource—against death alone he has no resource.

SOPHOCLES, *Antigone*

INTRODUCTION

Homo sapiens—the knowing, the wise, the sagacious animal—that is what the taxonomists of the eighteenth century named this remarkable—nay, this *transcendent*—species.

Of course, we are not quite so unique, or at least not unique in quite so many ways, as we once thought. We now know that other species can solve problems, make tools, learn, teach, and even experience "love" and "grief" (DeVore, 1965; Jay, 1968; Van Lawick-Goodall, 1968*a,b*; Hinde, 1970; Dolhinow, 1972; Jolly, 1972). Yet only man quests continually after new knowledge and accumulates it with accelerating efficiency. Only man has come to understand how he and all the other species evolved. And only man, through the knowledge he has so painfully acquired over the centuries, can superimpose his purposes on the blind "trial-and-success" process (in George Gaylord Simpson's phrase) by which he himself came to be. For man is more than a sapient animal; he is also a questioning and a purposive animal. He invents new ways to solve age-old problems (and, in the process, invents new problems as well).

Not only that, but he is a social animal, solving problems and pursuing goals in collaboration with his fellow human beings; he excels at devising, deploying, and playing "roles" in cybernetic behavioral systems—behavioral systems designed to achieve through collective action whatever ends he sets for himself.* And because "politics" is the process by which man goes about cybernating his behavior—the process by which he solves "public" problems, makes authoritative decisions, and organizes and coordinates the behavior of his fellows—man is distinctively a political animal.† In an important sense, therefore, Aristotle's way of classifying man (*zoon politikon*) is closer to the mark than is that of Linnaeus.

Indeed, politics is at once one of the most important of human activities and one of our least understood (or most misunderstood) terms. To quote the eminent political scientist, Robert A. Dahl:

> Whether he likes it or not, virtually no one is completely beyond the reach of some kind of political system. A citizen encounters politics in the government of a country, town, school, church, business firm, trade union, club, political party, civic association, and a host of other organizations, from the United Nations to the PTA. Politics is one of the unavoidable facts of human existence. Everyone is involved in some fashion at some time in some kind of political system. If politics is inescapable, so are the consequences of politics. That statement might once have been shrugged off as a rhetorical flourish, but today it is a horrid, brutal, and palpable fact. For whether mankind is to be blown to smithereens or whether political arrangements will be arrived at that enable our species to survive is now in the process of being determined—by politics and politicians. (Dahl, 1970, p. 1)

To Plato, Aristotle, and other Greek theorists of the fifth and fourth centuries B.C., the *polis* (or *polity*) connoted something more than a mere aggregation of individuals sharing a common language, territory, or culture. A political order was understood to be a very special kind of community—one that involved a division of labor, specialization of roles, and functional interdependence with respect to shared or complementary goals. A polity has synergy. It is an integrated, goal-oriented adaptive system—in

* The literature on cybernetics, and systems analysis in general, is extensive. See especially Demerath and Peterson (1967), Buckley (1968), von Bertalanffy (1968), and the yearbooks of the Society for General Systems Research, *General Systems*, from 1956 forward.

† In fact, the terms "cybernetics," "governor," and "government" all derive from the same Greek word, *kybernetes*, meaning steersman or helmsman. Thus the use by political theorists of the "ship of state" metaphor reflects an etymological as well as conceptual affinity between cybernetics and politics. As a matter of fact, more than a century before Norbert Weiner reintroduced the term, French scientist André Amperè had used "cybernetics" as a synonym for political science. A fuller discussion of politics as behavioral cybernetics, and of contrasting definitions of politics, will be presented later.

short, a cybernetic system—in which the whole is greater than the sum of the parts. And the activity of *doing* politics means the process of social cybernation.*

Aristotle employed the terms *polis* and *politics* with particular reference to the most inclusive of human organizations—those that are responsible for the common needs and goals of the community as a whole. Partly by convention, and partly as a consequence of academic specialization, modern political scientists tend to do likewise. But political behavior in the broad sense is not confined to specialized political institutions. It is a facet of any organized, interpersonal activity, insofar as that activity manifests the properties of a cybernetic system—insofar as it is goal-oriented, self-regulating, and has processes of communication and control (i.e., power, rule, or authority) among the individuals who comprise the "system."

In these terms, most complex human organizations have a political aspect, and, for analytical purposes, political scientists can abstract the "political system" out of total stream of social activity and treat it as a "subsystem" of the total process.† Indeed, numerous studies have been done of decision-making and control processes in large-scale organizations—from the American Medical Association to the federal antipoverty program.

By the same token, other species too could be said to exhibit certain aspects of a political way of life. Examples abound. There are, for instance, the ungulates that form defensive formations against predators when moving from one grazing area to another, the songbird flocks that mob potential avian predators, the emperor penguins, with their marvelous system of collective defense against the Antarctic winter, and the insect societies that display an elaborate division of labor. Among our primate relatives, the behavioral system of savanna-living baboon troops is often compared to that of *Homo sapiens*. However, the repertoire of "political" behaviors available to other species is quite limited and stereotyped. Other species cannot readily adapt their social behaviors to new goals.

* For a detailed and pellucid discussion of politics as social cybernation, see Deutsch (1966).

† Ralph P. Hummel has reduced political relationships to the following essential elements: (1) at least two people, (2) mutual recognition of a problem shared, and (3) design of an arrangement between these two or more people to engage in some sort of relationship for the purpose of solving the problem. Though perhaps not at first obvious, this "model" also includes conflict situations, as when two parties agree to fight out their "problem" in a court of law, or on the battlefield. Resolution of such conflicts involves, most importantly, a decision as to which party, or to what degree *each* party, shall control the outcome of the problem that gave rise to the conflict in the first place. This is what Clausewitz meant when he asserted that war is politics by other means.

What sets man apart from other species and makes him a distinctively political animal is the fact that he has learned how to elaborate on and adapt the basic principles of cybernetic behavioral organization—the basic social building blocks—to a breathtaking variety of human goals. Man alone is able to invent new and more intricate forms of organization, and the increasingly complex cybernation of human societies has been one of the most significant trends in cultural evolution.

Much is made by anthropologists of man's ability to invent tools and use language. A new generation of evolutionary anthropologists, searching for ways to characterize human "progress," single out "technoeconomic" development (Steward, 1955; White, 1959; Harris, 1969) or language development (see especially selections and references in Cohen, 1968), or simply increased complexity (Carneiro, 1972).

All this may have an element of truth. But, from an evolutionary perspective, what is most important is the way in which man puts language and tools to use in goal-oriented social contexts—in order to satisfy his needs and wants. Just as, at the morphological level, individual traits take on significance only as functioning parts of the total phenotype, so man's purposive behavioral systems—the systems by which means are fitted to ends—represent the cultural payoffs. This is what is implied when it is said that the partial autonomy of human behavioral adaptations introduces a Lamarckian element into cultural evolution.

However, the cybernetic character of human behavioral systems does not seem to have received sufficient attention from anthropologists—perhaps in part because their subject matter largely excludes modern industrial societies. But whatever the explanation, the "political" evolution of human societies needs to be brought into sharper focus, and, to that end, I shall consider here (1) the nature of politics as an evolutionary phenomenon, (2) the origins and development of politics (with an excursus on the causes of cultural evolution), and (3) the future of politics. The emphasis here will be on those specialized, macrolevel political organizations that have arisen to cope with the overriding problems of evolving human societies as collective entities.

ON THE NATURE OF POLITICS

Among the numerous advances that have been made in our understanding of the evolutionary process in the century or so since the publication of Darwin's *On the Origin of Species*, one of the most important has been the emergence during the past 25 years of what is commonly referred to as systems theory. Systems theory is grounded in the recognition that there is a fundamental difference between the organized, goal-oriented complexity of the biological realm and the unorganized complexity of the phenomena studied in classical physics. Accordingly, systems theorists have sought to identify and describe those properties that are peculiar to living systems. In the particular type of system with which we are concerned here (cybernetic systems), the following properties (though touched on above) may be singled out: (1) systemic goal-orientation, or "internal teleology," in F. J. Ayala's phrase; (2) interaction among functionally specialized structures, or "subsystems"; (3) hierarchical organization; (4) synergism; (5) interactions with the external environment across system boundaries; (6) internal control, or self-regulation; (7) communication; (8) feedback; (9) historicity (cybernetic systems are irreversible processes taking place through time); and (10) negentropy (cybernetic systems appear to violate the second law of thermodynamics because their processes do not lead inherently to a state of maximum disorder and energy dispersal but exactly the reverse).

That such properties can be the product of natural selection must strike many laymen as being utterly unbelieveable.* We like to think that purposive systems are a human invention. However, the inherently unstable and contingent nature of organic processes requires that *all* viable species be goal-oriented; the basic vocation of all life forms, including man, must be survival, or "system persistence" and reproduction. In a very real sense, the evolutionary process as a whole may be characterized as a series of explorations or experiments with hierarchical, purposive organization. There can be no doubt, certainly, that natural selection has favored the emergence of cybernetic systems.

It is equally clear, furthermore, that behavioral systems are also a product of the evolutionary process. Because the adaptiveness of an organism is largely dependent on its interactions with the environment (including other organisms of the same and different species), cybernated, purposive behavior is a very ancient evolutionary development. It has been shown that organisms with extremely primitive neural organization—army ants, for example (Schneirla and Piel, 1948; Topoff, 1972)—nevertheless may be capable of rather complex interactions with each other and their environments; social behavior is not at all dependent on a big brain and verbal or symbolic languages.

* In fact, it is still not entirely certain that cybernetic organization can be accounted for completely by the action of natural selection working on purely chance events, as most evolutionary biologists hold. Von Bertalanffy (1968) and others believe that, in view of the improbabilities involved, some as yet unrecognized "organizational forces" (though assumed to be of a purely physical sort) may have canalized the evolutionary process in ways we do not at present understand (see later).

In what way then do the social behaviors—particularly the macrolevel political systems—of *Homo sapiens* differ from those of other species? The answer is: Not in any absolute way. There are many differences of degree, which we shall explore, but the premise on which social scientists ought to be operating (one which may seem obvious to the evolutionary biologist) is that the political systems of human societies are but an amplification of behavioral principles that antedate—perhaps by several million years—the emergence of modern industrial societies. The political systems of modern nation-states are also products of the evolutionary process—just as man himself is an evolved species. A truism perhaps, but we are only beginning to apply it in a systematic way to the analysis of contemporary societies. Though I cannot go into very great detail here, I would like to consider now some aspects of politics as an evolutionary phenomenon.

Politics and Cybernetics

In the first place, the macrolevel political systems of human societies do indeed exhibit the properties of evolved cybernetic systems.* Until recently, it had been traditional among twentieth-century social scientists to assume that there is at best only a loose metaphorical relationship between political systems and organic systems. But, in point of fact, a precise and formal analogy exists, for there is a functional isomorphism between the two types of systems. Though each represents a different level of cybernetic organization, each is goal-oriented in relation to the same problem—the ongoing problem of survival and reproduction. Human societies are characterized by a division of labor, reciprocity, and a high degree of interdependence with respect to the satisfaction of basic survival and reproductive needs; in a very real sense, the social order is, *au fond*, a collective survival enterprise.

The heart of any political system—its *raison d'être*—is the process of decision-making and collective action, and peak-level political systems are the principal loci of societal integration and adaptation. David Easton (1965*b*, p. 21) defines politics as the process through which "values are authoritatively allocated for a society," and Karl Deutsch (1966, p. 124) speaks of politics as "the dependable coordination of human efforts and expectations for the attainment of the goals of the society." However, I would prefer to emphasize the fact that political decision-making implies

* There are, in fact, a number of recent works by political scientists that view politics from a cybernetic perspective. See especially Easton (1965*a,b*), Deutsch (1966), Almond and Powell (1966), and Fagen (1966). For some purposes, the polity may be treated as a distinct, autonomous system (as Easton does), but for the most part the practice is to view the polity as the steering and controlling "subsystem" within the social order as a whole.

selection in relation to the overall survival problem; *politics involves the authoritative selection and implementation of society's collective survival strategies.**

It has been pointed out that biological organisms are commonly endowed with "internal selection criteria"—mechanisms by which adaptive individual choices are made (Campbell, 1965). An almost banal example is the tastebuds, which serve to orient an animal toward selection of nutritious foods. Internal selection mechanisms of this kind both are the product of natural selection and are subject to further "editing" by natural selection in cases where such mechanisms lead an animal to make maladaptive choices—say, when there has been a significant change in the environment (see later).

What I am suggesting is that politics involves an analogous sort of intermediate, or internal, adaptation and selection process. In the course of hominid evolution, the evolving human brain became an increasingly powerful organ of behavioral creativity, enabling *Homo sapiens* to engage in complex cognition and problem-solving activities. Man has become capable of making increasingly potent conscious choices among alternatives that he himself has invented. The importance of this process has been greatly magnified, moreover, by the fact that it is operative at both individual and collective levels. One or more human brains may participate in the making of life-and-death choices for the larger social order and its members, utilizing criteria that may be partly learned (or culturally acquired) and partly derived from the motivational matrix that is inborn in each individual (albeit subject to wide individual variation). Furthermore, the particular choices in any given instance may involve new syntheses of previously developed cultural elements, so that novel forms of collective behavior may emerge from the decision-making process; social processes have partial autonomy. Yet it remains true that such choices are subject always to final review in the Supreme Court of the evolutionary process—by natural selection. There is thus potential feedback at all times between the behavioral system and the gene pool of the species. At the present time, we still know very little about this relationship, although I shall have something more to say on the subject later.

In small, primitive societies with few specialized activities and a relatively stable cultural system, macrolevel political activities may be intermit-

* Such holistic conceptions of politics are by no means universally accepted among political scientists. An alternative view is that individual self-interest is the primary stuff of politics and that the main function of political systems is to arbitrate among conflicting interests. The two points of view may not be as incompatible as they seem, but we will defer even an abbreviated discussion of this important issue until later.

tent and involve little or no functional specialization. But in complex modern societies, politics is a highly specialized affair conducted primarily by "politicians"—persons who make politics their principal social role. Political functions have also come to be performed in modern societies within the context of "institutions"—that is, explicitly organized and regularized behavior patterns involving well-defined "roles" that are designed specifically for the tasks at hand. In addition, modern politics is conducted within the framework of a cumulative body of codified behavioral rules (laws and lawlike regulations) of ever-increasing complexity.* (Ever since Plato, in fact, men have recognized the significance of the law as at once a repository of accumulated social experience and a form of depersonalized authority.)

It has been traditional among political scientists to characterize the steering and control activities of the policy in terms of the familiar Constitutional triad—legislative, executive, and judicial. But in recent years, as the volume of empirical studies of political processes has begun to mount, it has become increasingly evident that the traditional categories (which trace their origins to political theorists of the eighteenth century) are too restrictive and too closely associated with the particular kind of institutional arrangements found in Western democracies. Accordingly, several attempts have been made to improve on the conventional categories. Perhaps the most widely employed of the recent refinements is that of Gabriel A. Almond (Almond and Powell, 1966). Almond's categories are (1) interest articulation, (2) interest aggregation, (3) rule-making, (4) rule-application, (5) rule-adjudication, and, suffused throughout all the other categories, (6) communication (see later). Any such classification scheme is bound to be a bit procrustean; the dynamics of behavior in process do not always lend themselves to easy classification, and more than one function may be involved in the same act. Nevertheless, it is clear that political scientists are describing the same *kind* of functions that, in the terminology of cybernetics, fall under the headings of steering, regulatory, and control processes.

"Power" in Political Systems

One aspect of politics that is not included in Almond's list, though it is implicit and by no means overlooked by political scientists, is the role of power or authority. Indeed, the concept of power has been of central interest to political science, to the point that some consider power rela-

* Of course, macrolevel political systems are not the only loci of social selection with potential consequences for society at large. Partially autonomous subsystems at other levels of government or in the "private sector" also play a significant role.

tions—and the pursuit of power—to be the very essence of politics and a satisfactory way of delineating the political sphere. Harold Lasswell, an influential political scientist with a knack (at least in his earlier writings) for putting his thoughts into terse epigrams, summarized his view of politics in the title of his 1936 classic *Politics: Who Gets What, When, How.* More recently, Dahl (1970; p. 6) has defined politics as relations involving "power, rule, or authority." And, in their recent popularization *The Imperial Animal* (1971), anthropologists Lionel Tiger and Robin Fox gave the notion of politics as the struggle for power an ethological twist that brings the concept within the orbit of evolutionary biology, though in a misleading way. Since the issue is of central importance, it deserves to be discussed briefly.

What Tiger and Fox do—and with a certain relish—is equate politics in human societies with dominance competition in other species. Thus politics is "a world of winners and losers," and the "political system" is synonymous with the "dominance hierarchy." At first glance, it would seem Tiger and Fox are saying, as did O'Brien in George Orwell's masterpiece *1984*, that "power is not a means; it is an end . . . the object of power is power." But Tiger and Fox recognize that dominance competition is a product of the evolutionary process and a mechanism that generally has adaptive value. Because dominance is, in most species, linked with reproductive functions—with competition for food, nest sites, and mates—Tiger and Fox conclude that "the political system is the breeding system," and that politics *qua* dominance competition serves certain evolutionary functions.* Even in their terms, then, politics is not ultimately an end in itself, but a means to ends that transcend the individual and his motivations, although Tiger and Fox choose to emphasize the mechanism rather than its functions.

Having appropriated a word originally associated with human be-

* As an aside, Tiger and Fox have put forward what amounts to an interesting modern-day variant of Spencerian social Darwinism. It was the nineteenth-century social theorist Herbert Spencer who coined the phrase "survival of the fittest," and it was he who featured and extolled the role of interindividual competition in human evolution—although it was the self-serving pronunciamento of steel-magnate Andrew Carnegie that has been enshrined in the social histories of that era. Said Carnegie: "While the Law of competition may be sometimes hard for the individual, it is best for the race, because it ensures the survival of the fittest." Of course, in the nineteenth century the Capitalist economic system, and not governments and "politics," was presumed to be the major arena within which competition (and progress) took place. Spencer agreed with Marx on at least one point. The "state" was merely an epiphenomenon, a temporary strategem and a hindrance to progress that was destined eventually to wither away. Exactly the reverse has occurred of course, for reasons which a systems theorist could have predicted. In an era of big government, one can no longer ignore the state in formulating a theory of social evolution. So Tiger and Fox have co-opted it, putting politics at the center of the stage.

havior, and having redefined it in terms of breeding functions in other species, Tiger and Fox are then forced to concede that politics in human societies is quite a different kettle of fish. Dominance competition exists, of course, so there is still politics in that sense of the term, but protohominid hunting bands evolved a different kind of social structure—one involving a division of labor, specialization of skills, close cooperation, and reciprocity. Furthermore, dominance competition came to be dissociated, for the most part, from breeding functions—at least in mainstream human societies. So where does politics *qua* dominance competition fit into this new kind of social structure? How was it that dominance competition and dominance hierarchies survived the rigorous selection that must have taken place for collaboration and group-serving behaviors? Apparently dominance competition did not interfere with the progressive cybernation of protohominid societies. What functions could it have had? Because Tiger and Fox became transfixed with the idea of politics as the struggle for power and political systems as dominance hierarchies, they did not in the end bring this issue into focus.

It is not coincidental, I think, that among highly integrated group-living primates such as baboons and man, simple self-serving dominance–submission relationships evolved into "leadership" (and followership)—a well-defined set of "roles" with specific and limited functions for the group (Kummer, 1971). Though dominance competition might have had breeding functions for our remote ancestors, it has come to be adapted to new functions; social selection for breeding privileges has given way to a system of social selection for leadership.* (Such modifications of function are, of course, quite common in the evolutionary process, although we still do not know a great deal about how behavior evolves.) And regardless of what may motivate the individual to compete, leadership is a *function* that is related to behavioral cybernation. (It is a well-defined "office" in any formal institution.) As in any other kind of cybernetic system, the behavior of the parts must be coordinated to achieve the purposes of the whole. Yet behavioral "systems" are ephemeral things—dependent on the performance of individuals who are themselves partially autonomous cybernetic systems. The motivations and purposes of these individuals may or may not mesh with those of the larger systems of which they are a part.

The argument here is that dominance hierarchies *qua* leader–follower relations are structural members of political systems—a means by which steering and control functions are effectuated. Coercion—or rewards and

* We cannot, of course, be certain that our direct ancestors did in fact practice polygamy based on male dominance, although it is considered to be a reasonable hypothesis (Mayr, 1970, pp. 368–378).

punishments—may help to establish or reinforce the control structure, but leader–follower relationships are important both in pursuing common ends within existing political arrangements and in mobilizing collaborative efforts (social movements, revolutions, and counterrevolutions) to change or reestablish the polity itself. The use by social scientists of such terms as "legitimacy" and "authority"* reflects their appreciation of the fact that consensual and reciprocal relationships are an essential element of any durable political order.

Indeed, one of the things that is frequently overlooked by theorists who emphasize the struggle for power or the importance of individual self-interest in political processes is the ultimate dependence of the individual on a viable social order—and the need for constraint and reciprocity even in competitive situations. Although it would involve a digression to take up this issue at any length, it should be noted in passing that power-oriented theorists frequently make the error of taking the social order for granted—as though the basic problem of collective survival and an "agreement on fundamentals," as it is often euphemistically phrased, can be assumed as *given*. But if we start with the assumption of an inherent tendency to entropy, and if we take it as a working premise that collective survival is always potentially in jeopardy, then the struggle for power and the pursuit of self-interest must be fitted within the context of this larger problem. (I shall have more to say on this point later.)

In sum, while dominance competition and leadership are important aspects of the political process, to equate politics with the "power struggle" is to confine the term only to one facet—to one *means* among others† rather than to the *ends* of politics. In point of fact, means and ends are two sides of the same coin. Power is not some sort of disembodied entity. It is always relational and situational—a description of the relationship between means and ends in specific contexts. As Deutsch puts it, power is the *currency* with which political goals are attained.

Likewise, political systems are not simply static "power structures." They are dynamic, goal-oriented behavioral systems in process. As such, they involve the nexus of a number of behavioral components or elements. The "building blocks" of human politics include, among other things, operant conditioning, problem-solving activities (insight learning), modeling and imitation, memory, foresight, interpersonal and group

* Many social scientists have found useful sociologist Max Weber's threefold classification of legitimate authority: (1) traditional, (2) charismatic, and (3) rational-legal. All such classifications are "ideal types," of course, and may overlap with one another in the real world.

† Dominance hierarchies are not the only mechanism by which social control functions are exercised. In many group-living species—particularly among the insects—dominance competition is not in evidence (Wilson, 1971).

bonding, cooperative behaviors, leader–follower relations, authority-accepting behavior, language and various forms of nonverbal communication. Of course, the exact configuration assumed by these elements in any given instance depends on many factors—biological, cultural, and historical (or contextual). Not only are there wide variations between political systems at any given time, but also the same system may evolve through time. Diachronic differences can be observed in many systems, and we will consider some possible causes of political evolution shortly.

Communications and Politics

One other basic aspect of politics should be mentioned briefly, and that is political communication (including feedback). Patterned transactions of matter and energy are, of course, fundamental in cybernetic systems, but these processes are in turn directed and regulated by *information* transactions (or information "flows").* Information is essential to coordinating the behavior of the parts, or elements, of the system. In addition, most goal-seeking systems require feedback—information about the state of the system and about the consequences of steering and control processes. In a cybernetic system, the controllers must also be controlled.

This applies particularly to the kind of cybernetic system with which we are concerned here—the behavioral systems of social species; in recent years, there has been a convergence of thinking in many fields that communication is a functional requisite for all group-living animals (among the many references, see in particular Hockett, 1959; Lawson, 1963; Etkin, 1963; Wallace and Srb, 1964; Altmann, 1967; Sebeok, 1968; Cohen, 1968; Wilson, 1972). William Etkin has noted that "the existence of the social group presupposes that the members provide stimuli which evoke responses in other members of the group by which the group is held together" (Etkin, 1963; p. 149). This applies particularly to man, with his intricately organized "political" systems. There can be no doubt, certainly, that man has an unprecedented capacity for accumulating, storing, and transmitting behaviorally relevant information, and, as many scholars have noted, verbal language has played a major role in the evolution of human societies (Sapir, 1921; Cassirer, 1944; Pike, 1954; E. T. Hall, 1959; Whorf, 1956; Lawson, 1958; 1963; Berlo, 1960; McLuhan, 1962; Dobzhansky, 1962; Washburn and Shirek, 1967; Altmann, 1967; Lenneberg, 1967; Cohen, 1968; Lenski, 1970; Masters, 1970; Thorson, 1970; Gerbner, 1972; Washburn and McCown, 1972).

However, it is one thing to appreciate the role of language as an important "enabler," in Elman T. Service's phrase, and quite another to sug-

* See first footnote in this chapter.

gest that human culture can be reduced to language or that linguistic processes have "caused" cultural evolution, as some writers appear to do (Whorf, 1956; Cohen, 1968; Gerbner, 1972). This involves both a reification of language and a rather narrow definition of culture—and cultural evolution.

What the issue comes down to is this: To what extent can cultural processes and linguistic processes be equated with one another? There are, I believe, a number of reasons for questioning such reductionism. In the first place, it is becoming clear that human communications involve much more than what goes on at the conscious, verbal level. Explorations currently going forward in kinesics, the science of nonverbal communications (Birdwhistell, 1952, 1960, 1970; Hall, 1959), and sentics, the study of emotional communication processes in humans (Davitz, 1964; Clynes, 1969, 1970),* as well as the more extensively developed research among psychologists on modeling (see especially Bandura, 1971*a,b,* and references therein) all suggest that there are other important modes of information exchange in human societies besides symbolic language.† Indeed, we have only just begun to explore the extent to which direct observation of the *behavior* of others represents an important mechanism of cultural transmission, although some of the primate research, such as Menzel's (1971) important study of communications in chimpanzees, is very suggestive.‡ Menzel has shown that our primate relatives are able, without the use of specialized displays or symbolic language, to provide one another with considerable information about the environment (see also the reviews in K. R. L. Hall, 1963, and Lancaster, 1970).

Beyond the issue of nonverbal forms of communication, there is the more profound question of whether or not human mental faculties can be considered to be coextensive with symbolic language. Is culture (and politics) the product of language, or *vice versa*? At one pole, there are those who, like White (1944) and Whorf (1956), have argued that "meaning" for humans does not exist until words have been found to express it, and that a cultural artifact cannot be said to exist until it has been represented in the human "symbol pool." On the other hand, some hold, with Etkin (1963), that "The uniqueness of man must be sought not in the special characteristics of speech as a technique of communication but rather in the nature of the mind that makes use of speech as its chief instrument of communication. . . ."

* This parallels closely the work among ethologists and primatologists on "affective communication" in other species (Miller, 1967, and references therein).

† To a limited extent, humans appear to make use of all four modes of interindividual communication: auditory, visual, chemical (olfactory), and tactile.

‡ See also the account by Teleki (1973) of coordinated hunting behavior in chimpanzees.

In point of fact, much of the current argument over the relationship between language and man's mental capacities exists only because of our relative ignorance about what actually goes on inside the human brain. With the exception of Piaget's work (Piaget, 1963*a,b*; also Furth, 1969; Ginsburg and Opper, 1969), relatively little research was done on this particular issue until quite recently, although there has been an upsurge of activity in the past few years (see especially Vygotsky, 1962; Geschwind, 1964; 1972; Chomsky, 1965; Bruner, 1966; Furth, 1966; Lenneberg, 1964, 1967; Meadows, 1968; Pettifor, 1968; Lancaster, 1968; Reynolds, 1968; Premack, 1970). What is suggested by the limited evidence to date is that there is merit on both sides of the argument. Research on deaf persons (reviewed by Furth, 1966; also Pettifor, 1968) indicates that intellectual capacities, while possibly impaired, are not eliminated altogether by a person's difficulty in developing and using language skills. This is supported by neurological evidence of a partial independence of speech and higher intellectual centers in the human brain (Wooldridge, 1963; Thompson, 1967; Gazzaniga, 1971; Geschwind, 1972). In a similar vein, the rapidly accumulating literature on primate behavior indicates that fairly sophisticated intellectual and behavioral skills can be developed and sustained in the absence of symbolic language, as noted above. Learning, teaching, problem-solving, tool-making, information transmission, and complex behavioral coordination are all possible without verbal communication (Itani, 1958; Mirsky *et al.,* 1958; K. R. L. Hall, 1963; Marler, 1968; Jay, 1968; Van Lawick-Goodall, 1968*a,b*; Gardner and Gardner, 1969; Menzel, 1971; Kummer, 1971; Jolly, 1972; Dohlinow, 1972; Teleki, 1973). Furth (1966), Premack (1970), and others have argued that higher intellectual functions and the deep structure of language share a common neurological basis and that symboling and verbalizing abilities have served merely to increase the efficiency with which these skills may be utilized. There are, in fact, at least three major functional elements in the human language system—information-generating capacities, information-storing capacities, and information-transmitting capacities (*cf.* Premack, 1970)—and Peter Carlton Reynolds (1968) has provided us with a detailed analysis of what intellectual preadaptations might have been necessary for the emergence of human languages.

The issue remains unsettled, but one important conclusion can safely be drawn. It is not possible that language skills could have evolved for no apparent reason, after which our hominid ancestors learned how to engage in what we have defined here as "political" behaviors. Surely, the struggle to convey meaning in various social contexts and an increasing ability to do so must have evolved in a gradual feedback relationship to one another. Though some social scientists would seek to make language the prime mover of "political" evolution, politics—and the intellectual and linguistic skills on which modern political systems are based—must have evolved together in a relationship of reciprocal causation. Language has not caused cultural evolution to occur any more than mutations cause organic evolution to occur. (The fact that both primitive, homeostatic societies and rapidly evolving industrial societies share linguistic skills shows that the development of verbal language cannot in and of itself explain ongoing cultural and political evolution.) As with biological evolution, cultural evolution has been a multifaceted and relational process.

Finally, it involves a serious confusion of means and ends to equate language with culture or to equate cultural change with changes in the information system by which, in part, culture is transmitted.* The single most important characteristic of the biological realm is organization. Biological organization may involve many interdependent levels, each of which is erected on a distinct system of information storage and transmission. But these information systems—whether they be "gene pools" or "symbol pools"—are means to the end of biological organization, and it is biological organization on which natural selection acts.† It is not language that evolves but human organization as an extension of man's biological nature. Hence the importance of language and symboling lies, not in the fact that they have created cybernetic behavioral systems (which they have not), but that they have facilitated the exploitation of this level of biological organization as never before in evolutionary history.‡

ON THE ORIGIN AND DEVELOPMENT OF HUMAN POLITICS

One might assume that the evolution of goal-oriented collaborative behaviors in human societies represented an outgrowth of the larger process

* Indeed, if the term "culture" is confined only to what is carried in the symbol systems of human societies, then it would appear that culture can account for considerably less of the behavioral substrate of the species than was assumed only a few years ago.

† It has been fashionable of late to play on the analogy between the genetic code and human language (Teilhard de Chardin, 1959; McLuhan, 1962; Thorson, 1970; Masters, 1970). However, the analogy breaks down on one fundamental point. Evolution at the organic level is often initiated, or facilitated, by random changes in the information system which are then acted on by natural selection. In contrast, cultural changes do not, ordinarily, result from random changes in the symbol pool. They are the results of "purposive" adaptation and diffusion. As noted above, a Lamarckian process is involved in cultural evolution.

‡ In this regard, it should also be borne in mind that, despite its initial promise, information theory has so far failed to provide us with a calculus suitable for mathematical description of living systems (Johnson, 1970). In the biological realm, the *extensive* property of quantity (whether of information or energy) has no significance without reference to *intensive* (or qualitative) properties. Information must be modulated or structured so as to indicate its biological relevance or purposefulness.

by which our species evolved from small-brained Pliocene apes. But it is also possible that politics (in the sense that we have been using the term here) may have played a more crucial role (at least in the final stages) than is generally appreciated. Though we cannot yet be very confident about it, it appears possible that major changes in the behavioral system of evolving hominids (in the direction of behavioral cybernation) precipitated new selection pressures that led to the final emergence of *Homo sapiens*. If, as Mayr (1970, p. 363) has noted, "a shift to a new niche or adaptive zone is, almost without exception, initiated by a change in behavior," it is possible that the process of behavioral cybernation was not merely a product but also a cause of the biological evolution of man.

We may never be able to confirm the precise chronology or sequence, but the basic scenario of hominid evolution has been greatly clarified in the past few years by suggestive new studies of our primate relatives, by a number of new archeological discoveries, by more precise methods of dating fossil remains, and by some important refinements in our thinking. These have narrowed considerably the range of speculation.

Briefly, at the beginning of the Pliocene, some 13 million years ago, a rich variety of apes populated much of the Old World, including India, Africa, and China (Lancaster, 1970). Many of these species may have been unskilled tool-users and casual hunters (in much the same fashion as modern chimpanzees). Some millions of years into the Pliocene, but probably not less than 5 million years ago, pongid and hominid lines branched off from a common ancestor; a variety of morphological, immunological, molecular-level, and behavioral data (reviewed in Washburn, 1972) confirm a very close biological relationship between modern Hominidae and Pongidae.

For several million years thereafter and extending possibly into the Pleistocene, unspecialized protohominids succeeded remarkably well. Washburn (1972) suggests that these early ancestors of modern man may well have been ground-living, knuckle-walking tool users, and that the transition to bipedalism represented a response to behavioral changes associated with increased tool use. It was once thought that the genus *Australopithecus*, whose remains have been found in a number of deposits dating to the Lower Pleistocene (and perhaps even earlier), were transitional small-brained bipeds, but it now appears possible that they were instead contemporaries of larger-brained Lower Pleistocene hominids (Leakey, 1972; *cf.* Pilbeam, 1972). At any rate, until well into the Pleistocene, our hominid ancestors remained unspecialized hunters and employed crude (Oldowan) tools.

Yet tools and tool use do not tell the whole story. Though chimpanzees fashion tools and use them for a variety of purposes (Van Lawick-

Goodall 1968*a,b*), they do not appear to employ tools for hunting; knuckle-walkers can carry tools in their hands, but tool-carrying would interfere with chasing prey. Likewise, chimpanzees often hunt in groups, but they appear to be relatively poor at stalking and do not hunt larger, faster-moving game. Chimpanzees have also been observed to engage in collective defense against threatening agents (Kortlandt, 1967; Menzel, 1971), but chimpanzees do not appear to be exposed, for the most part, to systematic predation from large carnivores and pack-hunting animals.

Thus despite many suggestive similarities between human and chimpanzee behavior, there are many important differences, and adaptation to life on the savanna must have imposed new rigors, new selection pressures that led to the emergence of a more skillful bipedal hunter. We do not know why our hominid ancestors ventured out onto the savanna to seek a living. Changes in climate and vegetation have been the traditional explanations, but there are difficulties with this line of argument. Another possibility is migrations related to population pressures (Christian, 1970). In any event, we can formulate the issue of human evolution in this way: How might one expect natural selection to have modified an apelike ancestor resembling fairly closely the modern chimpanzee in order to adapt him for exploitation of an adjacent niche as a savanna-living predator? (Such adaptive radiations into underexploited niches are by no means uncommon.) The answer, of course, would be something like the genus *Homo*—though not necessarily the same in every detail.

With the emergence of the larger-brained *Homo erectus*, perhaps 500, 000 years ago, we begin to see evidence of a more complex social life, more varied and skillful tool-making (the so-called Aucheulian tradition), and more systematic and varied hunting. There seems little doubt that the morphological advances apparent in *Homo erectus* (and later) reflected positive selection to exploit an increasingly efficient set of behavioral adaptations. As William S. Laughlin (1968) has pointed out, hunting is not simply a subsistence technique but a demanding way of life and a complex system of social behavior. Efficiency as a hunter would require precisely the sort of mental and physical "improvements" that did develop in Hominidae.

Though *Homo erectus* apparently thrived for several hundred thousand years, the emergence of *Homo sapiens* traces to further evolutionary changes that occurred during the transition from Mousterian (or Middle Paleolithic) to Upper Paleolithic eras, and it has been proposed that the well-documented morphological changes of that period may have been preceded by a shift in subsistence patterns away from generalized hunting to specialized hunting of large migratory herd animals (Binford, 1968; also Pfeiffer, 1969). An ecological shift of this kind would have greatly increased positive selection pressures for large-scale task-oriented social organization. As Binford (1968, p. 714) notes: "Systematic exploitation of migratory herd mammals is a qualitatively different kind of activity, one that makes totally different structural demands on the human groups involved." Such a behavioral shift would have encouraged large-scale collaboration and an enlarged network of interpersonal relationships directed toward increasingly complex social activities. A premium would have been put on interband cooperation, more extensive planning and coordination of behavior, and the modification of supportive social relationships—from mating patterns to rituals and rule-systems.

In short, the capacities for complex behavioral cybernation in *Homo sapiens* were probably forged and perfected in the context of a progressive shift in subsistence patterns from small-scale, casual hunting to large-scale, specialized hunting. Once these skills had been developed, only a small additional step was required to apply them to new tasks—to warfare and the building of fortifications (Carneiro, 1970), to agricultural organization (Lenski, 1970; Boughey, 1971), to irrigation works (Wittfogel, 1957), and so forth. The process continues even today in such contemporary developments as conglomerate business enterprises, transnational corporations, and the European Common Market.

However, an important concomitant of this behavioral shift is the fact that the pack-hunting way of life must have represented a highly efficient new subsistence technique; the "economies" (the synergy) that resulted from shifting to this lifestyle must have produced positive reinforcement, or positive feedback. (Indeed, the wholesale extinction of large mammals during the Late Pleistocene suggests that evolving hominids became, if anything, *too* efficient at exploiting this niche; see Martin and Wright, 1967, and Boughey, 1971.) In effect, evolving hominids discovered (not all at once, of course, but by degrees) principles that economists, systems theorists, learning theorists, and, more recently, theoretical biologists (Trivers, 1971) would, many millennia later, formalize in their theories of social behavior. Early man must have found that specialization, division of labor, and cooperation can be "reinforcing"—especially in the harsh environment of the open savanna. That *hominidae* should have adopted such behaviors was not inevitable, but also not surprising considering the rewards to be gained for doing so and the penalties for not doing so.

Behavior as a Cause of Evolution

Indeed, the economies involved in the shift to hunting may help us understand the *mechanisms* by which such behavioral adaptations occurred. That behavioral cybernation can frequently have synergistic effects, rather than being disadvantageous to the individual, is a conclusion that an earlier generation of evolutionary biologists were reluctant to draw. In the past, extremely conservative assumptions were the rule among evolutionists; subordination to the group was presumed to involve self-sacrifice, by and large, with the result that the mechanisms of group selection (Williams, 1966, 1971) and kin selection (Hamilton, 1964) had to be invoked to account for the fact that group life does exist in nature nevertheless. But if we start, as Trivers (1971) does, with the more liberal assumption that a great many social behaviors are at very least pareto-optimizing (involving gains to at least some members of a group without loss to anyone) and often synergistic, then considerably more positive feedback, or positive "reinforcement," might be anticipated from experiments with group life.

Furthermore, the positive feedback from behavioral cybernation need not be confined to changes in gene frequencies (or in population size) through successive generations. "Adaptive responses" that are later reflected in the gene pool may be selected, in the first instance, through feedback involving internal mechanisms of behavioral modification within individual animals—the mechanisms of "learning." If so, these mechanisms may have significance as an evolutionary force that may not have been fully appreciated by either social scientists or evolutionists.

Over the years, learning theorists have demonstrated repeatedly that species as diverse as flatworms and man are endowed with a neural organization that permits them to "profit" from their experience in the environment. These species can "learn" which of the stimuli that impinge on them have "desirable" consequences and which "undesirable," and (within limits) they can orient their behavior accordingly. The most important of these learning abilities, for our purposes, are commonly classified as classical conditioning, operant or instrumental conditioning, insight learning (or problem-solving), and modeling or "vicarious reinforcement."

Unfortunately, though, there has been a tendency to reify these mechanisms, much as "natural selection" is sometimes given a life of its own. Often it is implied that these mechanisms exist independently of the particular organism involved—as though they were located in the environment, or inside a Skinner box (Skinner, 1965).* But in reality every learning situation involves a relationship between a particular organism—with its specific array of biological properties and survival "needs"—and the particular environment in which the animal must earn its living.

* This may be a reflection of the fact that an older generation of psychologists aspired to universal "laws" of behavior similar to the laws of Newtonian physics. That the so-called universal generalization paradigm may be inappropriate for dealing with biological phenomena, except at very high levels of abstraction with little or no predictive power, has only just begun to be appreciated.

Though the capacity for behavioral modification is innate, every instance of learning involves an interaction between the organism and its environment, and what is "aversive" or "rewarding" to one animal (or one species) may not be to another. Indeed, Seligman and Hager (1972) conclude that there may be systematic differences among species in their preparedness to learn in various situations, reflecting differences in the evolutionary history of each species and differences in the ways in which each is adapted to its environment (see also Rozin and Kalat, 1971).

Thus the mechanisms do not work the same in every case. And, instead of formulating learning theory in either/or terms—in terms either of "reinforcers" that exist independently of the animal or else "innate" behavior patterns—causation must be viewed in terms of the interaction between preprogrammed motivations and differential learning prepotencies, on the one hand, and the particular configuration of environmental influences, on the other.*

The psychologists' interest in learning theory has been concentrated on the mechanisms and processes of learning. They are interested in how an animal adapts to its environment during its own lifetime. For the most part, psychologists have not been concerned with the evolutionary consequences of learning abilities. Ethologists, likewise, have been interested primarily in how different animals learn their species-specific behaviors (to the extent that such behaviors are learned), and in the adaptive significance of learning for different species, while behavior geneticists have focused on the genetic bases of behavior and the selective consequences of behavioral differences—however these differences may occur. Some social scientists have shown an interest in operant conditioning in relation to the "cultural evolution" of human societies (Campbell, 1965; 1970; Alland, 1967) but have not, so far as I know, viewed it as a mechanism that could contribute significantly to biological evolution.

Evolutionists, on the other hand, have concentrated on transgenerational changes in gene frequencies. It is recognized that most behavior results from an "interaction" between organism and environment, and evolutionists often speak of "adaptive responses" or "opportunistic" behavioral adaptations, but they are in fact vague about the precise nature of these interactions.† Although it is now widely accepted that behavioral

* In fact, the causes of behavior cannot, properly speaking, be confined to the experience of an animal during its own lifetime. For when we consider how the biological substrate of an animal's behavior system came to be in the first place, we must introduce the effects of natural selection on previous generations of ancestors and ancestral species. Behavioral causation in the long view must be conceived in terms of a circular, time-bound feedback process in which natural selection "chooses" from the behavioral *phenotypes* of one generation the behavioral *genotypes* of the next.

† By the same token, systems theorists are often vague about precisely what mechanisms are involved in their all-important concept of feedback.

changes may often precipitate genetic change, for the most part evolutionists have focused on presumably "random" morphological or molecular-level changes (some of which result in individual differences in behavior, of course), while neglecting the potential role of capacities for adaptive behavioral modification that are found within many organisms. These capacities are the "selection criteria" to which Campbell (1965) referred in searching for an analogy between biological and cultural evolution. But what if the two forms of evolution are in fact *homologous*, at least in part? It is also possible that the neural mechanisms underlying learning capacities represent a significant force in biological evolution, serving to canalize evolutionary change to a degree that we have not as yet explored.*

Thus the fortuitous adoption and diffusion of a new subsistence pattern (such as occurred in hominid evolution) may often depend, in the first instance, on selective reinforcement for behavioral modifications that occur within the existing "reaction range" and learning capacities of a species. Perhaps initially only a few individuals (or populations) within a species perform a new behavior and do so successfully enough to be "rewarded." Or, alternatively, there may be individual differences in the *capacity* for being rewarded. In either event, the rewards will reinforce the behavior and, if the behavior confers an adaptive advantage on those individuals, it will subsequently be "rewarded" by natural selection. It will increase fitness and success in leaving progeny. Morphological changes to increase the efficiency with which such behaviors are performed (and to fully exploit its benefits) could then be expected to follow in succeeding generations.

In suggesting that the mechanisms of learning may have some potency in the evolutionary process itself, I do not, of course, ascribe to the behaviorist assumption that learning theory can fully explain the causes of behavior. My position is simply that classical conditioning, operant conditioning, and the like represent one set of mechanisms by which goal-oriented organic systems adapt to their environments, and some of these "purposive" adaptations (in the cyberneticists's sense) may have transgenerational effects on gene frequencies. If so, then the evolutionary process may have been shaped in part by the internal characteristics of evolved organic systems.†

It remains to be seen how significant a contribution, if any, these mechanisms make to the evolutionary process as compared to molecular-level changes, or changes in the environment. Presumably there will be considerable variation from one species to another. But, by any measure,

* See footnote on p. 257.

† In fact, the mechanisms of learning must ultimately be fitted within a cybernetic model of biological organization, if sense is to be made of them in evolutionary terms. A promising effort along these lines is the work by Cunningham (1972) (see also Boden, 1972).

Homo sapiens will rank at the very top—and this returns us to the main purpose of this discussion.

Toward an "Evolutionary" Theory of Cultural Evolution

Ever since Herbert Spencer's day, the cultural evolution of man has been treated as a thing apart, somehow, from the larger process of biological evolution. It was Spencer who coined the term "superorganic" to distinguish the cultural and biological realms, and though Spencer acknowledged an interaction between the two realms, some of his successors in the social sciences went so far as to deny any connection between the two (e.g., White, 1959; 1968). It became necessary for the biologists to remind us all that there are at the very least feedback relationships between the two "levels" of evolution (Roe and Simpson, 1958; Dobzhansky, 1962). Indeed, in recent years it has become apparent that cultural evolution (and, more specifically, man's transformation of the natural environment) is producing feedback effects not just for *Homo sapiens*, but for the entire biosphere.

If, in addition, it turns out that some of the mechanisms involved in cultural (and political) evolution have been independently influential in the evolution of a great many other species as well, then it would be more accurate to view cultural evolution as an amplification (and acceleration) of processes that are firmly embedded in the larger process of biological evolution. If we take the long view, cultural (and political) evolution should then be treated as different in degree, but not different in kind, from the process by which other species have evolved—and continue to evolve. In fact, if systematic behavioral adaptations independent of biological modifications are a common occurrence in the evolutionary process, then the very term "cultural evolution" loses its special meaning. The phenomena included in the evolution of human cultures become, quite simply, another aspect of *the* evolutionary process. Of course, we are far too anthropocentric and there are far too many vested intellectual interests presently associated with the concept of "culture" for us to abandon the term in the foreseeable future, even if it is becoming increasingly difficult to justify it on theoretical grounds.

Theorizing about the causes of "cultural" evolution in man is an ancient pastime. Some of the more euphoric theorists have portrayed the process as being goal-directed, or at the very least characterized by some overall trend (increased energy capture, increased homeostasis, or what have you). Yet even the most mechanistic- and materialistic-minded theorists have usually assigned an important causal role in cultural evolution to the workings of the human mind itself (Nisbet, 1969)—even though that role is sometimes only implicit. Thus, for example, Karl Marx's famous dialectic involved an interaction between successive states of human consciousness and the "mode of production of material life" (which, after all, consists of human artifacts); the dialectic, essentially, was between previously actualized human ideas and emergent new perspectives. More recently, anthropologist Leslie White (1959) espoused a "technological determinism" that, in the last analysis, involved the actualization of human mental processes, while Marvin Harris (1969) has put forward a more refined "probabilistic determinism" involving three sets of variables (economy, technology, and environment), two of which obviously represent the products of the human mind.

Many theorists have also recognized the multivariate and interactional nature of "cultural" evolution. Montesquieu, for instance, spoke of an interplay between basic human needs and such "variable factors" as climate, soil, geographical location, population size, occupations, commerce, and manners and customs. Of course, unilinear cultural evolutionism and "prime-mover" theories of social change have also been common. A recent example is Ester Boserup's (1965) model in which population growth is viewed as the major independent variable. There is also a new crop of quasi-social Darwinists who attribute cultural evolution primarily to human aggression (Bigelow, 1969; Ardrey, 1970).* However, most contemporary anthropologists favor multicausal models, with interest centering increasingly on the "reciprocal" relationship between population growth (or decline), changes in the means of subsistence, and social and political change (Dumond, 1965; Carneiro, 1967; Spooner, 1972).

The difficulty with such broad, configural approaches is that they have thus far tended to produce either global reifications or else unique and situation-specific "causes" with little or no general predictive value. At the same time, we remain unenlightened about the precise manner in which all of these variables interact with one another. What, finally, are the mechanisms by which cultures evolve or, equally important for a general theory of cultural evolution, achieve homeostasis or even devolve? Is there some analogue for natural selection in cultural life that can provide the basis for a general theory of cultural continuity and change?

The closest approximation is the model suggested by Campbell (1965). Building on the earlier work of Keller (1931), Campbell has proposed a frank analogy between "cultural" evolution and evolution at the biological level which he describes as a "blind-variation-and-selective-retention model" of societal change. His model involves three essential elements: (1)

* Though Carneiro (1970) hypothesizes that warfare may have had an important role in the emergence of the state, ecological factors—population increases and competition for scarce resources—are viewed by him as being prior in the chain of causation.

the occurrence of cultural variations that are analogous to mutations at the molecular level; (2) mechanisms (and consistent criteria) for selection among variations (the functional equivalent of natural selection); and (3) mechanisms for propagation, replication, and preservation of positively selected cultural variants.

There are several problems with this model. First Campbell's emphasis on the "blind," "random," and "haphazard" character of cultural variations is unfortunate. He meant to convey the idea that, in the first instance, the variations that provide the raw material for cultural selection need not be purposive and deliberate; variations are the important thing, however they occur. The point is well taken,* but it may be misleading to suggest that blind variations are the *rule*, or that they are literally random in nature. Just as molecular-level mutations are now recognized to be far from "random" in the strict sense of the word, so too cultural variations generally occur within a very narrow range. Furthermore, it may be inaccurate to assume that cultural variations in primitive societies are "blind" simply because they seem nonrational by our standards. They may reflect instead a process of hypothesis formulation and testing based on non-Western, nonscientific assumptions about the nature of causation in the phenomenal world. They may often be more "purposive" than we think. (Indeed, even many so-called trial-and-error processes may not be random; they may be oriented to predetermined goals.)

A more serious problem, one which Campbell acknowledges, is that he cannot specify with any precision the source or basis of the mechanisms by which cultural variations are selected. The inner principles elude him. He identifies at least six different ways in which cultural selection might occur: (1) the selective survival of complete social organizations, (2) selective diffusion or borrowing between social groups, (3) selective propagation of temporal variations, (4) selective imitation of interindividual variations; (5) selective promotion of individuals to leadership roles, and (6) rational or deliberate selection. But *how* does selection occur?

Reflecting the perspective of social scientists of the early 1960s, Campbell argues that cultural evolution is no longer "checked" for survival relevance in the strictly darwinian sense. Natural selection cannot be invoked to explain cultural evolution, in other words. On the other hand, while Campbell suggests the possibility that "internal selection criteria" might be operative in some instances, he does not develop the insight—perhaps because he proceeded from a crude pain–pleasure model of human

* As an example, Alland (1970) has found instances of hygienic cultural practices among primitive peoples (thorough cooking of food, washing of utensils, bathing, etc.) that are not consciously associated by the population with the formal medical practices and disease prevention techniques of the society.

motivation which seemed unable to account for the obviously complex process of societal evolution. As a result, Campbell's model of cultural evolution remains at the "pre-Darwinian" stage of development, so to speak; the specific mechanisms remain to be identified.

A third deficiency in Campbell's model is that he does not indicate what role, if any, is played by external variables—environmental constraints and opportunities. It is not at all clear that Campbell viewed these variables as having any causal efficacy whatsoever in cultural evolution.

Finally, Campbell's model presupposes an autonomous process of societal evolution that is disconnected from the larger process of biological evolution. This assumption grows out of a mode of thought that has been prevalent in the past two or three generations, but it is, as I have suggested, erroneous. It begs the question: How did our capacities for culture evolve in the first place? It also assumes that there is no signficant feedback between culture and our gene pool. However, the environmental crisis of the past few years—our "crisis of survival"—gives the lie to such an assumption. It has become abundantly clear that many cultural practices in modern societies are of direct consequence in terms of our continued viability as a species. Furthermore, it is not necessarily correct to assume that our gene pool is no longer "tracking" cultural changes.* As Wilson (1971) points out, microevolutionary changes in a population may occur over a much shorter time span (perhaps as few as ten generations) than we had previously assumed to be the case.

The model I would like to sketch briefly here may help us surmount the difficulties Campbell encountered. The keys lie in a proper understanding of how selection operates and in a more sophisticated understanding of the behavioral substrate of the human species.

In biological evolution, causation is circular and feedback dependent. Natural selection—the most important agent of transgenerational change—is not something external to the relationship between organisms and their environments. It is in reality a way of describing the outcomes of organism–environment interactions, and these outcomes, or "effects," are the "causes" of the more *systematic* changes between generations that we call evolution.

The textbook example is "industrial melanism" (Lerner, 1968). Until the industrial revolution, light strains of the peppered moth (*Biston betu-*

* This is a matter of considerable debate and speculation. Lerner (1968) has estimated that one-third of the current generation will contribute two-thirds of the genes to the second generation hence. In other words, fertility differentials do exist, and some assortative mating persists (Eckland, 1972). Glass (1972) has argued that the most systematic selection pressure today is probably for resistance to disease. But the fact is we really do not know.

laria) predominated in numbers in the English countryside over darker, melanic forms of the species. The latter were relatively rare. The reason was that when the moths were resting on lichen-encrusted tree trunks the light forms were all but invisible while the dark forms stood out. As a result, the light forms were far less subject to predation from insect-eating birds. They survived and reproduced in greater numbers. However, as industrial soot progressively blackened the tree trunks near factory areas, the effect was to alter the relative visibility of light and dark forms, and in time this change in the organism–environment relationship brought about a reversal in the relative frequencies of the two forms.

The same sort of relationship, I maintain, applies to cultural (and, needless to say, political) evolution. The causes of continuity and change in cultural processes are to be found neither in environmental factors nor in human mentation and action but in both—that is, in the interactions between human organisms and their various environments (cultural and natural). Changes either in the behavior of the human organism (or the collective behavior of a group of organisms) or in any of the environments in which the human organism is embedded (or changes in both, for that matter) become selectively relevant only insofar as the organism–environment *relationship* is significantly altered, and it is those alterations (or "variations") that are subjected to cultural selection.

Take the invention of the automobile, for instance. In the early days of automobile development, many variations on this theme were tried, but only those vehicles with a combination of characteristics that enabled man to enhance significantly (and reliably) his mobility and ability to transport goods were "selected" and further developed. Three-wheelers, naphtha-powered and steam-powered machines, and, in time, horsedrawn vehicles all were selected against—a selection process that involved human choices based on the relative economies or utilities involved.

What, then, are the mechanisms of cultural selection and evolution? Indirectly, natural selection. Natural selection has been responsible in the first instance for the emergence of the motivational and behavioral propensities of the human species and for shaping the biosphere within which man must earn his living. In addition, natural selection continues to pass final judgment on the various ways in which these propensities are expressed, as indicated previously. But the *immediate* source of cultural selection is a more proximate, intermediate mechanism that might be called "human selection."* By human selection, I am referring to the processes

* For generations, plant and animal breeders and geneticists have employed the term "artificial selection" with reference to deliberate efforts to modify the biological characteristics of a species of organisms through selective breeding. In contrast, my use of the term "human selection" refers to selection of cultural forms by individuals, groups, or "authorities"—selections that may or may not be conscious and deliberate.

by which Campbell's "internal selection criteria" "select" from ongoing behavior–environment interactions. Just as natural selection refers, not to the *initial* sources of change (say a new mutation), but to the *consequences* of that change, so human selection refers, not to human inventions or behavioral adaptations *per se*, but to selection based on the *effects* of those changes in relation to the satisfaction of human needs and wants. It is the perceived or actualized utility—the rewarding or reinforcing properties of the change—that counts. Artifacts and behaviors that contribute to the satisfaction of human needs and wants (needs and wants whose origins are biological and a product of the evolutionary process) will be selectively favored on balance (though, for a variety of reasons, favorable "variations" may sometimes fail to be incorporated into the cultural system), and those that are nonrewarding, or negatively reinforcing, will be disfavored.

The proposition, then, is that the variations (or pressures) that initiate cultural changes may arise from many sources: population increases, depletion of an essential resource, a climatic change, a human invention, or perhaps a new survival strategy developed and organized at the group level through the political process. However, it is the differential effects (or functions) of these changes in relation to the matrix of human needs that determine which cultural modifications will persist and contribute to the ongoing evolution of a society. Thus "human selection"—like natural selection—is a global abstraction that applies to a very large number of disparate phenomena. The reason such phenomena may be classed together is that they have a *functional equivalence* in terms of the problem of human survival as manifested in specific human needs.

Of course, if the matrix of human needs amounted only to a few rudimentary drives, or to the sort of crude pain–pleasure mechanism of naive behaviorist psychology, the mechanisms of human selection would be limited indeed, and it would be difficult to see how the human selection model could account for cultural evolution. However, the behavioral sciences have in the past few years gravitated toward a more complex model of human motivation that includes, not only the basic needs for self-preservation and sustenance, but also an array of derivative and *instrumental* needs in the social and psychological realm (for a comprehensive discussion of the subject, see Nash, 1970). Perhaps the most elaborate formulation of this revised view of "human nature" is Abraham Maslow's hierarchy of human needs (Maslow, 1954). Though flawed and a first approximation at best, Maslow's model does suggest strongly that the individual constitutes a set of preferences or selection criteria with considerable selective power.

In contrast to the earlier formulation of Campbell, therefore, I am proposing a model in which the human animal is viewed as a partially autonomous cybernetic system whose goal-directedness is the resultant vector, so to speak, of an array of internal needs and wants that have been acquired by the organism in the course of hominid evolution. These goals and value orientations are biologically based, even though they may be expressed in a variety of ways and be subject to considerable molding in the environment.

For the purposes of developing a formal model of cultural evolution, it might be useful to conceptualize the process as one involving interactions between two matrices. On the one hand, there is the matrix of human preferences, or selection criteria. (It may be necessary, in fact, to treat the process of human selection as being configural; needs and wants must often be weighed against one another, and the making of choices may require optimizing for several preferences at once.) On the other hand, the cultural and natural environments together may be viewed as a pattern of differential reinforcers (both positive and negative). This includes reinforcers that are created and maintained by different cultures for purposes of socialization and social control (Bem, 1970), as well as cultural practices that mediate the relationship between the individual and the natural environment.

In these terms, human selection is oriented to the "fit" between the individual's array of internal selection criteria and the pattern of reinforcers in the environment. Cultural continuity may be viewed as a reflection of an adaptive "fit" between these two matrices (though cultural stability may be maintained, even when maladaptive, by a conservatizing pattern of cultural sanctions and rewards—or by a lack of alternatives). Human selection will act in such cases to maintain that fit, in a fashion that is precisely analogous to stabilizing selection in Darwinian evolution. Conversely, either a degradation in that fit or an opportunistic improvement will set up "negative" or "positive" selection pressures (i.e., reinforcers or anticipations of reinforcement) for cultural change.*

Of course, the precise manner in which systematic cultural changes occur is complex, as Campbell suggested. Some instances of cultural evolution involve diffusion at the level of individual behavior, but many changes at the group or societal level involve "political" action. As Anthony F. C. Wallace (1956) noted in a seminal paper on the subject, the various kinds of "revitalization movements" that occur in human societies have in common the fact that they are organized, collective efforts to change a culture that is "unsatisfying." Wallace did not, however, deal in detail with the sources of dissatisfaction, whereas Ted Robert Gurr's empirically supported causal model of group violence (Gurr, 1968, 1970) singles

* Although there are parallels between the model sketched here and the formulations of gestalt theorists and field theorists in psychology, there are also obvious differences. What is involved here is a synthesis of configural, ecological, and behaviorist approaches within a cybernetic and evolutionary framework.

out "relative deprivations" as mediated by such factors as perceived benefits to be expected from political change and the perceived likelihood that collective violence can succeed.

By the same token, a sustained "trend" in cultural evolution suggests that a particular pattern of cultural change is reinforcing or rewarding in relation to human needs and wants. Yet all cultural trends are contingent. A trend will probably not continue long, once it ceases to be rewarding. (Much has been written on the phenomenon of cultural lag.)

There are, of course, many problems and many complexities relating to the model I have been describing, and I can only allude to a few of the more important issues here. For one thing, there is the thorny problem of specifying in greater detail precisely what the postulated substrate of human needs consists of and how these needs are integrated with one another. It is becoming increasingly obvious, for example, that there are biologically based individual differences in "personality" characteristics. There are also striking differences between cultures, and though there are very definite limits to the malleability of human nature, it is abundantly evident that cultures play an important role in molding the preference structures of individuals. By the same token, not all human needs are equally urgent and inflexible, as Maslow emphasized.

What the problem comes down to is the fact that our understanding of the mechanisms of human selection is limited at this juncture by the limits of our knowledge about the biopsychological bases of human motivation and learning. Of course, our understanding of the workings of *natural* selection is similarly limited; we are not therefore required to understand human selection completely in order to recognize its existence. Furthermore, as behavioral scientists increase our understanding of the substrate of human behavior, we will, at the same time, be illuminating the mechanisms of cultural evolution.

Some other complications that might be mentioned briefly:

1. Where, precisely, is the locus of human selection? Who does the selecting? Presumably, when one is given the opportunity to do so, the locus of selection is the individual. But, as Campbell suggested, individuals do not always choose for themselves. Much human selection may be imposed on an individual by parents, teachers, coaches, employers, and other "authorities." On the other hand, authorities may have considerable difficulty enforcing their choices. Many individuals may resist what they perceive or experience as aversive choices made by others in their behalf.

2. How are cultural changes diffused and retained in the cultural system? And does each succceeding generation in effect reassess the choices of previous generations? Although there are data available that bear on these questions, systematic analyses have not been done, so far as I know.

3. How do memory and foresight fit into the human selection model?

Intuitively, it would appear that *anticipated effects* are a frequent cause of human selection at both individual and collective levels, but the phenomenon will require further investigation.

4. The human selection model must also take into account the existence of feedback loops which affect the ongoing organism–environment relationship; man's behavior may alter the natural environment in such a way as to precipitate unanticipated and possibly aversive changes in the fit between man, culture, and natural environment. If necessity is the mother of invention, inventions are also the mother of necessity.

5. There is also the problem of defining "adaptiveness." Presumably, one can draw an analogy from the usage of evolutionary biologists. A cultural artifact is adaptive, or a behavior is adaptive, when it enables individuals to satisfy their needs. But what is adaptive for one individual's preferences may be maladaptive for others (or for the group) in the many cases where human preferences intersect. Equally important, what is "adaptive" in terms of the human selection model may not be adaptive in Darwinian terms—in terms of the problem of biological survival and reproduction through time.

6. Finally, there is the complicated question of how politics fits into the human selection model. As I have defined the term here, politics at the macrolevel is viewed as having primary responsibility for the collective survival strategies of human societies; it is a major contributor to the process of human selection at the collective level. Accordingly, the evolution of politics may be viewed as the progressive elaboration of specialized and routinized structures for the performance of these functions. And, like other cultural accretions, the emergence of formalized politics has involved a cumulative (though by no means linear) process of invention, testing for cultural "fitness," and subsequent modification.

Consider, for example, the evolution of a particular political artifact—say the U.S. Constitution. We need not invoke some sort of tortured explanation based on natural selection—except indirectly and over the very long run. In the shorter run, the incorporation of an invention of this kind can be accounted for in terms of evolved human capacities in interaction with accumulated cultural antecedents (or cultural "preadaptations," if you will), as well as selective modeling of other political systems and anticipatory problem-solving and planning. The result was a new organizational synthesis which was subsequently tested in relation to the political needs for which it was devised. In other words, political evolution is posited as being the result, in the main, of "positive selection" for cultural inventions which, *on balance*, contributed to the meeting of human needs—at both the individual and collective levels.* This is not to say that such de-

* Of course, "negative selection" is also part of the process. A prime example is the decline of absolute, hereditary monarchy.

vices are equally adaptive for everyone, that they are functioning as efficiently as they might, or that they will continue to do so in the future. Again, the utilities involved in such adaptations are always contingent. Of course, it is easy to describe in the abstract the evolution and functions of politics (or of a particular political institution), but it will be very difficult to translate such propositions into testable inferences.

THE FUTURE OF POLITICS

The argument here is that political behaviors are a product of evolution and an inextricable part of the process by which man himself evolved. I have postulated that the trend toward cybernated social behaviors was encouraged by the economies involved in a behavioral shift in this direction (for detailed discussions of the "economics" of collective action; see Arrow, 1951; Downs, 1957; Buchanan and Tullock, 1962; Riker, 1962; Olson, 1965).

To be sure, such a behavioral shift was possible only because early hominids were endowed with an appropriate combination of morphological and behavioral preadaptations. By the same token, had there not been underexploited resources present in the environment, this change in the basic hominid survival strategy would have gone unrewarded. Yet, in the final analysis, it was the utilities involved that gave direction to the progressive changes that occurred. In short, political behaviors proved to be adaptive for the group as a corporate entity and, on the whole, for the individuals who comprised the group.

If this interpretation is correct, then some important implications follow—implications relating to how macrolevel political systems are to be conceived and what role they should be assigned in society.

Western political theory has long been polarized around two opposing views of the political order (though there have also been bridge-builders who sought to reconcile the two camps). On the one hand, there are those who have espoused the view that societies have collective purposes—often termed the "general welfare" or "public interest"—that may transcend the individual and his needs. Society can thus be likened to an organism (or, in contemporary terminology, a cybernetic system), in which the whole is greater than the sum of the parts. This is the view I have been espousing here, but with some important qualifications, as I shall explain.

The origins of the so-called organismic analogy can be traced to Plato's *Republic*, although it enjoyed its greatest vogue at the turn of the present century (for examples, see Wilson, 1908; Lowell, 1910). Some theorists of that era went so far as to assert a detailed analogy between specific parts of the human body (heart, lungs, circulatory system, etc.) and the "body politic." However, more sophisticated theorists, such as Herbert Spencer (1891, 1899), fully appreciated that there was at best only a loose *functional* analogy between organisms and polities. Although the organismic analogy came under severe attack during the early 1900s (see especially Coker, 1910) and was eclipsed for a time, it has recently enjoyed a revival in a different guise within the theoretical frameworks first of structural-functionalism and then of systems theory.* And once again it has been subject to attack (Nisbet, 1969).

The other major perspective on politics derives from the view—traceable at least to the Epicureans—that the locus of human behavior is the individual and his preferences. A polity, therefore, is merely an arena within which individuals pursue their self-interests. Or, in the case of modern "interest group liberals," it is the arena within which organized interest groups pursue their group interests (Lowi, 1969). Thus one cannot assign higher purposes to politics than the express purposes of the individuals who engage in politics. And because people have so many different and often conflicting interests, the idea of a public interest, or general welfare, is a chimera. Often the public interest is invoked merely as a strategem to disguise one's self-interest—or so the argument runs. Accordingly, a sophisticated person should be cynical about any political claims made in the name of the public interest. As political scientist David B. Truman put it in one of the more influential texts of this genre, *The Governmental Process*: "Assertation of an inclusive 'national' or 'public interest' is an effective device. . . . However, [it does] not describe any actual or possible political situation within a complex modern nation. In developing a group interpretation of politics, therefore, we do not need to account for a totally inclusive interest, because one does not exist" (Truman, 1951, pp. 50–51).

Among other things, this viewpoint led to a posture of relativism and moral immobilism. Because no individual's preference could claim to have priority over any other, many otherwise compassionate men felt compelled on intellectual grounds to abjure the capacity for passing judgment on public policy or political conduct. Some liberal theorists became quite arrogant. Claiming that the liberal view of man was grounded in the "facts" about how people actually behave and how politics actually operates, these political "realists" fancied themselves as being more "scientific" than idealists who espoused supposedly "metaphysical" notions about global purposes (Brecht, 1959).

In my view, neither collectivist (or systems) theorists nor the individualist, liberal theorists are wholly right or wholly wrong. Both have hold

* Of course, socialist theory also fits within this rubric, as do National Socialism and other totalitarian ideologies. It is certainly true that "higher" collective purposes have been used by political leaders to justify political tyranny.

of a part of the truth, and a synthesis of both viewpoints best approximates the realities of the human condition.

The basis for reconciling "organismic" and individualist views of man and society is the fact that the ongoing problem of survival and reproduction is at once an individual problem and an aggregate or collective problem (especially in a group-living species such as man). Though the survival and reproductive success of the individual are not unrelated to the survival of the species, the larger problem transcends the individual and his needs. There is indeed a "general welfare"—the needs (which may often go unrepresented directly in macrolevel political processes) of the collective survival enterprise as a corporate entity and of future generations. These needs may not be consciously recognized or explicitly operative as an influence (or decision-criterion) in the political process, but the needs exist nonetheless. Accordingly, even if liberal theorists are correct about the locus of behavioral motivation, this does not mean that there are not "higher" purposes realized by the social order (what Robert Merton would have called "latent functions"). One must distinguish, in other words, between the proximate "causes" of political behavior and the "consequences."

Ten years ago, it was possible for many social scientists to assume that the problem of survival had somehow been "solved" and that the only "problems" associated with political processes were those relating to immediate and explicitly articulated "wants." Until quite recently, it was widely held that the most pressing problems confronting the human species were (1) how to avoid a nuclear war and (2) how to export to the underdeveloped countries the Western standard of living—or what Walt Rostow called in his influential book, *The Stages of Economic Growth* (1960), the stage of "high mass consumption."

We are sadder but wiser today. We now appreciate that the continued viability of our species is a pressing and immediate problem (even if one excludes the danger of a nuclear holocaust), and that the actions of macrolevel political systems may have a decisive influence over the outcome. In other words, we are now more conscious of the general welfare function and appreciate that it is not merely the sum (or product) of the self-interests of political actors.

The relationship between survival as an individual affair and as a collective enterprise can best be understood as a reflection of the hierarchical organization of biological phenomena. A social species such as man represents a hierarchy of partially autonomous—but also partially interdependent—systems, of which the individual organism is but one "level" (on this point, see Wilson, 1969, and references therein). As Koestler (1967) has noted, systems at every level of a system hierarchy tend to be Janus-faced. When facing "downward" (so to speak), they act as autonomous wholes, but when facing upward, they act as parts in more inclusive systems.

Thus any given individual may be at one and the same time a pursuer of his own system needs (or preferences) and a functional contributor to more inclusive systems. Indeed, more complex human societies exhibit several levels of behavioral organization, each of which is functionally specialized.*

Of course, the precise relationship between the individual and the collectivity depends on the context. Different individuals have different degrees of functional importance for the behavioral systems of which they are a part. Conversely, "higher" levels of social organization may have more or less relevance and value to different individuals in terms of satisfying or denying their individual needs. And because an array of needs is involved in the survival problem, different needs may be satisfied at different system levels.

But, in any event, the pursuit of self-interest may have functional consequences both for the individual and for the collective survival enterprise. One can distinguish, in fact, three different kinds of self-serving behaviors: (1) those that also contribute to the corporate needs of the collectivity, (2) those that are neutral in their effects, and (3) those that are maladaptive or detrimental. Accordingly, one of the most basic and persistent problems of any durable political order is to find and encourage (if not enforce) a suitable balance between individual self-interest and the "public interest." The relationship is a dynamic one and may require constant adjustment as conditions change.

In strictly "Darwinian" terms, an adaptive society is one in which the needs of future generations are given the highest priority; it is a society in which the self-gratifications of the present generation may be either encouraged or (as appropriate) discouraged, in the interest of the higher goal of collective survival over the long haul. By the same token, an adaptive political system is one which is able effectively to organize and coordinate individual and group behaviors so as to conform with higher, systemic goals.

But of course the imposition of such "higher" goals on a society might involve actions that are disruptive of social harmony and might well constrict the individual's pursuit of happiness for the present generation. And, in the real world, relatively few citizens may care about the long haul or be willing to make personal sacrifices for generations yet unborn. As a

* The modern nation-state is far more encompassing than any strictly biological deme; it is a superdeme. By the same token, the emerging international system—as yet only a tenuous entity but clearly impelled by the momentum of human evolution—may in time produce the ultrademe, that is, a level of biological organization encompassing (for certain purposes) the entire human species.

rule, political leaders are concerned with solving immediate problems and assuring their own political survival. Indeed, the "good of the state" has frequently been invoked to justify political oppression when in fact it is the survival of the regime rather than the collective good which is involved.

Thus after 20 centuries of theorizing and experimentation with political systems, we still have no definitive answer to the problem raised by Plato in *The Republic*—the fountainhead of Western political theory. Is it possible to engineer a political system that is capable of serving the general welfare and, at the same time, achieving social harmony by doing "justice" to individual citizens?

Plato's answer, one that is fully comprehensible to the modern systems theorist, was that individual talents should be matched to systemic roles, and that the whole should be coordinated and directed by a class of specially trained "guardians." Although the Platonic ruler was to be unfettered by law, tradition, or popular will (referred to contemptuously by Plato as "the great beast") and was to have absolute power to pursue the commonweal as he saw it, he would also be carefully schooled to become, in Sheldon Wolin's phrase, a "servant of truth untouched by his own subjective preferences or desires" (Wolin, 1960, p. 55). If the ruler's power would be unrestrained, it would also be yoked to absolute knowledge and to an absolute allegiance to the general welfare. In short, Plato had conceived a perfect steersman for the perfect cybernetic social system.

Plato's vision of the philosopher-king ruling over a perfectly ordered, totally mobilized society has proven to be one of the most compelling "models" in the history of political theory. For, in times of social crisis, men have often yearned for a charismatic leader who could reconstitute the social order and set it going on a new course. Cicero's "Lawgiver," Machiavelli's "Omnipotent Legislator," Hobbes's "Sovereign," Rousseau's "Legislator," and Hitler's "Fuehrer-Prinzip" were all echoes of the philosopher-king.

Unfortunately, though, Plato's model was fatally flawed. Human frailty is ineradicable even in the best of us. But more to the point, Plato's ideal state was, from a systems theorist's point of view, poorly engineered. It was deliberately designed to eliminate all "inputs" or "feedback" from the "system." Plato's perfectly harmonious order guided only by the ruler's higher vision would have had to be suspended somehow in a static, timeless equilibrium, whereas, in the real world, statesmen must deal with the incomplete. The real world is uncertain and changing, and the political art consists, quintessentially, of contingent solutions to both ongoing problems and new developments. Out of the myriad of particular interests in a society, the political leader must try to fashion areas of agreement for what are often trial-and-error responses to overriding public problems.

Late in life, Plato conceded these points when, in *The Laws*, he proposed to substitute "the golden cord of law" for the detached wisdom of the philosopher-king. Plato came ultimately to view the "wisdom" of the law, feeling its way from one limited precedent to another and building a structure of codified social experience, as the best real-world approximation of his ideal of "reason unaffected by desire"—as Aristotle put it. Yet even Plato's seasoned and more realistic view of politics was seriously deficient. While the idea of a philosopher-king had to be abandoned as impractical, Plato had failed—despite a lifetime of thought and teaching—to devise a satisfactory substitute. Conspicuously missing from his mature political vision was an answer to the cybernetics problem. Is it possible to design an instrument capable of achieving an adequate balance between, on the one hand, participation and control on the part of the citizenry, and, on the other, effective control and leadership from above?* If political leaders are given sufficient power and autonomy to *impose* the general welfare, then, in Juvenal's taunting phrase, *quis custodiet ipsos custodes* (who will guard the guardians)?

Since Plato's day, many other attempts have been made to cope with this problem. The Madisonian answer, embedded in the American Constitution, was to erect a set of institutions in which power was divided and shared and in which the political ambitions of one man were pitted against another's. It was a formula for stalemate, and it has had to be modified substantially in practice. Nor does anybody pretend that we have found the perfect solution.

Lenin's answer, by contrast, was to vest power in a disciplined and strongly led political party which would provide "collective" leadership for the "workers." The "Bolshevik" solution was implemented at the time of the Russian Revolution, but Stalin made a mockery of it and in subsequent practice it too has been greatly modified—with something less than ideal results.

Finally, modern liberals have argued that the public interest would emerge out of the clash of competing group interests, just as, in an earlier era, laissez-faire economists had asserted that free competition in the marketplace would maximize our collective economic well-being. But just as economic theory fell victim to the subversive effects of the real world, so the events of the past few years have dispelled the myth of the political "market." Contemporary political theory is in a state of disarray.

Is there any perfect solution? It could be that no system will ever be completely satisfactory (that the problem is inherently unsolvable in any

* Plato's council of elders, the Nocturnal Council, was apparently an afterthought and no more than a collective variant of the philosopher-king. It would have consisted of lifetime appointees who could veto any action of elected officials. The basic cybernetic problem remained unresolved.

definitive sense) but that almost any system will work after a fashion if there are good people running it. Some systems may make the job more difficult, or may be more likely to invite either oppression or stalemate. But perhaps the best we can hope for is that we manage to muddle through and cope with at least the most pressing of our collective problems.

One thing is certain, however. On at least a marginally satisfactory solution to this ancient conundrum depends not only the future of politics but, quite probably also the future of our species.

REFERENCES

Alland, A., Jr., 1967, *Evolution and Human Behavior*, Natural History Press, Garden City, N.Y.

Alland, A., Jr., 1970, *Adaptation in Cultural Evolution: An Approach to Medical Anthropology*, Columbia University Press, New York.

Almond, G. A., and Powell, G. B., Jr., 1966, *Comparative Politics: A Developmental Approach*, Little, Brown, Boston.

Altmann, S. A. (ed.), 1967, *Social Communication Among Primates*, University of Chicago Press, Chicago.

Ardrey, R., 1970, *The Social Contract*, Atheneum, New York.

Arrow, K. J., 1951, *Social Choice and Individual Values*, Wiley, New York.

Bandura, A., 1971*a*, *Social Learning Theory*, General Learning Press, New York.

Bandura, A. (ed.), 1971*b*, *Psychological Modeling*, Aldine-Atherton, Chicago.

Bem, D. J., 1970, *Beliefs, Attitudes and Human Affairs*, Brooks/Cole, Belmont, Calif.

Berlo, D. K., 1960, *The Process of Communication*, Holt, Rinehart and Winston, New York.

Bigelow, R., 1969, *The Dawn Warriors: Man's Evolution Toward Peace*, Little, Brown, Boston.

Binford, S. R., 1968, Early Upper Pleistocene adaptations in the Levant, *Am. Anthropologist* **70**(**4**):707–717.

Birdwhistell, R. L., 1952, *Introduction to Kinesics*, University of Louisville Press, Louisville, Ky.

Birdwhistell, R. L., 1960, Kinesics and communication, in: *Exploration in Communication*. E. Carpenter and M. McLuhan, eds., Beacon Press, Boston.

Birdwhistell, R. L., 1970, *Kinesics and Context*, University of Pennsylvania Press, Philadelphia.

Boden, M. A., 1972, *Purposive Explanation in Psychology*, Harvard University Press, Cambridge, Mass.

Boserup, E., 1965, *The Conditions of Agricultural Growth: The Economics of Agrarian Change Under Population Pressure*, Aldine, Chicago.

Boughey, A. S., 1971, *Man and the Environment*, Macmillan, New York.

Brecht, A., 1959, *Political Theory: The Foundations of Twentieth Century Political Thought*, Princeton University Press, Princeton, N.J.

Bruner, J., 1966, *Studies in Cognitive Growth*, Wiley, New York.

Buchanan, J. M., and Tullock, G., 1962, *The Calculus of Consent: Logical Foundations of Constitutional Democracy*, University of Michigan Press, Ann Arbor.

Buckley, W. (ed.), 1968, *Modern Systems Research for the Behavioral Scientist*, Aldine, Chicago.

Campbell, D. T., 1965, Variation and selective retention in socio-cultural evolution, in: *Social Change in Developing Areas: A Reinterpretation of Evolutionary Theory*, H. R. Barringer, G. I. Blanksten, and R. W. Mack, eds.), Shenkman, Cambridge, Mass.

Campbell, D. T., 1970, Natural selection as an epistemological model, in: *A Handbook of Method in Cultural Anthropology*. (R. Naroll and R. Cohen, eds.) Natural History Press, Garden City, N.Y.

Carneiro, R. L., 1967, On the relationship between size of population and the complexity of social organization, *Southwest. J. Anthropol.* **33**(**3**):234–243.

Carneiro, R. L., 1970, A theory of the origin of the state, *Science* **169**:733–738.

Carneiro, R. L., 1972, The devolution of evolution, *Soc. Biol.* **19**(**3**):248–258.

Cassirer, E., 1944, *An Essay on Man*, Yale University Press, New Haven.

Chomsky, N., 1965, *Aspects of the Theory of Syntax*, MIT Press, Cambridge, Mass.

Christian, J. J., 1970, Social subordination, population density, and mammalian evolution, *Science* **168**:84–90.

Clynes, M., 1969, Toward a theory of man, in: *Information Processing in the Nervous System* (K. N. Leibovic and J. Eccles, eds.), Springer-Verlag, New York.

Clynes, M., 1970, Toward a view of man, in: *Biomedical Engineering Systems* (M. Clynes and J. H. Milsum, eds.), McGraw-Hill, New York.

Cohen, Y. A. (ed.), 1968, *Man in Adaptation: The Biosocial Background*, Aldine, Chicago.

Coker, F. W., 1910, Studies in history, economics and public law, in: *Organismic Theories of the State*, Vol. 38, No. 2 (Whole No. 101), Columbia University, New York.

Cunningham, M., 1972, *Intelligence: Its Organization and Development*, Academic Press, New York.

Dahl, R. A., 1970, *Modern Political Analysis*, 2nd ed., Prentice-Hall, Englewood Cliffs, N.J.

Davitz, J. R. (ed.), 1964, *The Communication of Emotional Meaning*, McGraw-Hill, New York.

Demerath, N. J., and Peterson, R. A. (eds.), 1967, *System, Change, and Conflict: A Reader on Contemporary Sociological Theory and the Debate Over Functionalism*, Free Press, New York.

Deutsch, K. W., 1966, *The Nerves of Government: Models of Political Communication and Control*, 2nd ed., Free Press, New York.

DeVore, I. (ed.), 1965, *Primate Behavior*, Holt, Rinehart and Winston, New York.

Dobzhansky, T. H., 1962, *Mankind Evolving: The Evolution of the Human Species*, Yale University Press, New Haven.

Dolhinow, P. (ed.), 1972, *Primate Patterns*, Holt, Rinehart and Winston, New York.

Downs, A., 1957, *An Economic Theory of Democracy*, Harper and Row, New York.

Dumond, D. E., 1965, Population growth and cultural change, *Southwest. J. Anthropol.*, **21**(**3**):302–324.

Easton, D., 1965*a*, *A Framework for Political Analysis*, Prentice-Hall, Englewood Cliffs, N.J.

Easton, D., 1965*b*, *A Systems Analysis of Political Life*, Wiley, New York.

Eckland, B. K., 1972, Trends in the intensity and direction of natural selection, *Soc. Biol.* **19**(**3**):215–223.

Emerson, A. E., 1954, Dynamic homeostasis: A unifying principle in organic, social and ethical evolution, *Sci. Monthly* **78**(**27**):67–85.

Etkin, W., 1963, Communication among animals, in: *Psychology of Communication* (J. Eisenson, ed.), Appleton-Century-Crofts, New York.

Fagen, R. R., 1966, *Politics and Communication*, Little, Brown, Boston.

Furth, H. G., 1966, *Thinking Without Language: Psychological Implications of Deafness*, Free Press, New York.

Furth, H. G., 1969, *Piaget and Knowledge*, Prentice-Hall, Englewood Cliffs, N.J.

Gardner, R. A., and Gardner, B. T., 1969, Teaching sign language to a chimpanzee, *Science* **165**:664–672.

Gazzaniga, M. S., 1971, *The Bisected Brain*, Appleton-Century-Crofts, New York.

Gerbner, G., 1972, Communication and social environment, *Sci. Am.* **227**(3):153–160.

Geschwind, N., 1964, The development of the brain and the evolution of language, *Monogr. Ser. Lang. Linguist.* **17**:155–169.

Geschwind, N., 1972, Language and the brain, *Sci. Am.* **226**(4):76–83.

Ginsburg, H., and Opper, S., 1969, *Piaget's Theory of Intellectual Development: An Introduction*, Prentice-Hall, Englewood Cliffs, N.J.

Glass, H. B., 1972, The eugenic implications of artificial insemination and artificial inovulation, *Soc. Biol.* **19**(4):326–336.

Gurr, T. R., 1968, A causal model of civil strife: A comparative analysis using new indices, *Am. Pol. Sci. Rev.* **64**(4):1104–1124.

Gurr, T. R., 1970, *Why Men Rebel*, Princeton University Press, Princeton, N.J.

Hall, E. T., 1959, *The Silent Language*, Doubleday, New York.

Hall, K. R. L., 1963, Observational learning in monkeys and apes, *Brit. J. Psychol.* **54**(3):201–226.

Hamilton, W. D., 1964, The genetical evolution of social behavior, *J. Theoret. Biol.* **7**:1–52.

Harris, M., 1969, Monistic determinism: Anti-service, *Southwest. J. Anthropol.* **25**(2):198–206.

Hinde, R. A., 1970, *Animal Behavior: A Synthesis of Ethology and Comparative Psychology*, McGraw-Hill, New York.

Hinde, R. A., and Powell, I. E., 1962, Communication by postures and facial expressions in the rhesus monkey, *Proc. Zool. Soc. Lond.* **138**(1):1–21.

Hockett, C. F., 1959, Animal "languages" and human language, in *The Evolution of Man's Capacity for Culture* (J. N. Spuhler, ed.), Wayne State University Press, Detroit.

Itani, J., 1958, On the acquisition and propagation of a new food habit in the troop of Japanese monkeys at Takasakiyama, *Primates* **1**(1-2):84–98.

Jay, P. C. (ed.), 1968, *Primates: Studies in Adaptation and Variability*, Holt, Rinehart and Winston, New York.

Johnston, H. A., 1970, Information theory in biology after 18 years, *Science* **158**:1545–1550.

Jolly, A., 1972, *The Evolution of Primate Behavior*, Macmillan, New York.

Keller, A. G., 1931 (1915), *Societal Evolution*, Yale University, New Haven.

Koestler, A., 1967, *The Ghost in the Machine*, Macmillan, New York.

Kortlandt, A., 1967, Experimentation with chimpanzees in the wild, in: *Neue Ergebneisse der Primatologie*, (D. Starck, R. Schneider, and H. J. Kugh, eds.) Fischer, Stuttgart.

Kummer, H., 1971, *Primate Societies*, Aldine-Atherton, Chicago.

Lancaster, J. B., 1968, Primate communication systems and the emergence of human language, in: *Primates: Studies in Adaptation and Variability*, (P. C. Jay, ed.), Holt, Rinehart and Winston, New York.

Lancaster, J. B., 1970, On the evolution of tool-using behavior, *Am. Anthropologist* **70**(1):56–66.

Lasswell, H. D., 1936, *Politics: Who Gets What, When, How*, McGraw-Hill, New York.

Laughlin, W. S., 1968, Hunting: An integrating biobehavior system and its evolutionary importance, in: *Man the Hunter* (R. B. Lee and I. DeVore, eds.), Aldine, Chicago.

Lawson, C. A., 1958, *Language, Thought and the Human Mind*, Michigan State University Press, East Lansing.

Lawson, C. A., 1963, Language, communication and biological organization, *General Systems Yearbook* **8**:107–115.

Leakey, R. E., 1972, New fossil evidence for the evolution of man, *Soc. Biol.* **19**(2):99–114.

Lenneberg, E. H. (ed)., 1964, *New Directions in the Study of Language*, MIT Press, Cambridge, Mass.

Lenneberg, E. H. (ed.), 1967, *Biological Foundations of Language*, Wiley, New York.

Lenski, G., 1970, *Human Societies: A Macrolevel Introduction to Sociology*, McGraw-Hill, New York.

Lerner, I. M., 1968, *Heredity, Evolution and Society*, Freeman, San Francisco.

Lowell, A. L., 1910, The physiology of politics, *Am. Pol. Sci. Rev.* **4**(**1**):1–15.

Lowi, T. J., 1969, *The End of Liberalism*, Norton, New York.

Marler, P., 1968, Aggregation and dispersal: Two functions in primate communication, in: *Primates: Studies in Adaptation and Variability* (P.C. Jay, ed.), Holt, Rinehart and Winston, New York.

Martin, P. S., and Wright, H. E. (eds.), 1967, *Pleistocene Extinctions*, Yale University Press, New Haven.

Maslow, A. H., 1954, *Motivation and Personality*, Harper and Row, New York.

Masters, R. D., 1970, Genes, language and evolution, *Semiotica* **2**(**4**):295–320.

Mayr, E., 1970, *Populations, Species and Evolution*, Harvard University Press, Cambridge, Mass.

McLuhan, M., 1962, *The Gutenberg Galaxy*, University of Toronto Press, Toronto.

Meadows, K., 1968, Early manual communication in relation to the deaf child's intellectual, social and communicative functioning, *Am. Ann. Deaf,* **113**:29–41.

Menzel, E. W., Jr., 1971, Communication about the environment in a group of young chimpanzees, *Folia Primatol.* **15**(**3-4**):220–232.

Miller, R. E., 1967, Experimental approaches to the physiological and behavioral concomitants of affective communication in rhesus monkeys, in: *Social Communication Among Primates*, (S. A. Altmann, ed.), University of Chicago Press, Chicago.

Mirsky, I. A., Miller, R. E., and Murphy, J. W., 1958, The communication of affect in rhesus monkeys, *J. Am. Psychoanal. Assoc.* **6**(**3**):443–441.

Nash, J., 1970, *Developmental Psychology: A Psychobiological Approach*, Prentice-Hall, Englewood Cliffs, N.J.

Nisbet, R. A., 1969, *Social Change and History: Aspects of the Western Theory of Development*, Oxford University Press, New York.

Olson, M., Jr., 1965, *The Logic of Collective Action*, Harvard University Press, Cambridge, Mass.

Pettifor, J. L., 1968, The role of language in the development of abstract thinking, *Canad. J. Psychol.* **22**(**3**):139–156.

Pfeiffer, J. E., 1969, *The Emergence of Man*, Harper and Row, New York.

Piaget, J., 1963*a* (1947), *Psychology of Intelligence*, Littlefield, Adams, Patterson, N.J.

Piaget, J., 1963*b* (1952), *The Origins of Intelligence in Children*, Norton, New York.

Pike, K., 1954, *Language in Relation to a Unified Theory of the Structure of Human Behavior*, Institute of Linguistics, Glendale, Calif.

Pilbeam, D., 1972, Adaptive response of hominids to their environment as ascertained by fossil evidence, *Soc. Biol.* **19**(**2**):115–127.

Premack, D., 1970, A functional analysis of language, *J. Exp. Anal. Behav.* **14**(**1**):107–125.

Pringle, J. W. S., 1951, On the parallel between learning and evolution, *Behavior* **3**:174–215.

Reynolds, P. C., 1968, Evolution of primate vocal-auditory communication systems, *Am. Anthropologist* **70**(**2**):300–308.

Riker, W. H., 1962, *The Theory of Political Coalitions*, Yale University Press, New Haven.

Roe, A., and Simpson, G. G., 1958, *Behavior and Evolution*, Yale University Press, New Haven.

Rostow, W. W., 1960, *The Stages of Economic Growth: A Non-Communist Manifesto*, Cambridge University Press, Cambridge.
Rozin, P., and Kalat, J. W., 1971, Specific hungers and poison avoidance as adaptive specializations of learning, *Psychol. Rev.* **78(6)**:459–486.
Sapir, E., 1921, *Language*, Harcourt Brace, New York.
Schneirla, T. C., and Piel, G., 1948, The army ant, *Sci. Am.* **178(6)**:16–23.
Sebeok, T. A. (ed.), 1968, *Animal Communication*, University of Indiana Press, Bloomington.
Seligman, M., and Hager, J. L. (eds.), 1972, *The Biological Boundaries of Learning*, Appleton-Century-Crofts, New York.
Skinner, B. F., 1965 (1953), *Science and Human Behavior*, Free Press, New York.
Spencer, H., 1891, *Essays Scientific, Political and Speculative*, Appleton, New York.
Spencer, H., 1899, *Principles of Sociology*, Appleton, New York.
Spooner, B,. 1972, *Population Growth: Anthropological Implications*, MIT Press, Cambridge, Mass.
Spuhler, J., 1959, *The Evolution of Man's Capacity for Culture*, Wayne State University Press, Detroit.
Steward, J. H., 1955, *Theory of Culture Change: The Methodology of Multi-linear evolution*, University of Illinois Press, Urbana.
Teilhard de Chardin, P., 1959, *The Phenomenon of Man*, Harper and Row, New York.
Teleki, G., 1973, The omnivorous chimpanzee, *Sci. Am.* **228(1)**:33–42.
Thompson, R. F., 1967, *Foundations of Physiological Psychology*, Harper and Row, New York.
Thorson, T. L., 1970, *Biopolitics*, Holt, Rinehart and Winston, New York.
Tiger, L., and Fox, R., 1971, *The Imperial Animal*, Holt, Rinehart and Winston, New York.
Topoff, H. R., 1972, The social behavior of army ants, *Sci. Am.* **227(5)**:70–79.
Trivers, R. L., 1971, The evolution of reciprocal altruism, *Quart. Rev. Biol.* **40(4)**:35–57.
Truman, D. B., 1951, *The Governmental Process*, Knopf, New York.
Van Lawick-Goodall, J., 1968*a*, A preliminary report on expressive movements and communication in the Gombe Stream chimpanzees, in: *Primates: Studies in Adaptation and Variability*, (P. C. Jay, ed.), Holt, Rinehart and Winston, New York.
Van Lawick-Goodall, J., 1968*b*, The behavior of free-living chimpanzees in the Gombe Stream Reserve, *Anim. Behav. Monogr.* **1(3)**:161–311.
von Bertalanffy, L., 1968, *General System Theory: Foundations, Development Applications*, Braziller, New York.
von Foerster, H., (ed.), 1968, *Purposive Systems*, Spartan Books, New York.
Vygotsky, L. S., 1962, *Thought and Language*, MIT Press, Cambridge, Mass.
Wallace, A. F. C., 1956, Revitalization movements, *Am. Anthropologist* **58(2)**:264–281.
Wallace, B., and Srb, A. M., 1964, *Adaptation*, Prentice-Hall, Englewood Cliffs, N.J.
Washburn, S. L., 1972, Human evolution, *Evol. Biol.* **6**:349–360.
Washburn, S. L., and McCown, E. R., 1972, Evolution of human behavior, *Soc. Biol.* **19(2)**:163–170.
Washburn, S. L., and Shirek, J., 1967, Human evolution, in *Behavior-Genetic Analysis* (J. Hirsch, ed.), McGraw-Hill, New York.
White, L. A., 1944, The symbol: The origin and basis of human behavior, *Etc.* **1**:229–237.
White, L. A., 1959, *The Evolution of Culture*, McGraw-Hill, New York.
White, L. A., 1968, Culturology, *Int. Encycl. Soc. Sci.* **3**:547–550.
Whorf, B. L., 1956, *Language, Thought and Reality*, Wiley, New York.
Williams, G. C., 1966, *Adaptation and Natural Selection: A Critique of Some Current Evolutionary Thought*, Princeton University Press, Princeton, N.J.

Williams, G. C., (ed.), 1971, *Group Selection*, Aldine-Atherton, Chicago.

Wilson, D., 1969, Forms of hierarchy: A selected bibliography, *General Systems* **14**:3–15.

Wilson, E. O., 1971, Competitive and aggressive behavior, in: *Man and Beast: Comparative Social Behavior* (J. F. Eisenberg and W. Dillon, eds.), Smithsonian Institution Press, Washington, D.C.

Wilson, E. O., 1972, Animal communication, *Sci. Am.* **227**(3):53–60.

Wilson, W., 1908, *Constitutional Government in the United States*, Columbia University Press, New York.

Wittfogel, K. A., 1957, *Oriental Despotism: A Comparative Study of Total Power*, Yale University Press, New Haven.

Wolin, S. S., 1960, *Politics and Vision*, Little, Brown, Boston.

Wooldridge, D. E., 1963, *The Machinery of the Brain*, McGraw-Hill, New York.

8

Morphological Transformation, the Fossil Record, and the Mechanisms of Evolution: A Debate

Part I The Statement and The Critique

MAX K. HECHT
Queens College
Flushing, New York

Part II The Reply

NILES ELDREDGE
Department of Invertebrate Paleontology
American Museum of Natural History
New York, N.Y.

STEPHEN JAY GOULD
Museum of Comparative Zoology
Harvard University
Cambridge, Massachusetts

PART I. THE STATEMENT AND THE CRITIQUE

In a paper entitled "Punctuated Equilibria: An Alternative to Phyletic Gradualism," Eldredge and Gould (1972) properly criticize the naive concepts held by most paleontologists with respect to the interpretation of data from the fossil record. The model or "picture" (Eldredge and Gould, 1972) of the mechanism explaining morphological change used by most paleontologists is that of gradual change within a character or complex of characters which is expressed as shifts in the mean and associated statistics along a time axis. This type of evolutionary change has been described and illustrated by many paleontologists, such as Trueman (1922, Fig. 5), Simpson (1953, p. 387), Kurtén (1968, p. 239), and George (1971, p. 207). These authors and many others offer a picture of evolution as a progressive series of normal curves on a time axis in which there is a "long and insensibly graded chain of forms" (Eldredge and Gould, 1972). Basically, Eldredge and Gould (1972) criticize this simplistic model of paleontologists because it neither explains the important morphological stasis and gaps in the fossil record nor takes into consideration the major concept of modern systematics, the biological species concept. In their study, Eldredge and Gould (1972) deny the validity of phyletic gradualism and conclude that the best explanation of the paleontological data is the mechanism of geographic speciation expressed within their concept of "punctuated equilibria." It is the purpose of this essay to argue that neither phyletic gradualism nor speciation is in itself sufficient to explain evolution, but that both processes are important.

Eldredge and Gould summarize the basic tenets of phyletic gradualism as follows:

> (1) New species arise by the transformation of an ancestral population into its modified descendants.
>
> (2) The transformation is even and slow.
>
> (3) The transformation involves large numbers, usually the entire ancestral population.
>
> (4) The transformation occurs over all or a large part of the ancestral species' geographic range.
>
> These statements imply several consequences, two of which seem especially important to paleontologists:
>
> (1) Ideally, the fossil record for the origin of a new species should consist of a long sequence of continuous, insensibly graded intermediate forms linking ancestor and descendant.
>
> (2) Morphological breaks in a postulated phyletic sequence are due to the imperfection in the geological record. (Eldredge and Gould, 1972, p. 89)

Eldredge and Gould criticize the phyletic gradualistic concept, as they define it, because it does not comply with the modern biological concept of the species and its place in evolution. Furthermore, they believe there is little or no paleontological evidence that fits the model. Most paleontologists with a reasonably good fossil record as a basis for their evolutionary interpretations probably would believe that the transformation of the species in time is continual and should show a continuing morphological series. The basic tenets of phyletic gradualism as followed by many paleontologists (and stated in their textbooks) and summarized by Eldredge and Gould (1972) do not explain the rapid transformation within phyletic lineages and the morphological gaps in the fossil record. Therefore, Eldredge and Gould propose an alternative working hypothesis, the biospecies and punctuated equilibria.

The picture drawn according to the hypothesis of Eldredge and Gould is of a widely ranging biologically species, reproductively isolated from its nearest relatives and showing geographic variation in its morphology. Their model or picture requires small populations at the margin of the geographic range, and such populations are peripheral isolates. The requirement for peripheral isolates as a part of geographic speciation is an artificiality erected to answer why in any information there may be gaps in the continuous fossil sequence. Eldredge and Gould summarize their hypothesis as follows:

> In summary, we contrast the tenets and predictions of allopatric speciation with the corresponding statements of phyletic gradualism previously given:
>
> (1) New species arise by the splitting of lineages.
> (2) New species develop rapidly.
> (3) A small sub-population of the ancestral form gives rise to the new species.
> (4) The new species originates in a very small part of the ancestral species' geographic extent—in an isolated area at the periphery of the range.
>
> These four statements again entail two important consequences:
>
> (1) In any *local* section containing the ancestral species, the fossil record for the descendant's origin should consist of a sharp morphological break between the two forms. This break marks the migration of the descendant, from the peripherally isolated area in which it developed, into its ancestral range. Morphological change in the ancestor, even if directional in time, should bear no relationship to the descendant's morphology (which arose in response to local conditions in its isolated area). Since speciation occurs rapidly in small populations occupying small areas far from the center of ancestral abundance, we will rarely discover the actual event in the fossil record.
> (2) Many breaks in the fossil record are real; they express the way in which evolution occurs, not the fragments of an imperfect record. The sharp break in a local column accurately records what happened in that area through time. Acceptance of this point would release us from a self-imposed status of inferiority among the evolutionary sciences. The paleontologist's gut-reaction is to view almost any anomaly as an artifact imposed by our institutional millstone—an imperfect fossil record. But just as we now tend to view the rarity of Precambrian metazoans as a true reflection of life's history rather than a testimony to the ravages of metamorphism or the lacunae of Lipalian intervals, so also might we reassess the smaller breaks that permeate our Phanerozoic record. We suspect that this record is much better (or at least much richer in optimal cases) than tradition dictates. (Eldredge and Gould, 1972, p. 96).

The preceding statements may be summarized as follows:

> The model of evolution by speciation provides a mechanism to explain the morphological gaps and morphological stasis (or lack of continual change in morphology) as seen in the fossil record.

The picture developed by Eldredge and Gould is oversimplified and its criticism of phyletic gradualism is simplistic. Essentially they have set up phyletic gradualism as a straw man only to be knocked down. It therefore requires reevaluation.

First, there is a question of terminology. Speciation is that evolutionary process which results in the division of formerly interbreeding populations into reproductively isolated units. Phyletic gradualism as defined by Eldredge and Gould requires gradual overhauling of the phenotype of a species without reproductive isolation and with a uniform rate of differentiation and therefore is an oversimplified model. It is for this reason that I prefer the terms "phyletic evolution" (Simpson, 1953) or "phyletic transformation." The latter would perhaps be preferable because "phyletic evolution" has been misused as referring to a single selected lineage through time with a speciational mechanism. ("Phyletic speciation" is by definition invalid.) Furthermore, either "phyletic transformation" or "phyletic evolution" is acceptable because these terms do not imply the simplistic mechanism and tenets attributed to phyletic gradualism. The crux of the matter lies in the relative importance of phyletic transformation and of speciation in the phylogeny of organisms, and whether speciation and its consequences exclude phyletic transformation for major evolutionary changes.

Second, speciation as a model for evolutionary mechanisms is universally accepted and illustrated because it can be demonstrated by neozoological systematic studies and by biological experimentation, but it remains in most cases in the geological sense a single time level phenomenon. Furthermore, the initial steps of the process, the establishment of isolating mechanisms, are mostly behavioral and physiological features with no obvious morphological aspect to mark the evolutionary stage, the establishment of reproductive isolation—the final biological test being the establishment of sympatric populations or the reinvasion of an old portion of the geographic range by a formerly allopatric population. The level of resolution afforded by paleontological techniques in most cases does not allow such discrimination. In the neozoological situation, the neosympatric forms would generally not be distinguished on the morphological level. The picture of Eldredge and Gould requires the paleontological recognition of biological species, which they admit as not being paleontologically operational (Eldredge and Gould, 1972, p. 92–93). In fact, a new species that suddenly appears in a depositional series is most probably not the recent allopatric derivative of an ancestral species but a well-developed and more differentiated form which occupied a similar part of the adaptive zone in a slightly different ecological niche. If the original population had been partially displaced or replaced by its closest relative, it is doubtful that it could be recognized in most paleontological situations.

Third, Eldredge and Gould require that the population which will differentiate rapidly be derived from a peripheral isolate. Their model apparently construes that the major changes in biological species occur in demes that are peripheral, whereas the best examples of the speciational process (avian and lacertilian speciation in archipelagos and systematic reviews of continental vertebrates) actually are of species differentiation which is well within the major distribution of the species group or complex as a whole. This observation is easily demonstrated by Darwin's finches, Hawaiian honey creepers, Antillean anoline lizards in archipelagos, *Rana pipiens* complex on a mainland, etc.

Fourth, their biological species model of evolution does not necessarily contradict phyletic transformation of a lineage. In the frame of the biological species there are many examples of the morphological cline (Mayr, 1956), which is analogous to the phyletic lineages that are usually given as examples of phyletic gradualism, assuming the examples are correctly interpreted (Gould, 1972). Clines, whether step or gradual, show changes maintained over great distances and between many demes. Whether or not gene flow is important (Ehrlich and Raven, 1969) to maintain clines, clines indicate directional or orthoselectional trends (denied by Eldredge and Gould, 1972). Mayr (1956) stated that these clines are the result of adaptation to major aspects of the environment. They are expressed as generalizations as in Bergmann's and Allen's rules. There are major selective forces, often not biological in origin, which are determined by overriding physical aspects of the environment, such as climate. Certainly what occurs in a single time dimension, a geographical distribution of a morphological and perhaps genetic gradient, could occur more easily along the time axis because there are no counterbalancing forces of gene flow. Raven and Ehrlich's criticism of Mayr's evaluation of the importance of gene flow in a single time dimension has less significance during the entire life of a species and is based on cases of limited vagility. Endler (1973) emphasizes the importance of natural selection based on environmental gradients in the establishment of clines.

In contrast to speciation, which produces diversity, phyletic transformation can be defined as the change within a phyletic lineage without reproductive isolation or the separation of populations. It is not necessarily the transformation of entire populations of wide-ranging species as simplistically interpreted by many paleontologists. It obviously involves small populations responding to selection in a given direction. The origin of such selection is usually not biological, and the resulting adaptations are in response to conditions of the new environment being explored by the deme. Easily visualized examples of this type of change are invasions of a marine environment by a form of terrestrial origin, the development of cave-adapted troglobitic forms, and the invasion of the land by aquatic forms. For example, the physical conditions of the aquatic habitat place constraints on the genetic makeup of the evolving demes so that their morphology and physiology must adapt to the hydrodynamic characteristics of the new media. These selective factors will direct and override any biological aspects of selection which are associated with speciation.

Phyletic transformation can be seen in the *Drosophila* population cage and in the development of domestic types of animals and plants. Dobzhansky's *Genetics of the Evolutionary Process* (1972) is replete with examples of phyletic evolution which are essentially changes in the allelic frequencies through time. Bennett (1960) on DDT resistance in houseflies and Wright and Dobzhansky (1946) on gene frequency changes in *Drosophila pseudoobscura* are a few of the many experimental studies which essentially describe phyletic transformation in its earliest steps. In nature there are others showing, first, gene frequency changes in populations through time and without partitioning of populations in response to directional selection. The best example of this is Kettlewell (1956), where the increasing frequency of industrial melanism became widespread through the species *Biston betularia*. Ford (1964) elaborated on this theme and showed that in over 80 species of bark-dwelling Lepidoptera (scattered among several families) industrial melanism also developed independently. In *Biston betularia* and in other species of Lepidoptera, different populations produced high frequencies of industrial melanism independently and as a result of an overriding physical environmental change resulting in orthoselection. Directional selection or orthoselection can be easily demonstrated (as above) on the population level, but the convergence of character states among related or unrelated forms demonstrates that physical factors of the environment can override stabilizing and disruptive selective aspects of speciation.

Speciation is the only process leading to the multiplication of biological species. According to the allopatric speciation model, it requires reproductive isolation, sympatry, competition, and the resulting character displacement. This model used by Bock (1972) results in the adaptive radiation that leads to the rapid diversification and development of many closely related forms demonstrating diversity of type such as found in Darwin's finches, Hawaiian honey creepers, Hawaiian Drosophilidae, and Antillean anoline lizards. Closely related species, genera, and even families become diverse in a restricted geographic area. The selective factors in the original diversification are primarily biological such as behavioral features of isolating mechanisms, and not aspects of the physical environment such as climatic gradients.

The phyletic transformational model, on the other hand, usually involves adaptation to major aspects of the environment. It too must be based on populations which are subjected to directional selection resulting in continual morphological changes in a single adaptational trend. Many examples can be drawn from the fossil record to support such a model. Kurtén (1957, 1959, 1964) has clearly demonstrated a phyletic transformation in the *Ursus arctos* lineage. The polar bear, a species of recent origin, is derived from the ancestral biological species, the brown bear, from which it is not genetically isolated, and is completely allopatric in distribution. That the polar bear has not evolved any genetic isolating mechanisms from its closest relatives is indicated by the many hybrids and backcrosses in zoos. The morphological differences between the two forms are clear and are correlated with the more carnivorous habit and the ice pack habitat of the polar bear. Its breeding population is small because the total present world population is about 20,000 individuals (Norderhaug, 1972). Based on the Alaskan populations, about 48% are adult bears and slightly more than 60% of these would be females. The female bears breed in alternate years after their fourth to fifth year and average a single cub. The effective breeding population of females must be under 3000 individuals. The vagility of the bears has been studied by Larsen (1971, 1972*a,b*), and even within the small sample studied it is clear that the polar bears are highly nomadic, covering distances known to be as much as 3200 km within a single year. Such vagility indicates gene flow greater than that anticipated by Ehrlich and Raven (1969) and perhaps explains the insignificant geographic variation found by Manning (1971) in his analysis of the skull morphology. Kurtén (1964) clearly indicates that the early populations of *Ursus maritimus* from the Late Würm were characterized by dental variation more closely approximating *Ursus arctos* than the living populations of the polar bear. His chronosubspecies, *U. m. tyrannus*, emphasizes another trend of size reduction from the Early Würm to the present. To summarize, the phyletic trends within the polar bears show increasing carnivorous dentition and elongate skull form with a concomitant reduction in total body size for the last 15,000–20,000 years. As late as 8000–10,0000 years ago, the modern polar bear populations still had a high frequency of *arctos*-like molars. This condition exists in very low frequency in modern populations.

The history of the polar bear may be reconstructed as follows: An allopatric population probably of Asiatic brown bears invaded tundra and ice pack areas during Cromer-Mindel times. The isolation of these populations resulted in transformation of the brown bear morphology by changing limb proportions. By the Early Würm, a true ice pack bear had developed, ecologically isolated from the brown bears. The morphological trends within *U. maritimus* affected the entire population. At the greatest extent of the Würm or its North American equivalents (Bryson *et al.*, 1970), there could not have been more than twofold increase in the ice habitat and tundra. As a result, at no time were there more than twice the number of polar bears than during recent times. As a result, the changes within the polar bear are classical phyletic transformational changes involving small populations, lack of geographic isolation, directional selection derived from major effects of a new adaptive zone, and major morphological trends affecting an entire population. Whereas it is most probable that the polar bear will not make the adaptive breakthrough following its prey to a more marine existence, it is certainly on the threshold of a new and major adaptive zone for bears.

Another example demonstrating phyletic transformation is that of the dramatic changes within the Elephantinae (Maglio, 1972, 1973). Maglio (1973) interprets the transition from *Elephas ekorensis* to *E. recki* (all stages) to *E. iolensis* and the two series of European and North American *Mammuthus* (both derived from *M. meridionalis*) as phyletic transformation. In these two examples of the Elephantinae, the record is continuous and the morphological changes rapid, continuous, and functionally interpretable.

Therefore, phyletic transformation allows for rapid invasion of evolving populations into new major adaptive zones which are most probably characterized by radically different physical environmental factors. The initial phase is that of overhauling of gene frequencies with possible increase of polymorphism. This phase can be accomplished without dramatic or even evident phenotypic change, as shown by the polymorphism in the genus *Limulus* which certainly has the type of stasis required by Eldredge and Gould (Selander *et al.*, 1970). Further recent work on isozymes clearly indicates the hidden variability under the facade of low phenotypic variation (Hubby and Lewontin, 1966). The stasis or lack of morphological change, disrupted by rapid, sharp, nonintergrading change in the fossil series, that disturbs Eldredge and Gould can also be explained by rapid change in the phenotype after the major genotypic revolution has already taken place. After the development of "key" adaptations within a short period of time, the resulting adaptational breakthroughs allow for population expansions into new geographic areas. It is at this point that allopatric differentiation takes place, resulting in speciation, character displacement, and eventually adaptive radiation.

In conclusion, it appears clear that the fossil record does not support exclusively either of the two processes originally described by Simpson (1953). The model offered by Eldredge and Gould (1972) is in error because it is exclusionary in intent and does not allow for other demonstrated mechanisms. It appears to me that the major portion of the diversity of the present living world is the result of the speciational process but that the major breakthroughs into new adaptive zones (environments) are the result of phyletic transformation. On the other hand, there are cases of such breakthroughs which must be the result of the interaction of the two processes or primarily that of speciation. In any one case it will be difficult to determine which mechanism was involved and the most parsimonious explanation will not necessarily be correct. As Ghiselin (1972) and others have stated, evolution is not necessarily a parsimonious process.

ACKNOWLEDGMENTS

I wish to express my thanks to Drs. Marvin Wasserman, Michael Gochfeld, Walter Bock, and Vincent Maglio for their criticisms and suggestions. I also acknowledge the support of C.U.N.Y. National Institutes of Health Grant No. 5-SO5-RR-07064 and F.R.A.P. Grant No. 1589, which made this study possible.

PART II. THE REPLY

The main thrust of Hecht's remarks appears to be twofold: (1) we have identified our "own variant" of the neontological view of allopatric speciation, and in so doing have violated some aspects of biological theory, and (2) our depiction of "phyletic gradualism" is a straw man that does not accurately convey the biological notions commonly held by neontologists and paleontologists. Let us first address the issue of whether our notion of "punctuated equilibria" agrees with conventional speciation theory.

The main objection Hecht seems to raise is that we have insisted on using the expression "peripheral isolate" to denote a geographically discrete population that may or may not undergo speciation. His claim that we have done so merely to explain morphological gaps in the fossil record is a misinterpretation: to explain why one species succeeds its presumed ancestor in a local rock column, with no intermediate forms intervening, requires migration of the descendant from *anywhere* else. This area need not have been on the periphery of the ancestor's geographic range. Our use of the term "peripheral" alludes simply to the notion that (1) on probabilistic grounds, it is easier to effect genetic isolation in small populations near the periphery (see Mayr, 1963, p. 496, and his cited references on this point), and (2) the peripheral areas delimit the ecological boundary of a species' distribution. In such cases, it is more likely that within-population adaptations to extremes of ecological factors will yield radically "new" physiological and morphological adjustments to the environment. We readily admit that areas well within the ancestral range may also produce new species; at the present moment, this is more plausible than ever, given the recent argument (Ehrlich and Raven, 1969; Endler, 1973) that gene flow does not exert the homogenizing influence formerly attributed to it (Mayr, 1963). But surely the use of archipelagos to counter our preference for speaking of "peripheral isolates" is not germane, since "periphery" makes sense only when wholly contiguous areas, large portions of terrestrial continental habitats or marine environments, are under consideration. Speciation is easy to understand (now) in cave systems and island chains. We feel that peripheral isolates are more commonly involved in speciation events involving species living in more continuous and perhaps homogeneous environments. But, again, *peripheral* isolates *per se* are not crucial to our argument.

Hecht correctly points out that, in many cases, the initial stages of speciation are behavioral and physiological—with a genetic base to be sure, but commonly with little or no morphological expression. It is true too, that sibling species pose insoluble problems for paleontologists; furthermore, we fully agree with Hecht—in fact, we explicitly say so—that in geological terms, speciation can be considered as a virtually instantaneous event. But we do not concede (as Hecht claims) that the biological species concept is nonoperational in paleontology. We are faced with "unit taxa" which, at least in the case of marine invertebrates, frequently show (1) widespread geographic distribution and concomitant geographic variation, and (2) truly long periods of existence (5–10 million years is not uncommon) with virtually *no* concomitant morphological variation. How do we explain this? We feel that the biological species concept and its current understanding of speciation already contain the necessary ingredients to yield an explanation of these essentially, if not wholly, empirical observations. We believe that in all such examples we are dealing with true "biospecies" (always admitting, of course, that in these instances we are fortunate to have morphological clues for the definition of our taxa). In a nutshell, we have lifted the notion of allopatric speciation *virgo intacta*, and have not produced our own variant of it (though we abstracted only those aspects of the process relevant to interpretation of the fossil record). Indeed, we have merely added an argument based on developmental and genetic homeostasis to explain the truly astonishing *lack* of progressive change seen in most fossils, again an observation that we think is empirical and very general.

We characterized, then attacked, a notion that we called "phyletic gradualism." In essence, we claimed that most paleontologists, and their uncritical neontological colleagues who write chapters in texts dealing with evolution and the fossil record, have traditionally felt that evolution is predominantly a process of slow, steady transformation of entire lineages. We ascribed the source of this belief directly to Darwin (Eldredge and Gould, 1972, p. 87, for quotations and citations), but actually the philosophical roots lie deeper in early Victorian science (Simpson, 1970). We claim that this view is largely, if not wholly, false, for a plethora of theoretical and empirical reasons that we summarized in our paper. Now, issue may be taken with our characterization of this belief, though Hecht merely says it is a straw man not representative of actual notions held by modern scientists. In rebuttal, we simply refer the reader to any recent issue of the *Journal of Paleontology*.

What Hecht does criticize at length is our alleged rejection of phyletic transformation, which strictly speaking is *not* the issue we were addressing. A connecting point, however, is Hecht's argument that we have denied the validity of "orthoselection." This is true, but our understanding of this term seems to be different from Hecht's. We understand the term as a substitute for "orthogenesis," i.e., an attempt to explain long-enduring evolutionary trends in the fossil record in a selectional, rather than mystical, context. It is *not*, at least in our minds, synonymous with "directional selection" (Hecht, 1965). Surely we "believe" in regimes of natural selection which effect a transformation of the gene frequencies of populations, in a regular and progressive manner. How else can we explain adaptation? We know that selection studies on *Drosophila* in population cages clearly demonstrate the validity of directional selection—as if the work of animal breeders over the centuries hadn't amply done so already. We simply doubt that such regimes could persist over thousands or millions of years, through thousands of generations, without interruption by some more plausible event—such as local extinction or "migration." After all, every time an ecologist or physiologist or systematist examines a population of a species, we are told how thoroughly that population is adapted to its habitat. Are we to believe that a species can exist for a million years gradually improving its adaptation, very happy at the outset and presumably for a long time after that, with such imperfect original adaptation? Isn't it more likely that, faced with a linear change in environment extending through so long a period of time (in itself rather difficult to imagine), a species as a whole will change its area of residence, rather than sit there, grin and bear it, and "adapt"? Our argument is against truly long-term linear selection ("orthoselection"), not against the vitally necessary concept of short-term directional selection.

Again, we are faced with the dilemma of different uses and conceptions of time by two professions. When these are resolved, we detect very little difference between most of Hecht's remarks and our own. As paleontologists, we are arguing two essential points in our original essay: (1) that new species arise from a very small and geographically defined segment of the parental species i.e., the process is demic rather than panspecific, and (2) that the process happens very rapidly relative to the duration of a species. This combats the standard paleontological notion of slow, steady transformation of entire species. As a zoologist, Hecht cites against us several examples of directional selection (invasion of caves by terrestrial forms, directed transformation of Drosophilidae in population cages) that are long sustained by zoological standards. Yet these events are still but microseconds to a geologist; they are infinitessimal segments of a species' life that may last several million years. Both examples seem to support our ideas: the transformation occurs very rapidly in one small deme of an entire species. If *Drosophila willistoni* should "speciate" in its population cage, untold millions of its conspecifies are not affected; moreover, since the event can only have occurred sometime between T. H. Morgan and ourselves, it is essentially instantaneous.

Hecht compares geographic clines with stratigraphic character gradients (chronoclines of *some* authors). Geologists and paleontologists are steeped in the tradition that phenomena seen laterally are mirrored vertically. In many instances, this is true: geometrically, a sea encroaching over a continental area will leave a geographic pattern of contemporaneous but different nearshore-offshore sediments. With further encroachment, offshore sediments are deposited above nearshore sands. But is it legitimate to apply this rule to the evolutionary mechanism? To do so, one would *have* to uphold at least some version of orthoselection, while true geographic clines are merely the product of geographic variation requiring only short periods of directional selection, followed by much longer periods of "stabilizing selection" or perhaps simply of genetic homeostasis.

Finally, Hecht argues that it is his "orthoselection" (= our "directional selection") that accounts for the successful invasion of new "adaptive zones." We admit, certainly, that in some circumstances directional selection will be sustained longer than in others, and that Hecht has probably identified one such circumstance. We did not discuss this issue, *per se*, but would comment that, under the conditions Hecht specified (small populations, geographically if not fully genetically distinct from the parent species), such an all-or-nothing regime of directional selection might originate. We fail to see, however, how this is anything more than an extreme example (in terms of intensity) of normal processes of allopatric speciation involving, incidentally, peripheral isolates.

Again, we reiterate our preference for allopatric speciation (against gradual, wholesale phyletic transformation) as by far the most important process accounting for evolution and distribution of organisms in the fossil record. We cannot formally counter Hecht's citations of paleontological studies purporting to document phyletic transformation, for we have never claimed, contrary to Hecht's statment, that phyletic transformation couldn't happen or never happened. We have yet to resolve whether or not the studies cited by Hecht and others in the paleontological literature really represent this process. Further studies, keeping the two alternative models firmly in mind, will be necessary to lay the issue to rest.

REFERENCES

Bennett, J., 1960, A comparison of selective methods and a test of the preadaptation hypothesis, *Heredity* **15**:65–77.

Bock, W. J., 1972, Species interaction and macroevolution, *Evol. Biol.* **5**:1–24.

Bryson, R. A., Baerris, D. A., and Wendland, W. M., 1970, The character and late-glacial and post-glacial climatic changes, in: *Pleistocene and Recent Environments of the Central Great Plains* (W. Dort, Jr., and J. K. Jones, Jr., eds.), pp. 52–73, University of Kansas Press, Lawrence.

Dobzhansky, T. H., 1972, *Genetics of the Evolutionary Process*, 505 pp., Columbia University Press, New York.

Ehrlich, P. R., and Raven, P. H., 1969, Differentiation of populations, *Science* **165**:1228–1232.

Eldredge, N., and Gould, S. J., 1972, Punctuated equilibria: An alternative to phyletic gradualism, in: *Models in Paleobiology*, (T. J. M. Schopf, ed.), pp. 82–115, Freeman, Cooper, San Francisco.

Endler, J. A., 1973, Gene flow and population differentiation, *Science* **179**:243–250.

Ford, E. B., 1964, *Ecological Genetics*, 335 pp., Methuen, London.

George, T. N., 1971, Systematics in palaeontology, *J. Geol. Soc.* **127**:197–245.

Ghiselin, M., 1972, Models in phylogeny, in: *Models in Paleobiology* (T. J. M. Schopf, ed.), pp. 130–145, Freeman, Cooper, San Francisco.

Gould, S. J., 1972, Allometric fallacies and the evolution of *Gryphaea*: A new interpretation based on White's criterion of geometric similarity, *Evol. Biol.* **6**:91–119.

Hecht, M. K., 1965, The role of natural selection and evolutionary rates in the origin of higher levels of organization, *Syst. Zool.* **14**:301–317.

Hubby, J. L., and Lewontin, R., 1966, A molecular approach to the study of genic heterozygosity in natural populations. I. The number of alleles at different loci in *Drosophila pseudoobscura*, *Genetics* **54**:577–594.

Kettlewell, H. B. D., 1956, Further selection experiments on industrial melanism in the Lepidoptera, *Heredity* **10**:287–301.

Kurtén, B., 1957, The bears and hyaenas of the interglacials, *Quaternaria* **4**:1–13.

Kurtén, B., 1959, On the bears of the Holsteinian Interglacial, *Stockholm Contrib. Geol.* **2**:73–108.

Kurtén, B., 1964, The evolution of the polar bear, *Ursus maritimus, Phys. Acta Zool. Fenn.* **108**:1–26.

Kurtén, B., 1968, *Pleistocene Mammals of Europe*, 317 pp., Aldine, Chicago.

Larsen, T., 1971, Capturing, handling and marking polar bears in Svalbard (Spitzbergen), *J. Wildl. Manage.* **35**(**1**):27–36.

Larsen, T., 1972*a*, Norwegian polar bear hunt, management and research, *Int. Union Cons. Nat. Nat. Resour. Publ.* (*n.s.*) **23**:159–164.

Larsen, T., 1972*b*, Polar bear research in Norway, *Int. Union Cons. Nat. Nat. Resour. Publ.* (*n.s.*) **35**:60–65.

Maglio, V., 1972, Evolution of mastication in the Elephantidae, *Evolution* **26**:638–658.

Maglio, V., 1973, Origin and evolution of the Elephantidae, *Trans. Am. Phil. Soc.* (*n.s.*) **63**:3–149, (Part 3).

Manning, T. H., 1971, Geographical variation in the polar bear *Ursus maritimus, Phipps Canad. Wildl. Serv. Rep. Ser.* **13**:1–27.

Mayr, E., 1956, Geographical character gradients and climatic adaptation, *Evolution* **10**:105–108.

Mayr, E., 1963, *Animal Species and Evolution*, 797 pp., Belknap Press, Cambridge, Mass.

Norderhaug, M., 1972, Harvest and management of the polar bear in norway 1969–1971, *Int. Union Cons. Nat. Nat. Resour. Publ.* (*n.s.*) **35**:66–78 (suppl.).

Selander, R. K., Yung, S. Y., Lewontin, R., and Johnson, W. E., 1970, Genetic variation in the horseshoe crab (*Limulus polyphemus*), a phylogenetic "relic," *Evolution* **24**(**1**):402–414.

Simpson, G. G., 1953, *The Major Features of Evolution*, 434 pp., Columbia University Press, New York.

Simpson, G. G., 1970, Uniformitarianism: An inquiry into principle, theory, and methods in geohistory and biohistory, in: *Essays in Evolution and Genetics in Honor of Theodosius Dobzhansky* (A Supplement to *Evolutionary Biology*, M. K. Hecht and W. C. Steere, eds.), pp. 43–96, Appelton-Century-Crofts, New York.

Trueman, A. E., 1922, The use of *Gryphaea* in the correlation of the Lower Lias, *Geol. Mag.* **59**:256–268.

Wright, S., and Dobzhansky, T., 1946, Genetics of natural populations. XII. Experimental reproduction of some of the changes caused by natural selection in certain populations of *Drosophila pseudoobscura, Genetics* **31**:125–156.

Index